D0063389

ECOSYSTEM MANAGEMENT

**Edited by Mark S. Boyce
and Alan Haney**

Yale University Press
New Haven & London

Ecosystem Management

Applications for Sustainable

Forest and Wildlife Resources

Frontispiece: Prescribed fire to maintain prairie/savanna near Aldo Leopold's shack on the Leopold Preserve (courtesy of Nina Leopold)

Copyright © 1997 by Yale University. All rights reserved. This book may not be reproduced, in whole or in part, including illustrations, in any form (beyond that copying permitted by Sections 107 and 108 of the U.S. Copyright Law and except by reviewers for the public press), without written permission from the publishers.

Printed in the United States of America.

Library of Congress Cataloging-in-Publication Data

Ecosystem management : applications for sustainable forest and
    wildlife resources / edited by Mark S. Boyce and Alan Haney.
        p. cm.
    This volume is the result of a symposium held Mar. 3–5, 1994, at
the Univ. of Wisconsin–Stevens Point.
    Includes bibliographical references and index.
    ISBN 0-300-06902-2 (cloth : alk. paper)
        0-300-07858-7 (pbk. : alk. paper)
    1. Ecosystem management—Congresses.   2. Sustainable forestry—
Congresses.   3. Wildlife management—Congresses.   I. Boyce, Mark
    S.   II. Haney, Alan W.
    QH75.E295   1997
    333.95—dc20   96-34070

A catalogue record for this book is available from the British Library.

The paper in this book meets the guidelines for permanence and durability of the Committee on Production Guidelines for Book Longevity of the Council on Library Resources.

10 9 8 7 6 5 4 3 2

CONTENTS

FOREWORD

Congress has undertaken to undo important environmental legislation and is working on laws that would undermine the Endangered Species Act and other foundations for conservation in America. The current political climate is openly hostile toward proposals of yet additional environmental regulations, regardless of how necessary. All of this is happening during an era when environmental awareness is at an all-time high and most Americans view themselves to be environmentalists. How can this happen?

Some might argue that the problem is the American political process, but I suspect that the root is deeper. Many Americans, politicians included, perceive that we have many environmental regulations that do not make sense. Some laws appear to be inflexible and seem not to recognize the importance of humans as integral components of ecosystems. People whose livelihoods depend on natural resources perceive personal assaults on their very way of being.

But ecosystem management makes sense! Ecosystem management is about developing sound stewardship for our natural resources while recognizing that this is a complex task. Ecosystem management is about recognizing the value of such amenities as recreation, wildlife, and aesthetics. And ecosystem management ensures that we recognize the value of biological diversity in formulating strategies for use of natural resources. Indeed, ecosystem management is about putting people on the land and doing so in a way that doesn't degrade the very resources upon which we depend. Sustainable resource use is possible and necessary to ensure the future of life on this planet. This is all just good common sense.

But common sense doesn't mean knowing how to do it. Ecosystems are incredibly complex—perhaps the most complex entities in the universe, embracing all physical laws and all of biology to boot. Thus our personal experiences with various ecosystems shape our perceptions of key elements in ecosystem management. Each of the authors in this book puts a different spin on ecosystem management, but the commonalities are greater than one might expect.

Ecosystem management demands a tight relation with science. All too often we see science divorced from management. I maintain that good science makes good sense, but some practitioners of science don't use good sense— just as some resource managers don't use good sense. Science and resource management cannot be divorced. Sound ecosystem management involves strategies for incorporating science into management; adaptive management is how it happens.

Landscape ecology has become a highly popular discipline in recent years, and principles of landscape ecology have been embraced by ecosystem management. Throughout this book you will find references to scale and to how we need to expand the spatial scale at which we manage natural resources. This opens a veritable Pandora's box of controversial issues that will not go away, because ecological processes simply do not respect jurisdictional boundaries.

Issues of scale are not restricted to space. Ecosystem management requires also that we expand the temporal scale on which we manage. We must anticipate the long-term consequences of our management actions and ensure that our grandchildren's grandchildren will continue to benefit from the natural resources of today. And ecosystem management also requires that we expand the scale at which we consider the complexity of ecosystems.

Emergence of ecosystem management is a consequence of conflicts over how we manage our public lands. The American public will no longer tolerate commodity-production priority for managing our national forests and other public lands; neither can we afford to exclude commodity users from public lands and support a human population of 5.6 billion people. Scientists have been espousing the importance of ecosystem linkages in resource management for five decades. We have no choice but to embrace a broader vision.

Once, while I was testifying in front of a congressional committee, I was asked to explain—in simple and straightforward terms—just what was the purpose of "ecosystem management."

I asked the senator, "Did your momma ever tell you the fable of the goose that laid the golden eggs?" The senator allowed that he had heard the story. I then inquired, "Do you remember the lesson of that fable?" The senator replied that the point of the story was that "if you want golden eggs, it would be well to take good care of the goose." He looked at me in some puzzlement.

"Bingo!" I replied. "Such is the purpose of ecosystem management." That committee, and others, are frightened of "ecosystem management," which they perceive as some new and revolutionary concept. Ecosystem management is an evolutionary process in the application of science and philosophy to the husbandry of the good earth upon which all the inhabitants thereof depend for sustenance. Ecosystem management is a concept that must always be placed in context of time, space, and factors considered. We stand at a confluence of streams of science, philosophy, technological capability, and recognized human need, and the current is strong and compelling.

Ecosystem management is an idea whose time has come. There are arguments about just what ecosystem management is and is not. Let them continue. But there is no turning back against the current of the merged streams.

This book is a significant contribution by thoughtful scientists on how we may rationally begin to navigate the white water of those merged and merging currents. What lies ahead is an exciting journey. The concept of natural resource management has forever changed, and we practitioners are changed as well. Let's get on with it.

Jack Ward Thomas

PREFACE

This book explores ecosystem management, a crucial blend of science and management that is setting the stage for responsible stewardship of natural resources for the twenty-first century. We present the book in five sections. After an introductory chapter by the editors, the first section, Ecological Framework, tells us what ecosystem management is all about and establishes what we believe are its key elements. This leads into the next section, Disturbance, which highlights how natural disturbance drives many ecosystems, thus demonstrating that we cannot accomplish sound management without incorporating disturbance into the framework. The current revolution in resource management has been facilitated by the use of new techniques in computer technology that have further spawned advances in classification schemes for ecosystems, the topic of our third section, Techniques and Classification. And of course we must face the realities of implementation, as we do in our next section, Making It Happen, followed by the concluding section, Future Directions.

This volume is the result of a symposium on ecosystem management held at the University of Wisconsin–Stevens Point on 3–5 March 1994. Several people assisted us by reviewing manuscripts, including Eric Anderson, Robert Brush, Dan Coble, James Cook, Clive David, Robert Engelhardt, Jan Harms,

Robert Keiter, Evelyn Merrill, Robert Miller, Neil Payne, Robert Rogers, Stan Temple, and Linda Wallace.

We thank the Wildlife Management Institute for permission to adapt Chapter 11, and thanks to Charles Bradley and Nina Leopold for use of the frontispiece photo.

ECOSYSTEM MANAGEMENT

Chapter 1 Introduction

Alan Haney and Mark S. Boyce

Unrelenting demand for more and more commodities and services from global ecosystems raises questions of limits and sustainability. Staggering losses of topsoil each year from many of America's farmlands (Pimentel et al. 1995) demonstrate that these ecosystems are being exploited. Failure of certain tropical humid forests to rebound after clearfelling (Opler et al. 1977) dramatically illustrates their vulnerability to radical disturbance. Equally compelling evidence of ecosystem limits is seen in the altered flooding regimes, increased suspended loads, chemical contamination, and community structure changes in virtually every temperate river in the world (Johnson et al. 1995). The degradation of Earth's ecosystems is further signaled by the unprecedented decline of thousands of species, many of which have become extinct (Wilson 1992). Our purpose in this book is not to review how we are exploiting many ecosystems but to investigate how to apply good science so that we might manage ecosystems sustainably.

Examples of ecosystem degradation such as those mentioned here are usually the result of massive perturbations associated with human population growth and replacement of natural ecosystems by agriculture, cities, urban sprawl, and resulting fragmentation. In contrast, evidence of loss of ecosystem sustainability associated with long-term forest and wildlife management seems to be lacking or indirect. Timber harvests of old growth, shifts in the relative abundance of cover types, fragmentation by roads or clearcuts, and even loss of some species would not appear to interfere with forest productivity, for example. After all, we are reminded, most forests in eastern United States were heavily exploited less than one hundred years ago. Much of the forest land that was cleared and used for agriculture now supports productive forests again. Although soils were badly eroded and species were lost, available evidence does not suggest that sustainability of second growth forests has been reduced. Remarkable ecosystem resilience (rapid system recovery from perturbations) was further demonstrated in two forested watersheds in West Virginia. The areas were clear cut and rendered barren for five years through application of herbicides yet were found thirty years later to be similar to control watersheds in plant diversity and soil fertility (Patric 1995). We are reassured by such examples of ecological resilience that sustainable resource use will be possible in some forested systems even after abuse.

Sustainability is generally taken to mean that yield of goods and services from an ecosystem will not decline over time. Whether the priority is for commodity production, amenities such as recreation, or biodiversity (Knight 1995), all interests desire perpetuation. This book is about how ecosystem management tries to balance commodity production and amenity values while ensuring biodiversity protection and doing so sustainably.

ECOLOGICAL FRAMEWORK

Ecosystems are not only more complex than we know, they are more complex than we can know.
—Frank Egler

Regardless of how effective we think current management might be, we are witnessing an important change in forest and wildlife management, what Jerry Franklin (Chapter 2) calls a "paradigm shift." Perhaps careful forest and wildlife management can be carried on in many areas indefinitely without ecosystem degradation or other severe consequences, but ecosystem management demands more than sustainable yields of natural resources. Recreational users have become less tolerant of consumptive uses that reduce their enjoyment of forests. Conservation biologists are demanding that biodiversity be considered and principles of ecology be more critically applied in natural resource management. And courts of law have become arenas where resource management decisions are frequently litigated by nongovernmental organizations supporting environmental interests.

In the first section of the book we present five chapters that afford an overview of the seminal "big picture" issues of ecology in ecosystem management. Franklin, an ecologist from the University of Washington, begins with his overview of what is entailed in ecosystem management. Ecologists generally have been enthusiastic about the new focus on ecosystem management. Ecosystem management has spinoffs in ecological research, stimulating renewed interest in basic questions on the relation between ecosystem complexity and stability (Tilman and Downing 1994) and on the importance of biodiversity to ecosystem function (Naeem et al. 1994). Scientists are assertively challenging whether we are managing ecosystems sustainably (Ludwig et al. 1993, Rosenberg et al. 1993) and questioning how much we really know about ecosystem risks associated with losses of biodiversity (Baskin 1994). We are surprised at how little is known about the long-term effects of our management, and we share some of Fitzsimmons's (1994) concerns about basing management on such an immature science.

Yet resource managers must make decisions based upon incomplete and unsatisfactory sources of information and uncertainty of how ecosystems work. For example, global change associated with increasing greenhouse gases in the atmosphere could be devastating. We have little evidence that global change is occurring; instead, our predictions are based largely on global circulation models that can be wildly inconsistent (Schneider 1993). We cannot afford to ignore the possible consequences of global change, however, and we must make major commitments to reducing greenhouse gas emissions despite the uncertainty associated with our limited understanding of the system.

Weakness of understanding does not mean that we can ethically shirk responsibility for sustainable management. We must confront uncertainty head-on if ecosystem management is to be successful. Such ecological concepts as resiliency and persistence (continued existence) are central to what many people expect from "healthy" ecosystems. Specifically, the U.S. Forest Service has proposed that "the cumulative effects of human influences, including the production of commodities and services, should maintain resilient ecosystems capable of returning to the natural range of variability if left alone" (Kaufmann et al. 1994). Vague concepts like ecosystem health and integrity are usually poorly defined and can lead to confusion (Wicklum and Davies 1995). But ecological concepts of resiliency, resistance, stability, elasticity, constancy, and persistence are fundamental properties of ecosystems that must be distinguished if we are to anticipate ecosystem responses to management (Pimm 1984, 1991, 1994).

How ecosystems are configured on the landscape has enormous conse quences for how they should be managed. Connectivity, the spatial distribution of similar ecosystems, affects not only the way in which natural disturbances affect ecosystems (Turner et al. 1994) but also how well members of many populations and genes of some species are distributed over time (Crow and Gustafson, Chapter 3). For example, understanding regimes of natural fire disturbance may improve our ability to maintain diversity on the landscape (Bunnell 1995), but we seldom know the natural patterns of disturbance. Natural patterns of disturbance sometimes can be inferred from vegetation that has been reconstructed based upon pollen accumulated in wetlands, or fire history can be evaluated using charcoal bits from lake sediments (Whitlock 1993). As another example, we are only beginning to understand the ecological consequences of forest fragmentation. For some species, corridors can be used to facilitate movements among patches of habitat (Harris 1988). However, these well-meaning attempts to mitigate fragmentation by use of corridors may create avenues for spread of disturbances, diseases, insects, or exotics (Hobbs 1992, Simberloff et al. 1992, Mann and Plummer 1995a).

Managing at an appropriate spatial scale is fundamental to the accomplishment of conservation objectives. Ecological processes are seldom restricted by jurisdictional boundaries (Keiter 1994). Management on public lands, for example, may be insufficient to ensure population persistence if not coordinated with practices being performed on adjacent private property. The significance of large spatial scale of management became crucial in the efforts to manage the northern spotted owl *(Strix occidentalis caurina)* on fragmented landscapes in the Pacific Northwest (Boyce and Irwin 1990, FEMAT 1993) and likewise has been a key issue in managing the grizzly bear *(Ursus arctos horribilis)* that ranges over large areas in the Rocky Mountains (Servheen 1993).

Managing for biodiversity at too small a scale can be counterproductive. In general, diversity must be evaluated at a broad landscape scale. Consider the Kirtland's warbler *(Dendroica kírtlandii),* which requires sizable expanses of low-diversity 2–6 m tall jack pine *(Pinus banksiana)* forest for breeding habitat (Probst 1986). Managing to maximize diversity at a local scale would have disastrous consequences for the Kirtland's warbler. Accomplishing large-scale management requires often-difficult coordination across jurisdictional boundaries (Keiter 1994).

Some lands may be used primarily to retain biodiversity on the landscape while certain types of resource use are encouraged elsewhere (Seymour and Hunter 1992). Phillip deMaynadier and Malcolm Hunter (Chapter 4) suggest use of a keystone ecosystem concept in prioritizing patches in the landscape for protection, suggesting that certain ecosystems are of greater importance for a variety of reasons. Indeed, the status of various ecosystems also varies enormously; recently the Defenders of Wildlife has identified twenty-one of the most crucially endangered ecosystems in North America (Noss and Peters 1995).

Because the structures and functions of ecosystems depend on the organisms that are part of them (Schulze and Mooney 1994), it follows that maintenance of natural diversity is central to ecosystem management (Temple, Chapter 5). Familiar arguments for placing protection of biodiversity at the forefront of ecosystem management include ecosystem sustainability and the need to preserve genetic information. Whereas examples exist of rare strains of plants being found to contain important pharmaceuticals (Myers 1979) or becoming extremely valuable because they were resistant to disease (Wilson 1992), we have much to learn about the importance of biodiversity for ecosystem function. Recent research has demonstrated that diversity can increase productivity apparently because different species exploit resources of space and nutrients in different ways (Naeem et al. 1994). Marquis and Whalen (1994) reported that forest tree growth diminished when isolated from insectivorous birds, apparently because of effects of increased herbivory from leaf-chewing insects. Also, diversity can represent redundancy in ecosystem function that can provide resistance to perturbations. For example, the biomass of diverse grasslands in East Africa (McNaughton 1994) and Minnesota (Tilman and Downing 1994) decreased less after drought than similar grassland sites hosting fewer species.

Considerable confusion exists in the ecological literature due to inconsistent concepts and measures of stability (Botkin and Sobel 1975) and diversity (Palmer 1995). Contrary to common belief, for example, diversity does not necessarily confer stability to ecosystems. May (1972, 1973) demonstrated that complex model systems are actually less likely than simple ones to return to predisturbance state after being perturbed. Indeed, simply because a diverse ecosystem has more interactions and more rare species, it is easy to imagine that return after perturbation is less likely than in a simpler system with fewer species.

Managing ecosystems so as to maintain viable populations of natural species (see Boyce, Chapter 11) is one of the most widely suggested goals' of ecosystem management (Wilcove and Blair 1995), and population monitoring is arguably one of the best handles on sustainability. Although it is impossible to monitor all species of an ecosystem, persistent populations of sensitive and keystone species may indicate continuity of critical ecological processes (Frost et al. 1995, Schindler 1995).

In terrestrial ecosystems, the soil is the cornerstone of ecology. The last chapter in the Ecological Framework section examines the implications of ecosystem management on forest soil processes, with emphasis on macronutrient cycling (David, Chapter 6). Clive David argues that we have inadequate knowledge about many soil processes, but that if we manage to maintain the availability of macronutrients, plant productivity generally will be sustained. David also highlights the importance of managing woody debris in forests for both the retention of macronutrients and the preservation of structure in the forest. We do not know whether forests can be harvested sustainably over the long term, but how forests are managed likely will determine the outcome. For example, harvesting the boles of trees removes few nutrients from the system if the slash is left in place. On the other hand, fulltree removal—for example, whole-tree chipping—removes substantially more nutrients and may reduce the productivity of the next generation forest. Yet soil disturbance and associated erosion from logging may have greater impact on forest sustainability than removal of wood, especially on steeper slopes (Kimmins and Feller 1976).

DISTURBANCE

By chance, or nature's changing course untrimm'd.
—Shakespeare, Sonnet 18

Ecological process management gives primacy to retaining natural ecological processes, under the presumption that "nature knows best" (Keiter and Boyce 1991). Ecological process management is accomplished by either protecting natural ecological processes or mimicking these processes with management. Maintenance of ecological processes can be essential for maintaining biodiversity and, we hypothesize, for achieving ecosystem sustainability and resilience. For terrestrial ecosystems, the processes associated with the maintenance of populations and soils are arguably the most essential to ecosystem management. In addition, natural disturbance regimes are recognized as integral to the function of ecosystems (Pickett and White 1985).

Because environments are in a constant state of flux, either from natural or human causes, disequilibrium may be more the rule than the exception (Botkin 1990), thereby creating difficulties in defining dysfunction and attributing it to any specific cause. Mark Boyce and Neil Payne (Chapter 7) examine riparian habitats, providing some of the best examples of ecosystems maintained by disturbance regimes. But in riparian zones disturbance regimes have been greatly altered by human development and exploitation. In some cases, resource management can replace or mimic the natural disturbance regime (Hartshorn 1989), thereby preserving ecological processes while accommodating human needs.

Managing disturbances can be crucial to management for biodiversity (Budiansky 1995). Areas left without disturbance can be dominated by a few taxa, whereas some intermediate level of disturbance ensures that inferior competitors can coexist. Excessive disturbance, of course, results in few species because few can tolerate extreme levels or frequency of disturbance. Maximum species richness, therefore, is usually found in areas with an intermediate level of disturbance (Paine and Levin 1981). Leigh Fredrickson's essay (Chapter 8) highlights landmark work on managing disturbances for forested wetlands.

The recognition that disturbances can be important to the maintenance of biodiversity leads to the acknowledgment that for some rare species strict protection may be a poor conservation strategy. We protect rare species because of their scarcity and our concern over irreplaceability. For some rare species, disturbance may be crucial to providing appropriate habitat; for others protection against disturbance (especially human presence) may be crucial (leading to strict preservation). For example, oak and pine savanna habitat for the endangered Karner blue butterfly *(Lycaeides melissa samuelis)* in the eastern and midwestern United States depends upon disturbance management. Protecting a landscape from fire, mowing, and logging could eventually lead to the extinction of the species (Schweitzer 1994). In some systems grazing is a disturbance known to be important in maintaining floral diversity by reducing dominant plants (Reese and Brown 1992, McNaughton 1994, Sutherland 1995:9); grazing can also interact with fire (Hobbs et al. 1991).

TECHNIQUES AND CLASSIFICATION

Despite the complexity of ecosystem processes and resulting uncertainty, we are convinced that ecosystem management is important, necessary, and feasible. Moreover, we have sufficient knowledge to begin implementing programs of forest and wildlife management that are sustainable, and we have powerful new tools to help. Improved computer technology—for example, geographical information systems (GIS)—has improved our ability to organize and understand large quantities of spatially structured data, giving insight into how pattern affects processes and ultimately shaping new ways to manage landscapes (see D'Erchia, Chapter 10). Technology will help ecosystem management to improve the way natural resources are managed in the future.

Computers have enhanced our ability to derive classifications schemes using powerful numerical procedures. David Cleland and colleagues (Chapter 9) propose a land classification system based upon ecosystem principles and organized at a broad range of spatial scales. This hierarchical system, the National Hierarchical Framework of Ecological Units, is organized by ecological units irrespective of jurisdictional boundaries. Such classification schemes are necessary for the development of the sort of ecosystem-based GIS outlined by Frank D'Erchia (Chapter 10) and illustrated with resource management along the Upper Mississippi River.

Biogeography has played an important role in the applications of landscape ecology in ecosystem management; for example, the roots of conservation biology include ongoing debate over the design of systems of nature reserves (Shafer 1990). Similar considerations led to the formulation of gap analysis that uses GIS to find gaps in the representation of biodiversity among areas managed with a priority for conservation (Scott et al. 1993). The gap analysis project for a tri-state region in the upper Midwest also is explained by D'Erchia (Chapter 10).

In the spirit of "saving all the parts" (Barker 1993), the National Forest Management Act of 1976 requires maintenance of viable populations of vertebrates well distributed throughout their range. Obviously, managing for the sustainability of an ecosystem requires that we can ensure viability of the species contained in it. But despite decades of research on population processes, we are still struggling to understand what constitutes a viable population. Boyce (Chapter 11) takes a pragmatic view of population viability analysis (PVA) as a tool in developing conservation strategy for threatened and endangered species. Single-species approaches like PVA would seem antithetical to ecosystem management, with its focus on biodiversity and the recognition of the importance of all species. Yet species selected for PVA are often large and rare. Giving priority to "umbrella" species with large home ranges can be justified in the context of preservation of natural areas because preserving the extensive areas required by large or keystone species will ensure preservation of habitat for many organisms with smaller area requirements as well (Noss and Cooperrider 1994) and thus ensure ecosystem sustainability.

MAKING IT HAPPEN

Each of the principal federal land management agencies has openly embraced ecosystem management as a panacea for today's natural resource controversies (Keiter 1994). Because various land management agencies have different missions and mandates, each agency views ecosystem management differently (Stanley 1995). The U.S. Forest Service defines ecosystem management as "the use of an ecological approach to achieve multiple-use management of the national forests and grasslands by blending the needs of people and environmental values in such a way that the national forests and grasslands represent diverse, healthy, productive, and sustainable ecosystems." The reference to multiple-use management nicely coincides with the agency's congressional mandate. Likewise, the U.S. Bureau of Land Management uses language that emphasizes harvest of resources, defining ecosystem management as "an approach to sustain the integrity, diversity, and productivity of ecological systems while providing resource products for present needs and future generations." Mollie Beattie for the U.S. Fish and Wildlife Service reiterates her agency's charge by suggesting that ecosystem management "represents a new way of managing natural resources that takes into account the entire ecosystem and balances recreational use, economic development, and conservation of wildlife so each is sustainable." Franklin (Chapter 2) offers yet another interpretation that gives legislative context for ecosystem management by public agencies. In short, the current agency efforts to implement ecosystem management are occurring within the statutory and legal constraints governing each agency.

Many approaches will be necessary to make ecosystem management work. In most large landscapes, there is a mix of public and private lands and a variety of management objectives. Reed Noss and Michael Scott (Chapter 12) illustrate how diversity can be protected and maintained in a landscape approach to forest management that includes diversity protection areas to supplement more intensively managed zones.

On a much more local scale, the stand level, specific suggestions of a more diversity-sensitive approach to silviculture are suggested by John Kotar (Chapter 13). Modifying how we manage commodity extraction offers great opportunity for ensuring biodiversity preservation and can introduce more sensitivity to values of recreationists.

Implementation of ecosystem management on public lands in particular requires changes in policy. Douglas Norton and David Davis (Chapter 14) examine federal policy for implementation of ecosystem management through opportunities afforded under the Clean Water Act. The approach of using regulations to accomplish objectives of ecosystem management is often questioned because it can engender antagonism from those negatively affected by coercive regulatory requirements. For example, reactions to the Endangered Species Act sometimes have worked counter to the intentions of species protection (Mann and Plummer 1995b). Instead, creative incentives may work better. Yet we believe that it is unrealistic to imagine that ecosystem management is possible without regulations.

Stakeholders have demanded and received an increasing role in natural resource management through such laws as the National Environmental Policy Act (NEPA), the National Forest Management Act (NFMA), and the Federal Lands Policy Management Act (FLPMA). Yet they often lack adequate knowledge to understand the complex management decisions that they wish to influence. If ecosystem management is to be accepted by the public, if appropriate policies are to be developed and enforced, and if reserve managers are to apply effective ecosystem management practices, an appropriately integrated educational program is essential. This is particularly true in the case of ecosystem restoration. The need for education on ecological restoration is described, along with a suggested approach, by Steven Apfelbaum and Kim Alan Chapman (Chapter 15).

Because ecosystem management draws heavily on the complex science of ecology and conservation biology, some have proposed that ecosystem management designs should be prepared by scientists. Invariably, however, resource management decisions are value laden and must balance political, economic, and aesthetic considerations (Keiter 1994). Integrated planning and management are as essential as knowledge about the effects of human activities on the ecosystem (Slocombe 1993). Some would argue that "ecosystem management is more about people than anything else" (Salwasser 1994). But this does not imply that management and science should be divorced. Quite the contrary, it begs for the development of administrative structures that ensure close integration of science into resource management.

FUTURE DIRECTIONS

Uncertainty and complexity frustrate both science and management. Science should be the foundation for ecosystem management, but application of the scientific method in ecosystem studies has proven cumbersome because of the difficulty of conducting replicated experiments at large scales of time, space, and complexity (Carpenter 1990). Active adaptive management offers a solution by which science can be integrated effectively with ecosystem management. The complexity of ecosystems makes reductionist approaches to the scientific study of ecosystem management impractical. Instead, we must invoke the "learning by doing" paradigm of Walters and Holling (1990), which advocates: (1) the formulation of a model for the system, (2) manipulation of the system through active management, followed by (3) monitoring to document the response to management, and finally (4) a reassessment of model predictions and revision of the model or database before beginning the process anew. Successful management of ecosystems will require the development of active adaptive management, where intervention may occur to structure a range of alternative response models (Walters and Holling 1990). When possible, replication and treatment controls must be a part of such schemes, and we must develop effective monitoring programs (Goldsmith 1991).

We are concerned, however, that the "learning by doing" paradigm often is being used as an excuse for business as usual. Without the formal adaptive management structure including modeling, active management, and monitoring, we will not learn much from what we are doing. A similar sort of constant reevaluation of human concerns and values is also part of the adaptive management concept—the "social" as opposed to "scientific" discussions of it. Indeed, adaptive management can be a powerful approach for evaluating and harmonizing socioeconomic systems with ecological constraints (Gunderson et al. 1995). We do not believe that there should be a question about decision-making authority being placed with scientists or managers because adaptive resource management ensures that responsible management will develop based on scientific process and social considerations. Managers and scientists must work together closely to make it happen.

Ecosystem management has potential applications at a global scale, but the approaches evolving in North America may be poorly suited in much of the developing world. Indeed, the world's conservation community is making distressingly little progress at curbing devastating rates of tropical deforestation, population growth, and species extinctions, and we fear that our approaches are often misdirected, counterproductive, or ineffective (Boyce 1994). We often dwell on issues of the appropriate design for parks and nature preserves (Soulé and Simberloff 1986), yet in developing countries, parks and preserves are repeatedly shown to be ineffective and to alienate local people (Brockelman 1990, Alcorn 1991). We focus energies and invest conservation resources for the preservation of ancient forests in the Pacific Northwest (Boyce and Irwin 1990), yet this increases demand for tropical timber where consequences to biodiversity may be much more serious (Sprugel 1990). And we invest in detailed population viability analyses for single species (Boyce, Chapter 11) when it is clear that such efforts are hopeless given the enormous number of extinctions projected over the next thirty to fifty years (National Research Council 1992). We face an enormous challenge on a global scale: we must figure out how to check human population growth (Cohen 1995) and how to make ecologically sensitive management more responsive to the needs and interests of local citizens—by demonstrating the advantages (long- and short-term) associated with a sustainable management scheme.

Yet we share Norman Christensen's (Chapter 16) optimism that humans have the intelligence and adaptability to learn how to conduct ecosystem management. The fact that we are struggling with these issues is a step in the right direction. We are confident that the same Darwinian motivation that led us to develop efficient means to exploit natural resources can be redirected to develop efficient means for their stewardship. Adaptive management protocols provide a framework through which we can expect to learn how to accomplish such sound stewardship.

Is ecosystem management a panacea as touted by the federal land management agencies (Keiter 1994)? Perhaps not, but it is already causing an enormous shift in emphasis that is sorely needed in natural resource management. By its very nature, ecosystem management is a never-ending process. We will never complete it, and we will seem to never get it just right because doing ecosystem management right means that we are always learning, changing, and improving our management. We will have gotten it right if we find that we are able to do it sustainably. We are confident that ecosystem management is possible. Our future existence on this planet depends on it.

LITERATURE CITED

Alcorn, J. B. 1991. Ethics, economics, and conservation. Pages 317–349 *in* M. L. Oldfield and J. B. Alcorn, editors. Biodiversity: culture, conservation, and eco-development. Westview Press, Boulder, Colorado.

Barker, R. 1993. Saving all the parts: reconciling economics and the Endangered Species Act. Island Press, Washington, D.C.

Baskin, Y. 1994. Ecologists dare to ask: how much does it matter? Science 264:202–203.

Botkin, D. B. 1990. Discordant harmonies: a new ecology for the twenty-first century. Oxford University Press, Oxford, England.

Botkin, D. B., and M. Sobel. 1975. Stability in time-varying ecosystems. American Naturalist 109:625–646.

Boyce, M. S. 1994. Biodiversity at risk: deforestation. Pages 151–161 *in* S. K. Majumdar, F. J. Brenner, J. E. Lovich, J. F. Schalles, and E. W. Miller, editors. Biological diversity: problems and challenges. Pennsylvania Academy of Sciences, Easton, Pennsylvania.

Boyce, M. S., and L. L. Irwin. 1990. Viable populations of spotted owls for management of old growth forests in the Pacific Northwest. Pages 133–135 *in* R. S.

Mitchell, C. J. Sheviak, and D. J. Leopold, editors. Ecosystem management: rare species and significant habitats. Proceedings 15th Annual Natural Areas Conference, New York State Museum Bulletin 471.

Brockelman, W. Y. 1990. Wildlife conservation in Thailand: a strategic assessment. Pages 300–304 in J. C. Daniel and J. S. Serrao, editors. Conservation in developing countries: problems and prospects. Oxford University Press, Bombay.

Budiansky, S. 1995. Nature's keepers: the new science of nature management. Free Press, New York, New York.

Bunnell, F. L. 1995. Forest-dwelling vertebrate faunas and natural fire regimes in British Columbia: patterns and implications for conservation. Conservation Biology 9:636–644.

Carpenter, S. R. 1990. Large-scale perturbations: opportunities for innovation. Ecology 71:2038–2043.

Cohen, J. E. 1995. How many people can the earth support? W. W. Norton, New York, New York.

Egler, F. 1977. The nature of vegetation: its management and mismanagement. Aton Forest, Norfolk, Connecticut.

FEMAT. 1993. Forest ecosystem management: an ecological, economic, and social assessment. Forest Ecosystem Management Assessment Team, Government Printing Office, Washington, D.C.

Fitzsimmons, A. K. 1994. Federal ecosystem management: a "train wreck" in the making. Policy Analysis 217:1–33.

Frost, T. M., S. R. Carpenter, A. R. Ives, and T. K. Kratz. 1995. Species compensation and complementarity in ecosystem function. Pages 224–235 in C. G. Jeans and J. H. Lawton, editors. Linking species and ecosystems. Chapman and Hall, New York, New York.

Goldsmith, F. B., editor. 1991. Monitoring for conservation and ecology. Chapman and Hall, London.

Gunderson, L. H., C. S. Holling, and S. S. Light (editors). 1995. Barriers and bridges to the renewal of ecosystems and institutions. Columbia University Press, New York, New York.

Haney, A., and S. I. Apfelbaum. 1990. Structure and dynamics of Midwest oak savannas. Pages 19–30 in J. M. Sweeney, editor. Management of dynamic ecosystems. North Central Section, The Wildlife Society, West Lafayette, Indiana.

Harris, L. D. 1988. Landscape linkages: the dispersal corridor approach to wildlife conservation. Transactions of the North American Wildlife and Natural Resources Conference 53:595–607.

Hartshorn, G. S. 1989. Application of gap theory to tropical forest management: natural regulation on strip clear-cuts in the Peruvian Amazon. Ecology 70:567–569.

Hobbs, N. T., D. S. Schimel, and C. E. Owensby. 1991. Fire and grazing in the tallgrass prairie: contingent effects on nitrogen budgets. Ecology 72:1374–1382.

Hobbs, R. J. 1992. The role of corridors in conservation: solution or bandwagon? Trends in Ecology and Evolution 7:389–392.

Johnson, B. L., W. B. Richardson, and T. J. Naimo. 1995. Past, present, and future concepts in large river ecology. BioScience 45:134–141.

Kaufmann, M. R., R. T. Graham, D. A. Boyce, Jr., W. H. Moir, L. Perry, R. T. Reynolds, R. L. Bassett, P. Mehlhop, C. B. Edminster, W. M. Block, and P. S. Corn. 1994. An ecological basis for ecosystem management. U.S. Forest Service, General Technical Report RM-246.

Keiter, R. B. 1994. Beyond the boundary line: constructing a law of ecosystem management. University of Colorado Law Review 65:293–333.

Keiter, R. B., and M. S. Boyce. 1991. Greater Yellowstone's future: ecosystem management in a wilderness environment. Pages 379–412 in R. B. Keiter and M. S. Boyce, editors. The greater Yellowstone ecosystem: redefining America's wilderness heritage. Yale University Press, New Haven, Connecticut.

Kimmins, J. P., and M. C. Feller. 1976. Effect of clear-cutting and broadcast slash-burning on nutrient budgets, streamwater chemistry and productivity in western Canada. Proceedings of the XVI IUFRO World Congress, Oslo, Norway, Div. 1: 186–197.

Knight, R. L. 1995. Ecosystem management and Aldo Leopold. Rangelands 17:182–183.

Ludwig, D., R. Hilborn, and C. Walters. 1993. Uncertainty, resource exploitation, and conservation: lessons from history. Science 260:17, 36.

Mann, C. C., and M. L. Plummer. 1995a. Are wildlife corridors the right path? Science 270:1428–1430.

———. 1995b. Noah's choice: the future of endangered species. Knopf, New York, New York.

Marquis, R. J., and C. J. Whalen. 1994. Insectivorous birds increase growth of white oak through consumption of leaf-chewing insects. Ecology 75:2007–2014.

May, R. M. 1972. Will a large complex system be stable? Nature 238:413–414.

———. 1973. Stability and complexity in model ecosystems. Princeton University Press, Princeton, New Jersey.

McNaughton, S. J. 1994. Biodiversity and function of grazing ecosystems. Pages 361–383 in E.-D. Schulze and H. A. Mooney, editors. Biodiversity and ecosystem function. Springer-Verlag, New York, New York.

Myers, N. 1979. The sinking ark. Pergamon, New York, New York.

Naeem, S., L. J. Thompson, S. P. Lawler, J. H. Lawton, and R. M. Woodfin. 1994. Declining biodiversity can alter the performance of ecosystems. Nature 368:734–737.

National Research Council. 1992. Conserving biodiversity: a research agenda for development agencies. National Academy Press, Washington, D.C.

Noss, R. F., and A. Y. Cooperrider. 1994. Saving nature's legacy: protecting and restoring biodiversity. Island Press, Washington, D.C.

Noss, R. F., and R. L. Peters. 1995. Endangered ecosystems: a status report on America's vanishing habitat and wildlife. Defenders of Wildlife, Washington, D.C.

Opler, P. A., H. G. Baker, and G. W. Frankie. 1977. Recovery of tropical lowland forest ecosystems. Pages 379–421 in J. Cairns, Jr., K. L. Dickson, and E. E. Herricks, editors. Recovery and restoration of damaged ecosystems. University of Virginia Press, Charlottesville, Virginia.

Paine, R. T., and S. A. Levin. 1981. Intertidal landscapes: disturbance and the dynamics of pattern. Ecological Monographs 51:145–178.

Palmer, M. W. 1995. How should one count species? Natural Areas Journal 15:124–135.

Patric, J. H. 1995. Forest ecosystem recovery from induced barrenness. American Forest and Paper Association, TB 95-3, Washington, D.C.

Pickett, S. T. R., and P. S. White (editors). 1985. The ecology of natural disturbance and patch dynamics. Academic Press, Orlando, Florida.

Pimentel, D., C. Harvey, P. Resosudarmo, K. Sinclair, D. Kurz, M. McNair, S. Crist, L. Shpritz, L. Fitten, R. Saffouri, and B. Blair. 1995. Environmental and economic costs of soil erosion and conservation benefits. Science 267:1117–1123.

Pimm, S. L. 1984. The complexity and stability of ecosystems. Nature 307:321–326.

———. 1991. The balance of nature? ecological issues in the conservation of species and communities. University of Chicago Press, Chicago, Illinois.

———. 1994. Biodiversity and the balance of nature. Pages 347–359 in E.-D. Schulze and H. A. Mooney, editors. Biodiversity and ecosystem function. Springer-Verlag, New York, New York.

Popper, K. R. 1968. The logic of scientific discovery. 2d edition. Harper and Row, New York, New York.

Probst, J. R. 1986. A review of factors limiting the Kirtland's warbler on its breeding grounds. American Midland Naturalist 116:87–100.

Rees, M., and V. K. Brown. 1992. Interaction between invertebrate herbivores and plant competition. Journal of Ecology 80:353–360.

Regier, H. A. 1993. The notion of natural and cultural integrity. Pages 3–18 in S. Woodley, J. Kay, and G. Francis, editors. Ecological integrity and the management of ecosystems. St. Lucie Press, Delray Beach, Florida.

Rosenberg, A. A., M. J. Fogarty, M. P. Sissenwine, J. R. Beddington, and J. G. Shepherd. 1993. Achieving sustainable use of renewable resources. Science 262:828–829.

Salwasser, H. 1994. Ecosystem management: what's in store for wildlife? The Wildlifer 264:42.

Schindler, D. W. 1995. Linking species and communities to ecosystem management: a perspective from the experimental lakes experience. Pages 313–325 in C. G. Jones and J. H. Lawton, editors. Linking species and ecosystems. Chapman and Hall, New York, New York.

Schneider, S. H. 1993. Scenarios of global warming. Pages 9–23 in P. M. Kareiva,

J. G. Kingsolver, and R. B. Huey, editors. Biotic interactions and global change. Sinauer Associates, Inc., Sunderland, Massachusetts.

Schulze, E.-D., and H. A. Mooney. 1994. Ecosystem function of biodiversity: a summary. Pages 497–510 *in* E.-D. Schulze and H. A. Mooney, editors. Biodiversity and ecosystem function. Springer-Verlag, New York, New York.

Schweitzer, D. F. 1994. Recovery goals and methods for Karner blue butterfly populations. Pages 185–193 *in* D. A. Andow, R. J. Baker, and C. P. Lane, editors. Karner blue butterfly: a symbol of a vanishing landscape. Minnesota Agricultural Experiment Station, Miscellaneous Publication 84-1994. St. Paul, Minnesota.

Scott, J. M., F. Davis, B. Csuti, R. Noss, B. Butterfield, C. Groves, H. Anderson, S. Caicco, F. D'Erchia, T. C. Edwards, Jr., J. Ulliman, and R. G. Wright. 1993. Gap analysis: a geographic approach to protection of biological diversity. Wildlife Monographs 123:1–41.

Servheen, C. 1993. Grizzly bear recovery plan. U.S. Fish and Wildlife Service, Missoula, Montana.

Seymour, R. S., and M. L. Hunter, Jr. 1992. New forestry in eastern spruce-fir forests: principles and applications to Maine. Maine Agricultural Experiment Station, Miscellaneous Publication 716. University of Maine, Orono, Maine.

Shafer, C. L. 1990. Nature reserves: island theory and conservation practice. Smithsonian Institution Press, Washington, D.C.

Simberloff, D. S., J. A. Farr, J. Cox, and D. W. Mehlman. 1992. Movement corridors: conservation bargains or poor investments. Conservation Biology 6:493–504.

Slocombe, D. S. 1993. Implementing ecosystem-based management. BioScience 43: 612–622.

Soulé, M. E., and D. Simberloff. 1986. What do genetics and ecology tell us about the design of nature reserves? Biological Conservation 35:19–40.

Sprugel, D. G. 1990. Ancient forests: valuable data, questionable conclusions. Conservation Biology 4:461–462.

Stanley, T. R. 1995. Ecosystem management and the arrogance of humanism. Conservation Biology 9:255–262.

Sutherland, W. J. 1995. Introduction and principles of ecological management. Pages 1–21 *in* W. J. Sutherland and D. A. Hill, editors. Managing habitats for conservation. Cambridge University Press, Cambridge.

Thomas, J. W. 1995. Forest health: what it is, what we're doing about it. Transactions of the North American Wildlife and Natural Resources Conference 60:29–34.

Tilman, D., and J. A. Downing. 1994. Biodiversity and stability in grasslands. Nature 367:363–365.

Turner, M. G., W. H. Romme, and R. H. Gardner. 1994. Landscape disturbance models and the long-term dynamics of natural areas. Natural Areas Journal 14:3–11.

Walters, C. J. 1986. Adaptive management of renewable resources. Macmillan, New York, New York.

Walters, C. J., and C. S. Holling. 1990. Large-scale management experiments and learning by doing. Ecology 71:2060–2068.

Whitlock, C. 1993. Postglacial vegetation and climate of Grand Teton and southern Yellowstone National Parks. Ecological Monographs 63:173–198.

Wicklum, D., and R. W. Davies. 1995. Ecosystem health and integrity? Canadian Journal of Botany 73:997–1000.

Wilcove, D. S., and R. B. Blair. 1955. The ecosystem management bandwagon. Trends in Ecology and Evolution 10:345.

Wilson, E. O. 1992. The diversity of life. Harvard University Press, Cambridge, Massachusetts.

ECOLOGICAL FRAMEWORK

Ecosystem management must be based in ecological principles.
—*Hal Salwasser*

By ecosystem management, we mean an ecological approach will be used to achieve the multiple use management of our National Forests and Grasslands. It means that we must blend the needs of people and environmental values in such a way that the National Forests and Grasslands represent diverse, healthy, productive and sustainable ecosystems.
—*Dale Robertson*

Overleaf: Old-growth logging of white pine *(Pinus strobus)* in northern Wisconsin (Malcolm Rosholt collection, University of Wisconsin-Stevens Point, archives). Old-growth forests throughout the midwestern United States were largely eliminated by extensive logging and fires in the mid-1800s.

Chapter 2 Ecosystem Management: An Overview

Jerry F. Franklin

Forestry—and resource management in general—is undergoing a paradigm shift of massive proportions. Major changes have occurred in social perspectives and goals for forest management, and this is reflected in the contentious and complicated legal environment in which management activities are currently planned and conducted. A large body of new and significant ecological knowledge has accumulated during the past twenty-five years about the structure and function of forest and freshwater ecosystems and about biological diversity, including wildlife and fish species. Issues related to larger spatial scales—landscapes and regions—have become central elements in resource management.

Ecosystem management has emerged from a myriad of labels and concepts as the guide and organizing principle in meeting the technical and social challenges of this millennial paradigm shift. In this chapter I review some historical perspectives on the ecosystem concept and the evolution of management concepts that has culminated in ecosystem management. Next, I will provide my own definition of ecosystem management, followed by an identification and discussion of what I view as some of the elements critical to this concept. Finally, I will briefly review the development and application of ecosystem concepts in the forest policy controversies in the Pacific Northwest. My purpose is to provide a perspective on ecosystem management and its application rather than a comprehensive critical review of the massive body of literature which has contributed to the concept.

THE ECOSYSTEM CONCEPT

Ecological science has largely focused on organisms and the interactions among organisms at relatively small spatial scales through most of the twentieth century. Although the ecosystem concept emerged as early as 1935, only in the past three decades has it undergone extensive development and application. The ecosystem concept dramatically expanded the science of ecology by (1) focusing equal attention on abiotic and biotic components and (2) explicitly recognizing the potential of studying ecological processes at multiple scales.

An ecosystem is defined as "a spatially explicit unit of the Earth that includes all of the organisms, along with all components of the abiotic environment within its boundaries" (Likens 1992). The British ecologist A. G. Tansley is often credited with first identifying and labeling the concept in 1935, but the American ecologist E. P. Odum probably contributed as much as anyone to its popularization. The ecosystem perspective clearly brought the abiotic into a science that had previously focused primarily on the study of organisms and their interrelations.

In response to this more inclusive perspective (and to encourage a more integrated view of ecological science) Likens (1992) defined ecology as "the scientific study of the processes influencing distribution and abundance of organisms, interactions among organisms, and interactions between organisms and the transformation and flux of energy and matter," which clearly includes the ecosystem concept.

The ecosystem concept is open with regard to spatial scale. Conceptually, it is equally appropriate to view an aquarium, a forest stand, a watershed, a region, or the entire globe as an ecosystem. The scale that is selected is determined by the organisms or processes of interest. All that is required is attention to the interactions between organisms and the abiotic elements as per Likens's definition. The openness of the ecosystem concept with regards to scale is one of its great strengths, for it allows the human—manager, scientist, or layman—to select the scale that is appropriate to a particular objective or interest. The difficulties often arise in attempting to measure transfers of materials, energy, and organisms into and out of (across the boundaries) of a particular ecosystem. Hence, scientists often choose ecosystems with well-defined physical boundaries (for example, small watersheds);

the problem is measurement of fluxes, however, and not a conceptual definition of boundaries.

ECOSYSTEM MANAGEMENT: CONTRIBUTING CONCEPTS

Many topical areas of science and management have contributed significantly to ecosystem management. Some of these topical areas, particularly those with an "ecosystem" label, are sometimes mistaken for or equated with ecosystem management. Ecosystem science, the "greater" ecosystem concepts, landscape ecology, and a dynamic perspective on ecosystems all qualify as important contributors.

Ecosystem Science

Ecosystem science has been a major source of concepts and knowledge during the approximately thirty-five years that it has been recognized as an area of ecological research. The collaborative university–Forest Service program at Hubbard Brook is an early example, one which continues to make major contributions (see, e.g., Bormann and Likens 1979, Likens et al. 1977, and Likens 1985). Funded by the National Science Foundation, the U.S. International Biological Program—including, especially, the large-scale biome projects—produced a dramatic expansion in research and knowledge at the ecosystem level between 1968 and 1976 (Blair 1977). Much of the early IBP research focused on description and modeling of energy and material (carbon, nutrients, and water) cycles.

Ecosystem research topics broadened rapidly with the expansion of ecosystem research following the end of IBP. Emphasis was (and remains) on ecological processes or functions as opposed to species or biotic community issues. Examples of important processes are primary productivity and its controls, conservation of nutrients, interactions of nutrient regimes with productivity, and regulation of the hydrologic cycles. Major new areas of ecosystem science were opened by the IBP studies and developed in subsequent research projects. Ecosystem science resulted in a recognition of the importance of ecological systems previously very poorly known: the riparian or stream influence zone (see, e.g., Naiman 1992), belowground processes (for example, productivity and turnover) and community structure (see, e.g., Harris et al. 1979), the hyporheic zone associated with streams and rivers (Stanford and Ward 1992), and forest canopies (see, e.g., Carroll 1979). The scientific findings from this research have dramatically altered our view of forests and associated freshwater ecosystems and how they work (see, e.g., Edmonds 1982).

But ecosystem management is not just about material cycles or the body of science labeled ecosystem science.

"Greater" Ecosystem Concept and Landscape Ecology

Ecosystem management has often been associated with the need for large management units that would provide sufficient area for large, wide-ranging wildlife (see Agee and Johnson 1988), such as the Greater Yellowstone Ecosystem. This concept emerged from the recognition that many of our parks and other natural reserves are too small to provide for the full array of constituent species, especially for the charismatic megafauna (grizzly bear, elk, wolf, and others) on which environmentalists and conservation biologists have been so focused. The greater ecosystem concept has reinforced for ecosystem management the notion that ecosystems have to be defined and managed at appropriate spatial scales.

Landscape ecology also gained prominence in North America in the 1980s and has been equated with ecosystem management by some managers and scientists. Landscape ecology is concerned with ecological consequences of larger spatial patterns of abiotic and biotic resources; in a gross sense it can be characterized as the study of ecological effects of patches and their interactions, including collective effects. Landscapes that have lost their ability to fulfill important ecological functions—such as those subject to undesirable cumulative effects or fragmentation of forest habitat (see, e.g., Harris 1984)—have been of particular interest to managers and scientists. Landscape ecology has helped us to appreciate the spatial dimension of ecosystem management and to understand that many critical issues in ecosystem management must be dealt with at larger spatial scales than the individual forest stand (patch) or stream reach.

While it is easy to say that ecosystems have to be defined and managed at appropriate spatial scales, accomplishing this is often very difficult. The migratory boundaries of elk herds may provide us with a reasonable guide for that particular resource. But there are many animal species that roam much more widely—neotropical songbirds, whales, and sea turtles are excellent examples. Such resources require international agreements for management programs to be effective—although, as a part of these agreements, local agencies will have to provide appropriate ecosystem-based management programs for their portions of a species range.

Identifying and managing resources at appropriate spatial scales is clearly an important component of ecosystem management. However, ecosystem management is not defined by the "greater" ecosystem concept or even issues at larger spatial scales.

Ecosystem Dynamics

The increased emphasis on ecosystems as dynamic entities subject to periodic disturbance and continuing development has been a significant input to ecosystem management (e.g., Botkin 1990). The dynamic and nonequilibrium nature of ecosystems is certainly not a new concept—at least not among resource managers and applied ecologists—but much new information has been developed during the last two decades. There is greatly increased appreciation of the distinctions among disturbances with regard to impacts and subsequent recovery processes. Such major events as Mount St. Helens (Franklin et al. 1988) and the Yellowstone wildfires (Christensen et al. 1989) have greatly enriched our understanding of the importance of disturbance type, intensity, frequency, and size.

One emerging concept from disturbance research—biological legacies—is proving to have particular significance in understanding forest ecosystems (Franklin 1990). Biological legacies refer to living organisms (for example, seeds, spores, whole plants, and animals), biologically derived structures (for example, snags and logs), and biologically imposed patterns (for example, in soil communities) that are carried over from the predisturbance to the postdisturbance, recovering ecosystem. Such legacies strongly influence the structural, functional, and compositional complexity of the postdisturbance ecosystem. Natural forest disturbances typically leave high levels of structural (deadwood) and compositional legacies, whereas many human disturbances, such as clearcutting, typically leave very low levels.

Ecosystem management incorporates the concept of dynamic, nonequilibrium systems, but it is not defined by it.

ECOSYSTEM MANAGEMENT: RELATION TO SOME PAST MANAGEMENT PROGRAMS

A series of management concepts precedes the evolving philosophy of ecosystem management, particularly as it relates to management of federal lands. I will just consider a few of the more recent ones here.

Multiple use, the Forest Service management principle of the 1950s and 1960s, recognized the concept of managing the national forests for multiple values. Unfortunately, we did not yet have the ecological knowledge to practice truly integrative management approaches. We did not even consider biological diversity, focusing only on game fish and wildlife. Indeed, the occurrence, let alone importance, of nongame animals and commercially unimportant plants was essentially unknown. Furthermore—and, perhaps, most important—multiple use emphasized output of goods and services, rather than the stewardship of the ecosystem, as the objective of management. It was output oriented rather than sustainability oriented.

New Forestry emerged in the late 1980s and incorporated a greatly enlarged and continuously expanding knowledge of ecosystem processes and linkages, including many at larger spatial scales. New Forestry has often been viewed mainly as a "kinder and gentler" forestry—alternative silviculture or forest harvest practices—and this is certainly an important component (Hopwood 1991, Franklin 1992). Popular accounts of New Forestry typically emphasize structural retention in cutover areas. However, New Forestry also incorporates larger spatial components in its approaches: riparian protection as a key element, the need for and design of habitat reserves, and management approaches that minimize forest fragmentation.

New Perspectives emerged in the late 1980s as the Forest Service's tentative step—sort of a toe-in-the-water—toward New Forestry and Ecosystem Management (Salwasser 1991). Major changes in policy were clearly required with the emerging scientific challenges to dispersed management activities (e.g., Franklin and Forman 1987) and traditional clearcutting. New Perspectives tentatively embraced some of the ecological concepts but also incorporated many other agency programs in social interactions, recreation, and so on. New Perspective activities in the Pacific Northwest were often focused on entries into unroaded areas, which ultimately made it a very unpopular program with the environmental community. New Perspectives was probably most useful in moving the agency toward—and feeling more comfortable with—ecosystem management. Subsequently, in 1992 the Forest Service dropped the New Perspectives label as it adopted Ecosystem Management as its management paradigm.

The section on developments in the Pacific Northwest will provide some additional historical perspectives to this brief historical review of ecosystem management.

A CONCEPTUAL VIEW OF ECOSYSTEM MANAGEMENT

Many interpretations of ecosystem management exist, as suggested by the diversity—and contrasting goals—of stakeholders, professional groups, and decision makers who have adopted the label. Grumbine (1994) identified ten dominant themes in discussions of ecosystem management: a hierarchical context (multiple scales); recognition of ecological boundaries; ecological integrity; systematic research and data collection; monitoring; adaptive management; interagency cooperation; organizational change; humans as an ecosystem component; and human values as dominant in goal setting. While there are doubtless differences of opinion on a specific definition and interpretation of appropriate actions, most participants in the dialogue would agree that, fundamentally, ecosystem management is managing ecosystems so as to assure their sustainability.

Accepting this as a general principle, the issue then becomes one of defining sustainability. I offer my view with the proviso that there are many alternative perspectives. I view sustainability as the maintenance of the potential of our terrestrial and aquatic ecosystems to produce the same quantity and quality of goods and services in perpetuity (Franklin 1993). I emphasize *potential* because it makes implicit the option to return to alternative conditions and the need to avoid permanent loss of some capacity. Furthermore, the emphasis on potential makes clear that ecosystem management does not necessarily require maintaining an existing condition.

My concept of sustainability considers a full range of goods and services. For example, the services range widely from environmental processes, such as retention of the ecosystem's capacity to regulate streamflow and minimize erosional losses of nutrients and soil, to aesthetic and inspirational "services." Also implicit is the ability, either currently or at some future date, to provide habitat for the array of organisms native to the site and, of course, the continuing capacity to provide the same quantity and quality of products for human consumption. The requirement for sustainability may be met—on a specific site—by a variety of ecosystem conditions designed with different emphases, from commodities to amenities.

The basis for sustainability lies in maintaining the physical and biological elements of productivity. I suggest two guiding principles:

1. preventing the degradation of the productive capacity of our lands and waters—no net loss of productivity; and
2. preventing accelerated loss of genetic diversity (including species), recognizing that evolutionary processes will result in changes—no accelerated loss of genetic potential.

I make three observations about these principles. First, each has both an ecological and an ethical basis; that is, although they are human constructs, they can be objectively defined in ecological terms. Second, the principle of maintaining genetic potential is probably most fundamental because the physical elements of productivity can sometimes be restored to degraded ecosystems, whereas we have only very limited capacity to restore lost genetic potential. Third, these principles cannot be absolute or inviolable. There will be times when rational, even ecologically sensitive, human beings will knowingly violate a principle. Nature herself—through the process of evolution—can dictate the loss of existing genetic potential, including species, regardless of our efforts. Nevertheless, I think that the principles are a useful guide for sustainable management of natural resources—ecosystem management.

Haven't we already been providing for sustainability in management of forest and related resources? If you thought that existing management consistently passed the sustainability test, guess again. First, almost nowhere do we have good long-term assessments of how sustainable our current practices are; data are not sufficient to provide definitive answers, and numerous alternative hypotheses utilizing empirical and simulated evidence are offered up in the debate.

In some cases, we do know that proposed practices are not sustainable for at least some resources. For example, several plans approved in the early 1990s for national forests in the Pacific Northwest acknowledge that under the selected alternative there will continue to be further degradation in the quality of aquatic habitat. The threat to viable fish populations was a primary reason why most current forest plans rated so poorly in the assessment by the Scientific Panel for Late Successional Forest Ecosystems (Johnson et al. 1991).

Plans for management of the federal timberlands in the Pacific Northwest from the 1950s through mid-1980s provide another example. These plans called for virtual elimination of all of the remaining late-successional forest on federal lands outside of parks and wilderness preserves; only specimen groves and a few small reserves, such as Research Natural Areas, were to remain. Attainment of this objective would clearly have resulted in extirpation of much of the genetic diversity which we now know is associated with late-successional forests, possibly eliminating hundreds of species and reducing the genetic variability of hundreds of others. Whether we still have the resource base, management skills, or societal commitment to fulfill sustainability of either the aquatic or late-successional ecosystems is a major element in the debate over the Forest Ecosystem Management Assessment Team (1993) report. It would appear that we have not been consistently fulfilling either of the sustainability principles in current forest management activities.

The proposition that you do not begin ecosystem management by stipulating outputs is a corollary of basing it in sustainability. If sustainability is the basic premise or goal of ecosystem management, you do not begin by stipulating some level of output whether it is wood, deer, fish, or recreational visitor days. The capacity of the ecosystem determines the output levels that are consistent with sustainability. Allowable cuts are not appropriate inputs into the design of an ecosystem management plan, nor are they appropriate constraints on a plan that will meet the test of sustainability. It is equally clear that ecosystem management should not be funded primarily through timber sale programs, as has been attempted on the federal timberlands throughout the United States.

Again, if you think that we have consistently respected this corollary, we have not. The government of the Province of British Columbia indicated that it wants sustainable, state-of-the-art management in the Clayoquot Sound region on the west coast of Vancouver Island. However, simultaneously the government mandated an annual timber yield of 600,000 cubic m of wood.

In sum, sustainability is the underlying premise or goal of ecosystem management; the physical and biological elements of productivity must be maintained, and products (goods and services) must be outputs from management, not inputs or constraints. Finally, I would note that others have proposed that ecosystem management explicitly involves acceptance of an alternative philosophical viewpoint regarding environmental values (e.g., Grumbine 1994: 34) and people's relation to nature; however, this point-of-view is not widely accepted among participants in the ecosystem management dialogue.

ELEMENTS OF ECOSYSTEM MANAGEMENT

If sustainability is the strategic goal of ecosystem management, what are some of the tactical elements that are an overall part of this strategy? Let me begin by listing and providing a perspective on some of the tactical elements that are often identified with ecosystem management:

Ecosystem management is not about actively or intensively manipulating all locations in the landscape—harvesting or silviculturally manipulating every forest stand within a landscape, for example.

Ecosystem management is not just about ecological structure and functioning at larger spatial scales, but it does incorporate the science and perspectives of landscape ecology.

Ecosystem management is not just about alternative silvicultural practices or other stand- or stream reach–level manipulations, but it does consider the ecological consequences of conditions and manipulations in stands and stream reaches.

Ecosystem management is not just about technology, although it uses such modern technologies as geographic information systems (GIS).

Ecosystem management is most definitely not just about individual species, but it does incorporate species and their viability and functional roles as part of a comprehensive view of the ecosystem and what we want to achieve as a society.

Ecosystem management does address a full range of spatial scales. For example, strategies to conserve biological diversity go from the level of small to large (fig. 2.1). In ecosystem management scales are selected that are appropriate for the particular set of processes or species of concern; for example, we will not conserve all elements of biological diversity everywhere but will achieve various goals at various scales. Ecosystem management does incorporate both habitat "reserves" and a "matrix" within which more active management for goods and services occurs.

Furthermore, ecosystem management does integrate ecological concepts at a variety of spatial scales. For example, it recognizes the interactions between stands and landscapes, as in the case of connectivity (Harris 1984, Thomas et al. 1990, Franklin 1993). Ecosystem management does take an integrated view of terrestrial and aquatic ecosystems, a common example being the linkage of a forest with associated streams and rivers; however, it also includes interactions between continental and marine ecosystems, such as impacts of terrestrial activities on coral reef ecosystems.

Ecosystem management does take a very long temporal perspective and recognizes the dynamics of ecosystems, including that associated with global change.

Ecosystem management does vary in its application with landowners. Ecosystem management does not mean the same prescription everywhere, only that the prescriptions must meet the principle of sustainability.

Ecosystem management *is* evolutionary. It is based upon the concepts of adaptive management. Practitioners of ecosystem management recognize that *all* management decisions are based upon limited information, with significant degrees of uncertainty as to outcomes and, further, that new knowledge will continue to accumulate, altering basic assumptions and modifying predicted outcomes. Ecosystem management recognizes that *all* management prescriptions are, effectively, working hypotheses, with substantial levels of uncertainty regarding the outcomes. Recognizing this, ecosystem managers design their activities so as to contribute to the learning process—managing to learn and learning from management. This may involve a highly formalized process (e.g., Holling 1978, Walters 1986) or less formal approaches. In all cases, a scheme that provides for the systematic collection of information and its feedback into the decision-making process is required.

Scientifically designed, socially credible, sustained monitoring is an essential part of the management program to objectively measure the effectiveness of the management programs in achieving the desired goals—along with clear mechanisms for feedback to and adjustments in management practices. This is an immense challenge for both management agencies and society; credible

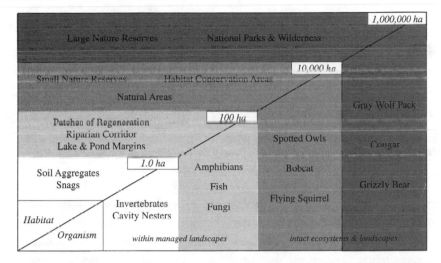

Fig. 2.1. Habitat protection involves a broad range of spatial scales, depending upon the species or process of interest, not simply provision of large natural preserves as is sometimes proposed. Maintenance of intact ecosystem conditions and provision of habitat may involve very large habitat preserves or, at least, areas managed in ways that are consistent with the needs of specific organisms or processes. The habitat needs of many other organisms can and must be met at much smaller spatial scales within managed landscapes.

monitoring programs are expensive, which is one reason why they are so rarely encountered in resource management. Technical choices of what and where to monitor to obtain efficient and effective assessments are also difficult. Nevertheless, recognition of the hypothetical nature of our management activities and adoption of adaptive management must make monitoring a major part of future management programs.

Does ecosystem management incorporate the human element? It most assuredly does. I will make only a few comments here, for exploring the social elements of ecosystem management would require another entire volume. Beginning with the most basic, "goods and services" are human constructs; we define the objectives of ecosystem management in terms of our own needs. Beyond this fundamental principle, ecosystem management must serve and inform society. Local communities and expertise can contribute in many ways, including as major elements in stewardship—the management and monitoring programs. Who better to carry out designed monitoring programs and ecosystem restoration than individuals who live with and depend upon the resource? Appropriate educational and training programs are essential, of course, but this brings additional benefits to the local communities. Such concepts illustrate how ecosystem management can substantially assist in the creation and maintenance of viable rural societies as a source of both (in the broadest possible senses) education and employment. For example, ecosystem management can help assure that rural populations are technologically current and well educated. Residents may be among the best possible candidates to assist in assessing the effectiveness of ecosystem management.

I would like to consider several of these elements in greater detail: ecosystem management as an activity involving multiple spatial scales and the importance of adaptive management.

ECOSYSTEM MANAGEMENT AT MULTIPLE SPATIAL SCALES

Ecosystem management incorporates concepts, plans, and activities at a full range of spatial scales—from stands and stream reaches to landscapes and larger drainage basins. It must also consider the interactions between patch-level conditions and processes and those at larger spatial scales, such as the effect of structural conditions in a landscape matrix on movement of organisms—that is, on landscape connectivity.

Silviculture at the Stand Level

Stand-level silvicultural prescriptions in ecosystem management recognize the importance of structural complexity in maintaining ecosystem function, including provision of a variety of habitats or niches (Franklin 1992, Hansen et al. 1991).

The functional role of the dead tree and its structural derivatives (logs and other coarse woody debris) often receives special recognition in managing forests as ecosystems because of the importance of these structures (e.g., Franklin et al. 1989, Harmon et al. 1986, Maser et al. 1988, Trappe and Maser 1984). The consideration of snags, logs, and other coarse woody debris often distinguishes ecosystem management from traditional silvicultural prescriptions. However, many foresters appear not yet to recognize or acknowledge the ecological importance of a dead tree, perhaps because it is so foreign to traditional forestry training. As a specific example, traditional silvicultural and mensurational textbooks contain little or nothing about the function of dead trees, let alone management practices which incorporate the dead tree and its derivative structures. A recent and extensively utilized textbook entitled *Forest Stand Dynamics* (Oliver and Larson 1990) is devoted almost exclusively to the live tree as though it were the only structural element in a forest stand.

Acknowledging the roles of dead trees requires drastic revisions in basic silvicultural premises, and this is probably a more serious problem than the lack of exposure in educational programs and textbooks. Forest management is simplified by assumptions that the dead tree has limited ecological value or that only very small numbers or small sizes of such structures are required to achieve management objectives. Such assumptions can be very misleading, however; for example, they can lead to equating ecological conditions after clearcutting and after wildfire. Such assumptions also encourage managers to believe that essentially all of the large wood can be removed with limited ecological impact. A recent *Journal of Forestry* article by four American silviculturalists (O'Hara et al. 1994)—which did acknowledge the fundamental differences between a clearcut and wildfire—identified the landscape perspective as the major change affecting silviculture; I think that the newly recognized ecological roles of the dead tree are at least as important in challenging traditional silvicultural practices. Incorporating dead wood into forest management is a challenge that is still not fully recognized or accepted by many foresters.

At the stand level, silviculture under ecosystem management emphasizes prescriptions that will retain and/or quickly restore structural complexity (Franklin 1992). Recognition of the role of biological legacies—organisms and organic structures carried over from a predisturbance to a postdisturbance ecosystem—is an important contribution by studies of natural ecosystem recovery following catastrophic disturbances (Franklin et al. 1986, Christensen et al. 1989). Silvicultural approaches directed toward structural complexity invest some of the merchantable wood back into the site to achieve ecological objectives. Many silvicultural variants are possible, most of them involving retention of at least some of the green trees from the harvested stand for incorporation into the regenerating stand (see fig. 2.2). For example, varying quantities of green trees can be retained and retention can occur as either dispersed individual structures or in aggregates (small patches within the harvested unit).

A "variable retention" forest harvest system (utilizing the entire retention gradient, fig. 2.2) could be completely substituted for traditional regeneration harvest systems and terminology. Such a shift could free foresters from the constraints of these formalized but limiting systems while greatly improving our ability to communicate what we are doing and why. For example, it does not facilitate understanding to describe a harvested stand with twenty-five large trees per acre as a "clearcut with reserves" because the residual trees were not retained for regeneration purposes; the harvested area is obviously not a clearcut by any nontechnical understanding of the word. Using a variable retention system, harvest prescriptions would be designed to meet a set

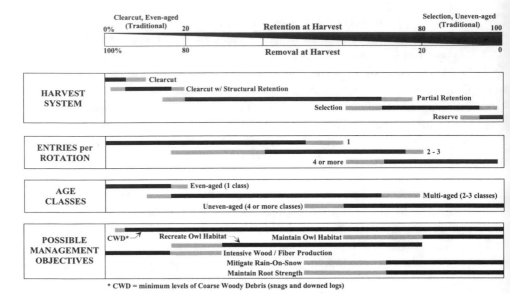

Fig. 2.2. Forest regeneration harvest systems should span the gradient of structural retention (or, conversely, the gradient of removal). Traditional regeneration harvest systems have focused on the extremes of little of no retention (clearcutting) and, less often, on low levels of removal (selection cutting). An array of regeneration harvest prescriptions exists between these extremes.

of defined management objectives for the stand in question. These prescriptions can be simply and clearly described by stipulating the (1) kind and (2) density of structures to be retained, as well as (3) the spatial pattern for retention—whether, for example, they are to be evenly dispersed over the harvest unit or retained in small forest aggregates or patches.

Regardless of terminology, silvicultural approaches under ecosystem management recognize the need to maintain much of the biological diversity on the harvested site in order to preserve its productive capacity (Franklin 1993). Examples include many elements of the belowground community, such as a diversity of fungi capable of forming mycorrhizae. Such diversity can be critical for optimal, as well as long-term, productivity of the site. Such retaining structures as live green trees and down logs are often necessary to achieve the desired in situ survival of affected organisms and processes.

Silvicultural approaches under ecosystem management also recognize the importance of intensive land management in overall landscape function—the role of the matrix, for example, in controlling connectivity and the movement of organisms and materials across the landscape or in regulating such hydrologic regimes as the frequency and severity of flooding.

Landscape-Level Management

As important as stand-level manipulations are, larger spatial scales—typically referred to as landscape scales—are essential. It is necessary to view individual stands or stream reaches in a broader spatial context so that we can assess the collective effects of our activities and plan management at appropriate spatial scales. This involves integrated landscape-level planning. Issues which are dealt with at these larger scales include riparian protection, identification and management of habitat reserves, and consideration of the relative merit of dispersing or aggregating management activities and impacts in space and time.

A riparian-based strategy is often an appropriate centerpiece or starting point for ecosystem management in many heavily forested regions that are situated on steep mountain topography with a high density of streams and rivers and that support important fisheries. For example, in places like the northwestern coast of North America, an uncompromising commitment to maintaining the integrity of the freshwater ecosystems might well achieve the majority of our objectives for terrestrial and upland habitats as well. FEMAT (1993) provides one example of such a strategy and an assessment of its potential effectiveness.

A riparian-based management strategy involves much more than just protection of the riparian zones, although this is clearly an important element. It also involves such issues as the transportation and logging systems, particularly the design, density, and permanence of roads, the greatest single source of human impact on forested streams. The choice of transportation systems affects, in turn, the choice of logging technologies, possibly increasing the emphasis on aerial and long-span cable systems. Transportation systems also constitute an important factor influencing the spatial pattern of management activities.

Ecosystem management includes the identification and protection of "habitat reserves." We will not manipulate all of the landscape in most cases. Reserves—areas involving low levels of human manipulation, at least for significant time periods—will almost certainly be necessary for a variety of reasons. For example, they are necessary to protect scarce, high-quality habitat for particular organisms or functions. Currently, provision of late-successional, including old-growth, forests is of particular concern because of the rapid decline in habitat of this type. Ecosystem management does recognize, however, that maintenance of diversity cannot be accomplished simply by set-asides or reserves; the need to maintain many elements of biological diversity in the matrix is also recognized. Furthermore, ecosystem management also recognizes the dynamic nature of ecosystems and the need to plan for replacement or refreshment of "reserved" habitats over the longer term.

There are many difficult and important issues regarding ecosystem management at the landscape level that we do not fully understand. For example, we cannot describe with certainly the most effective management strategies to achieve high levels of connectivity in a landscape. What are the relative merits of investments in corridors versus modifications of matrix management? Another issue is that of appropriate patch sizes and intensity of edge effects (or, conversely, the importance of interior forest conditions) for various organisms and processes. The question of dispersing versus concentrating management activities in time and space—especially forest harvest—is an important issue (Franklin and Forman 1987). Past management has generally opted for dispersing activities throughout a forest property requiring extensive permanent road systems and producing chronically elevated levels of human disturbance to aquatic systems. Concentrating management in portions of the landscape for periods of time might better fit several objectives and better match the intense but infrequent pattern of disturbance characteristic of many stream ecosystems.

Resource managers in the lake states have been leaders in developing and applying landscape-level approaches in forest management. Scientific studies are providing the critical underpinning. Important examples include the research on presettlement and current forest patch patterns (Mladenoff et al. 1993) and the application of this information in managing old-growth forests (Mladenoff et al. 1994), as well as studies of the impacts of forest edges on deer populations and plant populations (e.g., Mladenoff and Stearns 1993). The Chequamegon National Forest in Wisconsin has been one of the leaders in putting such information into practice in developing management practices which utilize appropriate spatial scales. This includes the use of larger patch sizes for late-successional and other biological reserves and concentrating timber harvest activities in time and space rather than dispersing them (e.g., USDA Forest Service Eastern Region 1991).

ADAPTIVE MANAGEMENT: MANAGEMENT ACTIVITIES AS WORKING HYPOTHESES

Ecosystem management is—and always should be—evolutionary in nature rather than a static set of prescriptions. Ecosystem managers recognize that their prescriptions and plans are, in fact, working hypotheses, based on limited information and with outcomes predicted but not assured. Hence, ecosystem management incorporates the philosophy, if not the formality, of adaptive management (Walters and Holling 1991, Holling 1986, Walters 1986). One major component of that adaptive management approach is the development and implementation of credible monitoring programs, something that has rarely even been attempted, let alone achieved, by management agencies.

I would like to briefly explore this notion of management activities as working hypotheses because of its importance as a philosophical underpinning for ecosystem management. Professional myths are always a major impediment to incorporation of new knowledge in management activities. One of the greatest myths, shared by both managers and scientists/academics, concerns the way that scientific information is incorporated into resource management.

A part of this myth is the common notion that science is gradually incorporated into resource management through a systematic process. Supposedly, the sequence of steps begins with creation of new knowledge, construction of hypotheses, and experimental testing of those hypotheses. After a series of such experiments with associated revision of hypotheses, proposals for management application are developed and subject to small-scale testing (pilot trials). Eventually, after the body of knowledge is fully developed and appropriately extrapolated, extensive adoption of the new management practices occurs.

Incorporation of new scientific information into resource management may sometimes occur this way, especially when it involves incremental modifications of generally accepted practices. But it does not work that way during major paradigm shifts, when basic assumptions and traditional concepts are being challenged by new goals, information, and conditions. Paradigm shifts in resource management are driven by shifts in societal goals, including economic factors, agency and stakeholder agendas, and only to a limited degree by new knowledge. New knowledge can certainly contribute to pressure for paradigm shifts but is probably insufficient in itself; I would judge that to be the case with the old-growth forests of the Pacific Northwest, for example.

When forced to make changes we—professional resource managers and society at large—have always had to operate, to make decisions, with inadequate information. We resource managers have never known as much as we thought, nor as much as we have led the public to believe, about our natural resources and how they will respond to a proposed management treatment. But lack of knowledge has never prevented us from moving ahead in making decisions and implementing management programs.

Therefore, the idealized systematic approach to the incorporation of scientific knowledge into major management decisions may occur only rarely. Certainly the adoption of the dispersed patch clearcutting system for harvest of Douglas fir forests in the Pacific Northwest in the 1940s offers an important example which did not follow the idealized concept.

Dispersed Patch Clearcutting: A Learning Experience

The "staggered-setting system of clearcutting," as the U.S. Forest Service termed it, dispersed clearcuts of some 10 to 25 ha through a forested matrix (Smith 1986). It was believed that this system would provide for prompt natural regeneration of Douglas fir and associated conifers and optimal conditions for growth of the trees. Further, it both justified and provided the financial basis for rapid development of road systems. Once this approach to harvest was adopted by the agency, all others, including any systems involving retention (for example, shelterwood) or selection, were excluded from consideration.

Was the dispersed patch clearcutting system a proven system at the time it was adopted? No, it was not; in fact, there had been no experiments or tests of any kind with this system of management. Dispersed patch clearcutting was a working hypothesis. A number of factors—societal and agency agendas, logging technology, and the current state of scientific knowledge—were responsible for the expansion of timber harvest on the national forests of the Pacific Northwest in the 1940s and for the adoption of the dispersed patch clearcutting system. On society's agenda was the need to provide for large amounts of inexpensive wood to feed the expanding post–World War II economy.

The U.S. Forest Service had several important agency objectives. One motive was the development of the timber growing potential of the national forests, beginning with conversion of the "decadent" old-growth stands to rapidly growing young forests (Clary 1986, Hirt 1994). Another important objective was to develop access to the largely inaccessible national forestlands so that they could be protected and managed; dispersing the cutovers would provide the basis for a much more rapid expansion of the road system than concentrating initial management activities.

The Forest Service also wanted to differentiate its management approach from approaches being used on industrial forestlands. The agency had been campaigning since Gifford Pinchot took over in 1905 for authority to regulate timber cutting on private and state lands—that is, to become the sole regulatory authority over timber harvest in the United States. Obviously, harvest practices on national forestlands had to look significantly different from those on the large private industrial forestlands in the Douglas fir region, so large continuous clearcuts—which would also have reduced the rate of road development—were not appropriate.

An additional Forest Service concern was to avoid the application of single-tree selection harvest and uneven-aged management in Douglas fir forests. The development of the Caterpillar tractor and log truck as effective harvest equipment had led to serious proposals in the 1930s for single-tree selection in old-growth Douglas fir forests. Such proposals had advocates even within the agency. However, many scientists and managers in the Forest Service—including such leaders as Thornton T. Munger and Leo A. Isaac—were opposed to this approach on the basis that it was not appropriate to the massive forests and to an intolerant crop tree like Douglas fir (e.g., Isaac 1943). Opponents of selection harvest believed that it would result in "high grading"—removal of the most valuable trees—and deterioration in the growth and composition of the stands. Hence, any management approach had to involve even-aged harvest systems, with clearcutting preferred.

Biological knowledge was, of course, a significant consideration in the selection of the harvest cutting system. Any system had to be consistent with the available ecological knowledge about Douglas fir, including its regeneration and growth habits (Isaac 1943). However, contemporary knowledge of Douglas fir silvics actually provided a rather broad biological envelope within which many management approaches, including dispersed patch clearcutting, were possible (Isaac 1943).

So the Forest Service adopted the dispersed patch clearcutting system as the approach that fit the biological envelope and fulfilled the agency's political and managerial agendas. As noted, it was implemented extensively as the only approved approach to timber harvest in the Douglas fir region. No regional or local site variations were anticipated. No experimental trials were conducted. No monitoring program was established.

Dispersed patch clearcutting did not work out exactly as hypothesized by the managers and scientists. It took more than twenty-five years for the agency's foresters to figure out, largely by trial and error, how to attain dependable prompt tree regeneration on cutover areas. It was about ten years before it became clear that natural regeneration could not be depended upon. Good seed crops of Douglas fir were too infrequent, and environmental conditions on the cutover areas were often so severe that young seedlings succumbed to heat and drought (Silen 1960). The emphasis shifted to artificial regeneration, but, again, it took more than a decade to develop the techniques—including provision of healthy planting stock—to achieve a high level of dependability.

Agency foresters also learned that some sites in the Douglas fir region could not be regenerated following clearcutting, even with the most intensive planting efforts. Exemplary problem sites were areas on gentle topography at moderate to high elevations, where intense frosts defeated regeneration efforts, or steep southerly exposed slopes in southwestern Oregon, where heat and drought were severe problems. Eventually, in the 1970s, shelterwood harvest systems were adopted for many of these difficult sites. In the Douglas fir region we still live with a legacy of nonstocked and understocked cutovers from the 1950s and 1960s that were a part of that learning experience.

To recapitulate, a harvest cutting system was devised, based on societal and agency objectives and the best available silvical knowledge, and was extensively implemented without trial. Problems with the system were identified by a crude monitoring program (regeneration surveys on cutovers) and limited (but very relevant) research projects. Adjustments occurred gradually over a thirty-year period and were confined almost entirely to regeneration; there was essentially no assessment of the impact of dispersed-patch clearcutting on other resource values until the 1980s. The system was a working hypothesis, an ad hoc approach to adaptive management, but it was never clearly recognized as such by either the resource managers or the public.

Management prescriptions as working hypotheses have always been the situation in resource management and will remain so in the future. The fact is that we do not know, will not know, cannot know, everything that we would like to know and maybe should know before we move ahead. But this cannot and will not deter us from making those decisions and proceeding.

However, the lack of knowledge is not a reason to continue using approaches that clearly do not achieve our societal and agency objectives, although that is what many foresters have proposed that we should do. Nor is it a reason to stop everything until we have experimentally proven that new approaches will work, which is what many environmentalists and academics would have us do.

As before, faced with a paradigm shift in forest resource management, we must (1) take our best current scientific information, (2) craft some new approaches that offer an improved likelihood of achieving our new objectives, (3) implement those approaches with a very strong dose of humility, and (4) *learn* through a systematic process of monitoring and research. This is what adaptive management is about—design, implement, monitor, learn, and adjust.

Ecosystem management must be based upon adaptive management principles so that it can evolve. And we obviously can manage in ways that will facilitate that learning process.

APPLICATIONS OF ECOSYSTEM MANAGEMENT IN THE PACIFIC NORTHWEST

We have been going through a very rapid learning process regarding ecosystem management in the forests of the Pacific Northwest. This is focused primarily upon concerns about: (1) late-successional (including old-growth) forests and habitats and the associated species; and (2) riparian habitats and organisms, particularly anadromous fish.

Some important elements in the learning process on federal timberlands have included:

- the National Forest Management Act (NFMA) and associated forest planning and biodiversity requirements;
- the Interagency Scientific Committee for the Recovery of the Northern Spotted Owl (ISC, sometimes known as the Thomas Committee; Thomas et al. 1991);
- the Scientific Panel on Late-Successional Forests (sometimes known as the Gang of Four; Johnson et al. 1992);
- the Scientific Analysis Team or SAT (1993); and
- the Forest Ecosystem Management Assessment Team (FEMAT) and its report (FEMAT 1993).

Although I am using the Pacific Northwest as my major example of the evolution of ecosystem management, the reader should be aware that similar developments are occurring in forested regions throughout most of the temperate and much of the boreal forest world. For example, Sweden, which has perhaps represented a national commitment to traditional tree farming, has undertaken major legal and technical changes in forest policy in order to maintain a broader array of ecological values (Sweden National Board of Forestry 1992). At the opposite side of the globe, major changes in policy and practice have occurred in Australia as a result of concerns over forest-related biological diversity (see, e.g., Lunney 1991).

The National Forest Management Act

NFMA has been a major impetus to development of ecosystem management on federal timberlands in the Pacific Northwest, primarily by (1) requiring development of comprehensive national forest plans and (2) mandating the maintenance of biological diversity. The planning requirement directed each national forest to develop plans that incorporated interdisciplinary approaches and public involvement. Criticisms of the planning process have been numerous and often justified; putting it into action has taken far more time and money than anyone expected.

Development of interdisciplinary perspectives was one major contribution from the planning process. For the first time, the Forest Service began to address all resources on the same systematic basis that had previously governed only timber management planning. However uneven, the development and use of interdisciplinary teams gave a regular voice to such other resource values as soils, wildlife, and water. Factual information on all resources and effects of various alternatives were reported much more systematically, sometimes even when this information did not support the selected alternative. Inclusion of balanced and complete information was further stimulated by stakeholder groups that became ever more proficient in identifying errors of omission and commission in the draft plans. Even some of the failures in the planning process have been extraordinarily instructive—most notably, recognition of the need for spatial information in both planning and implementing activities.

Development of an interdisciplinary workforce within the Forest Service was a necessary requirement for fulfilling the planning mandate. It significantly altered both the professional and gender mix within the agency, moving it from an organization dominated by white male foresters and engineers to one with a broad mix of professional disciplines—and varied professional and personal perspectives, as well.

The task of implementing ecosystem management on the federal timberlands would be much more difficult without the previous development of the interdisciplinary workforce, the experience of the planning efforts, and the associated exponential expansion in knowledge and analytic capabilities.

The legal commitment to biological diversity provided by NFMA and associated regulations has also been an important element in shifting management paradigms for the national forests (see, e.g., Caldwell et al. 1994). Regulations promulgated under NFMA in 1979 and 1982 provide a high standard for national forest lands; the 1979 regulations, for example, require that "fish and wildlife habitats will be managed to maintain viable populations of all existing native vertebrate species in the planning area" (Caldwell et al. 1994).

NFMA and other environmental laws, notably the National Environmental Policy and the Endangered Species Act, have also been responsible for providing a central role for scientists in the development and implementation of forest policy. Legal challenges to agency plans and proposals, based on these laws, ultimately produced a gridlock in which agencies had lost technical and scientific credibility with the courts and a majority of Congress (Yaffee 1994). This led the agencies, Congress, courts, and, ultimately, the White House to turn to independent scientific bodies to develop scientifically credible plans and policy alternatives (Franklin 1996).

The Interagency Scientific Committee

The Interagency Scientific Committee to Address the Conservation of the Northern Spotted Owl (ISC) was created by four federal agencies to develop a scientifically credible conservation strategy for the northern spotted owl (Thomas et al. 1990). This effort was unique in many respects, including its exclusive use of a scientific team and scientific criteria in addressing a major forest conservation issue. The committee broke important new scientific ground in its proposals on (1) size and distribution of habitat reserves and (2) modified treatment of the matrix.

The conservation strategy developed by the ISC called for protection of large blocks of suitable habitat, called habitat conservation areas (HCAs), in place of earlier proposals for a system of one- to three-pair spotted owl habitat areas (Thomas et al. 1990). Proposals for such large reserves were actually part of the earlier dialogue on owl conservation strategies (Yaffee 1994), but they were not widely circulated. The HCAs proposed by ISC, based upon empirical and modeled data, were delineated and mapped so as to include a minimum of twenty pairs of owls in each, with a maximum distance between HCAs of twelve miles. More than 2 million ha of federal forest was included in the HCAs, and ISC recommended no harvest in those areas, at least during an interim period.

Modified management of the forest matrix between HCAs was the second major innovative element of ISC. The issue was how to facilitate dispersal of northern spotted owls among HCAs; mortality of dispersing birds is typically very high in clearcut areas. Corridors between the forest reserves were considered and rejected in favor of a matrix management strategy. Specifically, the ISC recommended that at least 50% of the federal forest landbase outside HCAs be maintained in stands of timber with an average diameter at breast height of 11 inches or greater and at least 40% canopy cover; this recommendation is popularly known as the 50–11–40 rule.

The recommendations for the reserve and matrix strategy by ISC represented major intellectual and practical contributions to the problem of maintaining biological diversity in Pacific Northwest forests. It became the primary basis for the recovery plan developed by the U.S. Fish and Wildlife Service for the northern spotted owl. However, ISC was directed to address only the northern spotted owl, not to develop strategies for protection of late-successional forests and associated species or for protection of aquatic habitats and fish.

Scientific Panel on Late-Successional Forest Ecosystems

In May 1991 the Scientific Panel on Late-Successional Forest Ecosystems was established by several committees of the U.S. House of Representatives to provide objective and comprehensive information concerning the federal timberlands within the range of the northern spotted owl (Johnson et al. 1991). The charges to the panel included (1) mapping and ecologically grading all of the late-successional and old-growth (LS/OG) forests, (2) developing a range of alternatives for protection of LS/OG ecosystems and associated organisms—including northern spotted owl, marbled murrelet, and sensitive fish stocks—and (3) conducting a biological risk analysis and economic evaluation of each alternative.

The scientific panel completed its activity and submitted its final report within six months. Fourteen major alternatives involving various levels of LS/OG preservation and—for ten of the alternatives—three levels of matrix management were presented and analyzed. The analyses included biological risk ratings (each assessing the probability that a specific organism or ecosystem will survive to the next century) and calculations of sustainable allowable timber harvest and associated employment and income in the region (Johnson et al. 1991).

The recommendations of the scientific panel (often referred to as the Gang of Four or the Gang of Four Plus Two, to acknowledge the critical role of two aquatic biologists, James R. Sedell and Gordon H. Reeves) were partially built on ISC but went beyond it in developing a regional strategy for conserving biological diversity. As in the case of the ISC report, the panel took a regional approach in developing alternative LS/OG reserve systems. Large blocks involving thousands of hectares were identified rather than a fine-scale pattern of reserves involving tens or hundreds of hectares.

The Gang of Four took a multiple-species, ecosystem approach—a significant advance over ISC. The scientific panel addressed all late-successional forest ecosystems and all of the organisms believed to be associated with them, including habitat for sensitive fish stocks. A comprehensive plan of this type allowed for much greater efficiency than is possible when species and issues are addressed on an individual basis, as was clearly evident in an analysis of the alternatives (Johnson et al. 1991). For example, reserve systems were developed with the multiple objectives of providing for viability

of the late-successional forest ecosystem, including several hundred constituent species, and of aquatic ecosystems, not just of habitat for northern spotted owls and marbled murrelets, although viable populations of these species
were a consideration.

The recommendations of the scientific panel also recognized the critical
role of the forest matrix, the lands surrounding any system of LS/OG
reserves. As with the ISC, altering management of the matrix was viewed
as a strategy superior to delineation of habitat corridors in providing for
both movement of organisms between reserves and for conservation of organisms and processes within the matrix. The 50–11–40 rule of ISC was
viewed as inadequate for the larger, multiple-species objectives of the
scientific panel, however; for example, ISC did not require maintenance of
large live trees, large snags, and large downed logs in the matrix. Hence,
the scientific panel developed a minimal matrix management recommendation that prescribed both 50–11–40 and retention of an average of at least
fifteen large live trees, five large snags, and five large downed logs on each
hectare.

The incorporation of aquatic ecosystems and sensitive fish stocks in the
Gang of Four exercise was a major step toward ecosystem management.
Integration of terrestrial and aquatic objectives had never previously been
attempted on this scale. This initial effort became the basis for greatly expanded approaches to conservation of aquatic ecosystems and fisheries by
the Scientific Analysis Team (1993) and its current culmination in the FE-
MAT alternatives (FEMAT 1993).

The development, presentation, and assessment (both ecologic and economic) of multiple alternatives was an additional original contribution of the
scientific panel. Instead of providing a single plan, the panel furnished thirty-
four alternatives ranging from high levels of timber harvest (comparable to
those during the 1980s) to the protection of essentially all remaining LS/OG
forests. The ecological and economic costs and benefits of each alternative
were presented without a specified preference. Having presented the factual
information, the panel (appropriately) left it to societal institutions (specifically, Congress) to select among the alternatives.

Scientific Analysis Team

The U.S. Forest Service ultimately developed a plan for the northern spotted
owl based on ISC and submitted it to Judge William Dwyer. By this time
the ISC report (Thomas et al. 1990) was two years old. Additional demographic data had accumulated for the northern spotted owl, suggesting a faster

decline in populations than had been predicted. Furthermore, the Bureau of Land Management had rejected the ISC plan and adopted its own approach to management of owl habitat; the ISC strategy was based on joint adoption and implementation by both Forest Service and Bureau of Land Management. Finally, the Gang of Four report had raised issues about other old-growth-related species. Hence, Judge Dwyer returned the plan to the Forest Service with a series of questions related to these issues.

The Scientific Analysis Team (SAT), composed entirely of federal agency scientists, was convened to develop a response to Judge Dwyer's questions and issued its report in early 1993 (Scientific Analysis Team 1993). After considering the implications of the additional demographic data and BLM's deferral from adoption of the ISC strategy, the SAT identified and assessed the potential significance of the plan for 667 species that had a "high likelihood" of being associated with late-successional forests. SAT was the first exercise that systematically applied an expert panel approach in developing viability assessments for various groups of species, an approach that was refined and extensively applied in FEMAT (Forest Ecosystem Management Assessment Team).

Development of an aquatically oriented conservation strategy was probably the most important contribution of SAT, however. The SAT riparian proposals built on the efforts initiated in the Gang of Four exercise, but with substantial refinement and expansion. Most notable were dramatically expanded prescriptions for riparian buffer zones, proposed to extend across the flood plains of major streams, for example. Even more significant was the extension of riparian buffers to the smallest—but most extensive—elements of a drainage system: intermittent and perennial first-order streams. Since these smallest streams are the majority of the stream mileage in any riparian network, may have extremely high densities in wet coastal regions, and occur high on slopes (and even close to ridgetops), management implications of the SAT riparian proposals were immense.

The SAT aquatic conservation strategy also developed the concepts of critical watersheds and watershed restoration, initiated in the Gang of Four exercise. Also proposed was a formal process of watershed analysis, designed to provide for sophisticated geomorphic and hydrologic analysis of watersheds prior to designing site-specific riparian buffers.

The SAT report was largely overshadowed by developments associated with President Clinton's Forest Conference and the subsequent scientific analysis known as FEMAT. Nevertheless, it contributed important new information and analyses, experienced team members, and, particularly, an aquatic conservation strategy, to the FEMAT process.

Forest Ecosystem Management Assessment Team

The FEMAT activity is the most comprehensive scientific analysis of resource management and conservation issues attempted to date in the Pacific
Northwest (Forest Ecosystem Management Assessment Team 1993, Thomas
1994). This analysis involved more than six hundred scientists, managers,
and support personnel. It was adopted as the basis for an Environmental
Impact Statement (Interagency SEIS Team 1993) and, ultimately, for a record
of decision (Espy and Babbitt 1994) adopting a specific management plan
for federal timberlands within the range of the northern spotted owl. The
FEMAT process and conclusions are presented in considerable detail in the
April 1994 *Journal of Forestry,* along with critical commentaries representing
a diversity of viewpoints.

FEMAT contributed significantly to development of ecosystem management approaches as applied in the Pacific Northwest; indeed, it is seen by
many as a blueprint for initiating the evolutionary process known as ecosystem management. Major contributions include:

- development and adoption of a strong riparian-based conservation strategy;
- integrating terrestrial and aquatic and multiple-species objectives design of
late-successional reserve systems; and
- emphasis on adaptive management approaches, including identification of
specific locales (Adaptive Management Areas) as focal points for adaptive
management programs.

FEMAT is clearly the culmination of the series of scientific analyses which
began with the ISC effort in 1989.

FEMAT is designed primarily as an ecosystem-based strategy, incorporating both stand- and landscape-level perspectives in the analysis and recommendations (Franklin 1994). A major objective was the development of a
system of forest reserves that integrated aquatic and terrestrial objectives, as
well as one that provided for late-successional forest ecosystems and multiple
species. This was achieved in Option 9, the FEMAT alternative ultimately
adopted, in modified form, by President Clinton.

A strictly ecosystem-based approach was not possible, however, and this
is likely to be the case in other forest policy analyses, given the current legal
mandates (Franklin 1994, Raphael and Marcot 1994). Laws and regulations
require assessments for individual species, such as the northern spotted owl
and marbled murrelet. FEMAT did illustrate the difficulty of addressing multiple species, however, and identified the types of problems that can arise

when conservation plans are directed to a single species. For example, the system of Dedicated Conservation Areas (DCAs) developed (and nearly adopted as part of the recovery plan (USDI Fish and Wildlife Service 1992)) for the northern spotted owl did not provide for optimal representation of old-growth ecosystems, marbled murrelets, and watershed protection (FE-MAT 1993). On the Umpqua National Forest, the DCAs encompassed less than half of the best remaining old-growth forest ecosystems as identified in the Gang of Four report.

The FEMAT exercise also makes clear the immense difficulties associated with a species-based approach to conservation (Meslow et al. 1994, Raphael and Marcot 1994). Viability assessments were carried out for 82 terrestrial vertebrates, 21 groups of fish, 102 mollusks, 124 vascular plant species, 157 lichen species, 527 fungal species, 106 species of Bryophytes, and 15 functional groups of arthropods (involving 7,000 to 10,000 species). These assessments quickly exhausted the scientific knowledge base; the majority of species identified in the study were ultimately classed as being of unknown status with regard to endangerment and level of association with or dependence upon late-successional forests.

The program for conservation and restoration of aquatic ecosystems reached full development from its Gang of Four and SAT progenitors as a part of FEMAT (Sedell et al. 1994). It includes four elements: riparian reserves (based upon the SAT model); identification of, and more restrictive management procedures for, key watersheds; watershed analysis incorporating geomorphic and ecologic analyses; and watershed restoration.

Adaptive Management Areas (AMAs) are a key concept introduced for the first time in Option 9 of FEMAT. Identification of these areas provides explicit recognition of the importance of the adaptive management approach in ecosystem management. The AMAs are intended to provide locales where new approaches to technical, administrative, and social issues could be developed and tested. They represent a cross-section of the physical, biological, and social conditions within the FEMAT region. A specific objective is to develop alternative approaches to achieving the management objectives set by the plan. Two essential elements of the technical programs in AMAs are (1) scientifically credible experiments and (2) development and testing of major monitoring programs. Suggestions for funding the essential research, educational, and monitoring functions are also provided.

Criticism of FEMAT and the selected Option 9 have represented a wide variety of sociopolitical and scientific perspectives (see commentaries in April 1994 *Journal of Forestry* for a sampling). For example, the extensive system of late-successional reserves proposed under FEMAT has been criticized by some scientists and managers as unnecessary, potentially dangerous to the continued existence of these forests, and based upon an "equilibrium" view of ecosystems. Other scientists have argued that various ecological factors—such as continued population declines in northern spotted owls and lack of definitive information on other late successional species—require reservation of all remaining late-successional forests.

Following public comment and revision, Option 9 from FEMAT (1993) was incorporated into a final Environmental Impact Statement and Record of Decision (Espy and Babbitt 1994). Modifications of Option 9 were designed to meet environmental concerns with both aquatic environments and terrestrial species. Judge Dwyer has ruled that the plan does meet all legal tests and has given the Clinton administration permission to implement it.

With this judicial approval, the ongoing process of developing and applying ecosystem management in the forests of the Pacific Northwest has now returned to the field where resource managers are implementing the plan. Legal challenges do remain from both environmental and industry organizations. In addition, the 104th Congress has the potential to modify or eliminate the legal requirement for such a plan.

Ecosystem management is an emerging management paradigm that is holistic in its view of natural and human resources and has sustainability as its central premise. Since it is an integrative or synthetic approach, many concepts and scientific disciplines are significant contributors to ecosystem management. Among these are landscape ecology, including the "greater" ecosystem concepts, and ecosystem science, with its traditional focus on ecosystem function. Ecosystem management is not defined by such concepts, however. Ecosystem management does incorporate activities and their consequences at a full range of spatial scales, including much larger scales than managers have typically addressed. Ecosystem management does take an integrated approach to terrestrial and aquatic ecosystems. Ecosystem management is adaptive in character, recognizing the hypothetical nature of essentially all management decisions. Consequently, credible and focused research and monitoring programs, as well as mechanisms for information feedback to managers, are essential components of ecosystem management. Experiences with forest issues in the Pacific Northwest illustrate progressive efforts to shift from traditional economic and species-oriented programs to ecosystem management.

LITERATURE CITED

Agee, J. K., and D. R. Johnson (editors). 1988. Ecosystem management for parks and wilderness. University of Washington Press, Seattle, Washington.

Blair, W. F. 1977. Big biology: The US/IBP. US/IBP Synthesis Series 7. Dowden, Hutchinson & Ross, Stroudsburg, Pennsylvania.

Bormann, F. H., and G. E. Likens. 1979. Pattern and process in a forested ecosystem. Springer-Verlag, New York, New York.

Botkin, D. B. 1990. Discordant harmonies. Oxford University Press, New York, New York.

Caldwell, L. K., C. F. Wilkinson, and M. A. Shannon. 1994. Making ecosystem policy: three decades of change. Journal of Forestry 92(4):7–10.

Carroll, G. C. 1979. Forest canopies: complex and independent subsystems. Pages 87–108 *in* R. H. Waring, editor. Forests: fresh perspectives from ecosystem analysis. Oregon State University Press, Corvallis, Oregon.

Christensen, N. L., J. K. Agee, P. F. Brussard, et al. 1989. Interpreting the Yellowstone fires of 1988. BioScience 39:678–685.

Clary, D. A. 1986. Timber and the Forest Service. University of Kansas Press, Lawrence, Kansas.

Edmonds, R. L. 1982. Analysis of coniferous forest ecosystems in the western United States. US/IBP Synthesis Series 14. Hutchinson Ross Publishing, Stroudsburg, Pennsylvania.

Espy, M., and B. Babbitt. 1994. Record of decision for amendments to Forest Service and Bureau of Land Management planning documents within the range of the northern spotted owl. USDA Forest Service and USDI Bureau of Land Management, Portland, Oregon.

Forest Ecosystem Management Assessment Team. 1993. Forest ecosystem management: an ecological, economic, and social assessment. Report of the Forest Ecosystem Management Assessment Team. USDA Forest Service, Portland, Oregon.

Franklin, J. F. 1990. Biological legacies: a critical management concept from Mount St. Helens. Transactions of the North American Wildlife and Natural Resources Conference 55:216–219.

———. 1992. Scientific basis for new perspectives in forests and streams. Pages 25–72 *in* R. J. Naiman, editor. Watershed management: balancing sustainability and environmental change. Springer-Verlag, New York, New York.

———. 1993. The fundamentals of ecosystem management with applications in the Pacific Northwest. Pages 127–144 *in* G. H. Aplet, N. Johnson, J. T. Olson, and V. A. Sample, editors. Defining sustainable forestry. Island Press, Washington, D.C.

———. 1994. Ecological science. A conceptual basis for FEMAT. Journal of Forestry 92(4):21–23.

———. 1996. Scientists in wonderland: experiences in development of forest policy. BioScience (in press).

Franklin, J. F., and R. T. T. Forman. 1987. Creating landscape patterns by forest cutting: ecological consequences and principles. Landscape Ecology 1:5–18.

Franklin, J. F., P. M. Frenzen, and F. J. Swanson. 1988. Re-creation of ecosystems at Mount St. Helens: contrasts in artificial and natural approaches. Pages 1–37 *in* J. Cairns, Jr., editor. Rehabilitating damaged ecosystems. Volume 2. CRC Press, Boca Raton, Florida.

Franklin, J. F., H. H. Shugart, and M. E. Harmon. 1987. Tree death as an ecological process. BioScience 37:550–556.

Grumbine, R. E. 1994. What is ecosystem management? Conservation Biology 8:27–38.

Hansen, A. J., T. A. Spies, F. J. Swanson, and J. L. Ohmann. 1991. Conserving biodiversity in managed forests. BioScience 41:382–392.

Harmon, M. E., J. F. Franklin, F. J. Swanson, et al. 1986. Ecology of coarse woody debris in temperate ecosystems. Advances in Ecological Research 15:133–302.

Harris, L. D. 1984. The fragmented forest. University of Chicago Press, Chicago, Illinois.

Harris, W. G., D. Santantonio, and D. McGinty. 1979. The dynamic belowground ecosystem. Pages 119–129 *in* R. H. Waring, editor. Forests: fresh perspectives from ecosystem analysis. Oregon State University Press, Corvallis, Oregon.

Hirt, P. W. 1994. A conspiracy of optimism. University of Nebraska Press, Lincoln, Nebraska.

Holling, C. S. (editor). 1978. Adaptive environmental assessment and management. John Wiley and Sons, London.

Hopwood, D. 1991. Principles and practices of new forestry: a guide for British Columbians. British Columbia Ministry of Forests, Land Management Report 71: 1–95.

Interagency SEIS Team. 1993. Draft supplemental environmental impact statement on management of habitat for late-successional and old-growth related species within the range of the northern spotted owl. USDA Forest Service Pacific Northwest Region, Portland, Oregon.

Isaac, L. A. 1943. Reproductive habits of Douglas fir. Charles Lathrop Pack Forestry Foundation: Washington, D.C.

Johnson, N., J. Franklin, J. Gordon, and J. W. Thomas. 1991. Alternatives for management of late successional forests of the Pacific Northwest: a report to the Agriculture Committee and the Merchant Marine Committee of the U.S. House of Representatives. College of Forestry, Oregon State University, Corvallis, Oregon.

Likens, G. E. 1992. The ecosystem approach: its use and abuse. Ecology Institute, Oldendorf/Luhe, Germany.

——— (editor). 1985. An ecosystem approach to aquatic ecology and its environment: Mirror Lake. Springer-Verlag, New York, New York.

Likens, G. E., F. H. Bormann, R. S. Pierce, J. S. Eaton, and N. M. Johnson. 1977. Biogeochemistry of a forested ecosystem. Springer-Verlag, New York, New York.

Lunney, D. (editor). 1991. Conservation of Australia's forest fauna. Surrey Beatty & Sons: Chipping Norton, New South Wales.

Maser, C., R. F. Tarrant, J. M. Trappe, and J. F. Franklin. 1988. From the forest to the sea: a story of fallen trees. USDA Forest Service General Technical Report PNW-GTR-229.

Maser, C., and J. M. Trappe. 1984. The seen and unseen world of the fallen tree. USDA Forest Service General Technical Report PNW-164.

Meslow, E. C., R. S. Holthausen, and D. A. Cleaves. 1994. Assessment of terrestrial species and ecosystems. Journal of Forestry 92(4):24–27.

Mladenoff, D. J., and F. Stearns. 1993. Eastern hemlock regeneration and browsing in the northern Great Lakes Region: a re-examination and model simulation. Conservation Biology 7:889–900.

Mladenoff, D. J., M. A. White, T. R. Crow, and J. Pastor. 1994. Applying principles of landscape design and management to integrate old-growth forest enhancement and commodity use. Conservation Biology 8:752–762.

Mladenoff, D. J., M. A. White, J. Pastor, and T. R. Crow. 1993. Comparing spatial pattern in unaltered old-growth and disturbed forest landscapes for biodiversity design and management. Ecological Applications 3:293–305.

O'Hara, K. L., R. S. Seymour, S. D. Tesch, and J. M. Guldin. 1994. Silviculture and our changing profession: leadership for shifting paradigms. Journal of Forestry 92(1):8–13.

Oliver, C. D., and B. C. Larson. 1990. Forest stand dynamics. McGraw-Hill: New York, New York.

Raphael, M. G., and B. G. Marcot. 1994. Key questions and issues. Species and ecosystem viability. Journal of Forestry 92(4):45–47.

Salwasser, H. 1991. New perspectives for sustaining diversity in the U.S. National Forest ecosystems. Conservation Biology 5:567–569.

Scientific Analysis Team. 1993. Viability assessments and management considerations for species associated with late-successional and old-growth forests of the Pacific Northwest: the report of the Scientific Analysis Team. USDA Forest Service Pacific Northwest Region, Portland, Oregon.

Sedell, J. R., G. H. Reeves, and K. M. Burnett. 1994. Development and evaluation of aquatic conservation strategies. Journal of Forestry 92(4):28–31.

Silen, R. R. 1960. Lethal surface temperatures and their interpretation for Douglas fir. Ph.D. thesis. Oregon State University, Corvallis, Oregon.

Smith, D. M. 1986. The practice of silviculture. John Wiley & Sons: New York, New York.

Stanford, J. A., and J. V. Ward. 1992. Management of aquatic resources in large catchments: recognizing interactions between ecosystem connectivity and environmental disturbance. Pages 91–124 *in* R. J. Naiman, editor. Watershed management: balancing sustainability and environmental change. Springer-Verlag, New York, New York.

Sweden National Board of Forestry. 1992. A richer forest. National Board of Forestry, Jönköping.

Thomas, J. W. 1994. Forest ecosystem management assessment team. Objectives, process and options. Journal of Forestry 92(4):12–17, 19.

Thomas, J. W., E. D. Foreman, J. B. Lint, et al. 1991. A conservation strategy for the northern spotted owl. USDA Forest Service, Portland, Oregon.

USDA Forest Service Eastern Region. 1991. Final environmental impact statement for the Sunken Camp Area Management Area 351 Washburn Ranger District Chequamegon National Forest, Park Falls, Wisconsin.

USDI Fish and Wildlife Service. 1992. Recovery plan for the northern spotted owl. Volume 1. USDI Fish and Wildlife Service Pacific Region, Portland, Oregon.

Walters, C. J. 1986. Adaptive management of renewable resources. McGraw-Hill, New York, New York.

Walters, C. J., and C. S. Holling. 1990. Large-scale management experiments and learning by doing. Ecology 71:2060–2068.

Yaffee, S. L. 1994. The wisdom of the spotted owl. Policy lessons for a new century. Island Press, Washington, D.C.

Chapter 3 **Concepts and Methods of Ecosystem Management: Lessons from Landscape Ecology**
Thomas R. Crow and Eric J. Gustafson

In their book *Landscape Ecology: Theory and Application,* Naveh and Lieberman (1984) stated that the science of landscape ecology was almost unknown outside of Europe and that reference to this science was virtually absent in the North American literature at that time. In the years since their book was published, developments in landscape ecology have been both rapid and substantial.

Recent theoretical developments have emphasized how pattern and process are related and how changes in spatial scale affect this relation (Addicott et al. 1987, Brown and Allen 1989, de Bruin and Jacobs 1989, Wiens 1989, Lord and Norton 1990, O'Neill et al. 1991, Cornell and Lawton 1992, Fahrig 1992, Holling 1992, McLaughlin and Roughgarden 1992, Pahl-Wostl 1993). Considerable attention has been given to the nature of patch dynamics in landscapes and the effects of habitat fragmentation on biological diversity and population dynamics (Turner 1987, Lord and Norton 1990, May and Southwood 1990, Kolasa and Pickett 1991). Landscape patches are defined by boundaries or ecotones, and characteristics of these boundaries are of research interest because they represent a spatial and temporal discontinuity in structural properties and probably in functional properties as well (Forman and Moore 1992, Wiens 1992).

Many of the common issues confronting resource managers—forest fragmentation, loss of biological diversity, edge effects, integrated resource management, cumulative effects, old-growth forests, endangered species and endangered ecosystems, sustainable management—are in fact spatial problems or at least have some spatial component. For example, widespread concern about environmental quality in the United States resulted in the passage of the National Environmental Policy Act (NEPA) in 1970, and with it came the legal requirement to consider and evaluate the ecological consequences of such human activities as road building, pesticide use, and timber harvesting. To understand the cumulative effects of human activities in space and time requires a landscape perspective. While landscape ecology may not be integrated with land-use planning and decision making in North America to the degree that it is in Europe, it is increasingly viewed as fundamental to a broad range of resource management issues (Crow 1989, 1990). Incorporating spatial analysis into planning and management activities and considering the implications of pattern on biotic and abiotic processes at multiple spatial and temporal scales remain major challenges to resource managers.

The following important concepts are beginning to define a theoretical framework for the science we call landscape ecology. We consider the significance of these concepts and principles to managing natural resources and present a case study.

SCALE AND HIERARCHICAL ORGANIZATION

How we perceive an object or a phenomenon is greatly influenced by the scale, both in space and time, at which we view it. This rather obvious fact has important implications for both science and resource management (Hoekstra et al. 1991). Addressing a question at the wrong scale often leads to a failure of explanation and to the wrong conclusions (Allen and Hoekstra 1986, Turner 1989). Scientists commonly view biological systems as being so complex that they can be fully understood only by reducing them to their smallest, and presumably their simplest, components. Indeed, the biological sciences are dominated by fine-scale approaches. This common reductionistic approach to science is firmly entrenched in our academic institutions and is reflected in their departmental organization. Such subject areas as molecular and cellular biology dominate many funding programs, most prestigious awards, and many scientific journals. Yet most critical environmental issues are broad-scale problems. Witness concerns about global warming, ozone depletion in the upper atmosphere, deforestation in the tropics, global loss of biological diversity, long-distance transport and dispersion of toxins, and regional forest decline caused by interactions of drought, pathogens, and pollution.

At least two assumptions implicit in reductionism are questionable. First, that the components of a whole operate independently and in isolation from one another; and second, that the whole is merely the sum of its components. Basic research using holistic approaches is needed at all levels of organization in physical and biological systems; for example, carbon budgets or energy budgets can be developed for entire regions or biomes just as they can be developed for an individual leaf, a whole plant, or a forest stand.

How do we take fundamental information about ecological processes obtained at small scales, such as the physiological response of leaves to elevated levels of ozone, and apply it to a landscape, a region, or even a global scale? Understanding the properties of complex hierarchical organizations might be useful in addressing this question (Pattee 1973, Simon 1973). For example, the transfer of information from one level of ecological organization to another is not a simple additive process. Vital information from lower levels in a hierarchy may be extraneous information at higher levels of organization. Furthermore, levels in an organizational hierarchy can be isolated from one another because they operate at distinctly different rates. That is, high-frequency signals tend to attenuate as they pass up the hierarchy (Allen and Starr 1982). A leaf, for example, is sensitive to changes in light conditions that can be measured in seconds and minutes. A forest stand integrates this information over weeks, months, and growing seasons, and its growth responses are measured in these time frames. At each level in the organizational hierarchy, new organizing principles apply and new properties emerge, so properties of whole systems cannot effectively be predicted from the properties of simpler subsystems (Allen and Starr 1982, O'Neill et al. 1986).

Our understanding of complex hierarchical systems is still rudimentary at best. There are a number of hierarchical structures, both nested and non-nested, that can exist in nature (Allen and Hoekstra 1992). Hierarchy theory, however, provides an organizing framework to search for common properties across broad classes of complex systems, including physical, biological, social, and artificial systems.

The emphasis in landscape ecology is primarily but not entirely on large temporal and spatial scales. Landscapes vary in size, but they commonly encompass tens to hundreds of square kilometers in which mosaics of ecosystems are considered collectively and interactively. An important property when considering landscapes is the convergence of time and space. Large spatial scales often equate to long response periods.

Many tools for applying our large-scale science are already available. Geographic information systems (GIS), remote sensing, database managers, and computer simulations that are spatially explicit are all tools of the trade.

Technologies for applying landscape ecology are less a problem than is the lack of recognition that phenomena operating at large scales are important. It is not a question of having better science at lower levels of biological organization but rather of needing good science at all levels of biological organization, with more emphasis on working across scales.

CONTEXT AFFECTS CONTENT

Ecosystems are not closed, self-supporting systems, but rather parts of larger interacting systems. Serious problems can arise from treating each ecosystem as an isolated entity without regard for the context created by the broader landscape matrix. Surrounding lands can significantly affect the biological character of a habitat patch (Janzen 1986, Noss and Harris 1986). Direct impacts can result from such things as pesticide drift from agricultural fields into a small forest patch, movement of materials by wind or water, such as runoff from agricultural land to surface water (Sharpley et al. 1993), the invasion of exotic species from surrounding disturbed lands (Harris 1984), or changes in the microclimate due to increased exposure near the edges of a patch (Chen et al. 1993). Other effects may be more subtle. Increased levels of competition, predation, herbivory, or parasitism by open-land species on forest species, for example, can profoundly affect the community composition of forest islands. One such example is the increased levels of predation near habitat edges (Harris 1984, Wilcove et al. 1986).

Increasing the extent of our perspective helps in our interpretation of local phenomena. The species diversity of a local community, for example, is influenced by a number of factors operating at landscape and regional scales. Local changes in population size may merely reflect changes in the regional population of a species and not be due to local conditions. Likewise, habitat selection by individuals of a species may be determined not only by the characteristics of a given site but by the densities of populations in other habitats over a larger area. High regional populations may force a species into marginal habitats at the local level.

A "big picture" approach to resource management immediately brings us to the challenge of coordination and cooperation among ownerships and across political boundaries (Crow 1991). Management across boundaries brings with it issues relating to private property rights, proprietary information, conflicting legislated mandates, and general lack of trust among the participating parties—issues that can be, either individually or collectively, major obstacles to success. Perhaps the best-known example of cooperative management is the attempt to manage Yellowstone National Park as part of a larger landscape. The Greater Yellowstone Ecosystem includes the park, surrounding national forests, wildlife preserves, and private lands. As this attempt at cooperative management has demonstrated, collaborative effects are difficult, but the need will only increase as demands for relatively scarce natural resources continue to grow (Patten 1991). While no single model will be successful in all places, there are a few common elements that will increase the potential for success. For example, owners of adjoining properties are more likely to cooperate if firm guidelines exist to direct transactions and cooperation interactions (Schonewald-Cox et al. 1992). The experiences gained from the Greater Yellowstone Ecosystem will be valuable to others searching for models to promote cooperative management.

Obviously, not all decisions in forest management can be made at the stand level. Resource planning, as well as plan implementation, requires an integrative view that incorporates several spatial and temporal scales. An expanded perspective that includes not only within-boundary conditions (content), but also the broader landscape (context) is essential to better understand many ecological processes and interactions and to better predict the consequences of management actions.

BOUNDARIES AND SPATIAL HETEROGENEITY

Patches, boundaries, and spatial heterogeneity are inextricably linked (Forman and Moore 1992, Wiens 1992). Boundaries define patches, and patchiness is what produces heterogeneity. Boundaries may be sharp or fuzzy, linear or convoluted, and they may define patches that are small or large. These differences affect such things as the pattern of landscapes and the temporal dynamics of the patches. They also affect how organisms that follow a particular patch-foraging strategy respond to patchiness (Pyke 1984, Wiens 1976), how populations that are subdivided among patches stabilize or suffer local or regional extinction (Wiens 1976, Pulliam 1988), and how disturbances such as fire or insect outbreaks spread over an area (Forman 1987, Roland 1993).

Boundaries occur when structural or functional properties of ecological systems change (Wiens 1992). Because of this, ecologists often seek to minimize the effects of this heterogeneity on the phenomena of interest by confining investigations to areas that are internally relatively homogeneous. For example, ecosystem ecologists have used such boundaries as watersheds to define units within which they can conduct their studies (e.g., Dillon and Kirchner 1975). Another approach focuses on the boundary itself. These are studies of edges with both their positive and negative effects. Neither of these approaches gives much attention to the dynamics that may occur across boundaries.

If boundaries between patches or landscape elements are absolute, no exchanges take place. In this case, patches exist as closed systems, isolated from their neighbors. Such situations rarely occur in nature. It is equally rare, though, that boundaries are neutral and completely permeable to all fluxes. Most typically, landscape boundaries act as filters, with the structural characteristics of the edge and the context in which the edge exists playing key roles in determining its permeability (Buechner 1987, Schonewald-Cox 1988, Wiens et al. 1986, Forman and Moore 1992, Johnson 1973). For example, penetration into forest stands by cowbirds in their searching for host nests depends in part on the density of cowbirds in the broader agricultural matrix (F. Thompson, pers. comm.). The vertical structure of vegetation at the edge and the abruptness of boundaries also influence permeability (Forman and Moore 1992). Abrupt, high-contrast edges are likely to be less permeable than gradual, low-contrast edges.

To summarize, patches are linked at some level by movements of organisms or flows of materials across boundaries. These flows influence the structure and function within the patches as defined by their boundaries. Patches, their boundaries, and the heterogeneity that patches and their boundaries create are sensitive to the scale on which they are viewed. Detail that defines a complex mosaic of patches and boundaries at one scale may disappear at either finer or broader scales of resolution.

PATTERN AND PROCESS

The effect of pattern on process has long been a central theme for ecological research. Classic studies dealing with the spatial distribution of species in relation to environmental factors can be found in both the European and North American literature (e.g., Watt 1947). Historically, however, much of the emphasis was on describing the processes that created patterns observed in the biota. The explicit effects of spatial patterns on ecological processes have not been well studied (Turner 1989).

Characterizing landscape structure is fairly straightforward. Various techniques are available for quantifying landscape structure (see O'Neill et al. 1988, Turner 1990, Baskent and Jordan 1995). Understanding the effects of landscape structure on ecological processes is more difficult because the broad spatial-temporal scale makes experimentation and hypothesis testing challenging. And, as always, the patterns and related processes are scale dependent.

Dunning et al. (1992) presented four types of ecological processes relating to species and populations operating at the landscape level. First, when a species requires nonsubstitutable resources found in two different habitats, a landscape where both habitats are relatively close will support more individuals than landscapes where one habitat is relatively rare. In the second process, habitat requirements are substitutable. In this case, a local population may increase if the habitat is located in a portion of the landscape that contains additional available resources. It is conceivable that under this scenario a population can be maintained in a marginal habitat (for example, one that is too small) to sustain the population based solely on the resources found within the patch itself. The third landscape process, sources and sinks, is somewhat related to the second process. Here, relatively productive patches serve as sources of emigrants that disperse to other patches that alone may not be able to maintain viable populations (that is, sinks). Relatively large populations may be maintained in patches that are sinks due to close proximity to a patch that is a strong source (Pulliam 1988). The fourth process, called neighborhood effects, is based on the geographic principle that objects in close proximity are more likely to interact than objects that are distant. The corollary to this principle for populations is that the abundance of a species in a patch is more strongly affected by characteristics of adjacent patches than by patches that are more distant in the landscape.

Changes in the composition and structure of landscapes have profound implications for a variety of issues, ranging from climatic change to biological diversity and water pollution. Extensive deforestation, for example, can result in reduction in surface net radiation, evaporation, and precipitation and in higher surface temperature (Franchito and Rao 1992). In terms of plant and animal species, changes in patterns from natural to human-dominated landscapes tend to favor generalists and common species at the expense of specialist and rare species (Probst and Crow 1991). Conversion of wildlands to urban and agricultural uses can have significant impacts on both surface water and groundwater quality (Charbonneau and Kondolf 1993). Urbanization adds various point sources of pollution, including from septic systems, underground storage tanks, industrial and municipal wastewater effluent. Greater soil erosion, intensive fertilizer and pesticide usage, nonpoint source pollution from surface runoff contaminants and toxins are associated with conversion of wildlands to agricultural production.

The use of remote sensing technology offers opportunities for studying scale-dependent patterns of productivity, water balance, and biogeochemical cycling. For example, evapotranspiration from forested landscapes has been estimated using remotely sensed data (Luvall and Holbo 1991). Remote sensing is also a valuable tool for analyzing landscape change.

APPLICATIONS

Timber harvesting affects the composition, structure, and function of land-scapes. Changes in composition and structure can be easily quantified, using GIS along with remotely sensed imagery. When these tools for generating, organizing, storing, and analyzing spatial information are combined with mathematical models, resource planners and managers have the means for assessing the impacts of alternative management practices on pattern and process.

Forest cover in a 23,600 ha portion of the Pleasant Run planning unit on the Hoosier National Forest (HNF) in southern Indiana is shown in figure 3.1a. This map was derived from a digital land cover map produced from a LANDSAT Thematic Mapper satellite image collected on April 26, 1988. Land cover was classified using PC-ERDAS image processing software and supervised classification techniques (Lillesand and Keifer 1987), as described by Gustafson and Crow (1994). For purposes of this example, the map de-lineates forest interior (more than 210 m from an edge) in red, forest edge in blue, and nonforest areas comprising primarily agricultural lands or re-cently harvested areas (within 20 years) in white. Forest interior conditions provide critical habitat for a number of species (Solheim et al. 1987). Large blocks of forest with closed canopy are relatively rare in southern Indiana because of ownership patterns, increased urbanization, and abundance of ag-ricultural lands.

Given this regional context, a management goal for the HNF is to maintain existing forest interior conditions wherever possible within the forest (USDA Forest Service 1991). To determine the effects of variation in harvest size and total area harvested on forest interior for this planning area, a timber harvest allocation model (HARVEST) was applied that allows the input of specific rules to allocate harvest units in space and time (Gustafson and Crow 1994, 1996). The model runs within a PC-ERDAS GIS environment and was developed using ERDAS Toolkit routines for input and output. ERDAS is a grid-cell (raster) GIS that allows flexible display and manipulation of digital maps. Timber harvest allocations were made by HARVEST based on a digital stand map with values for grid cells reflecting the age of each forest stand. Only even-aged management was considered. The model allows control of harvest intensity, size of harvest unit, and rotation length to produce forest patterns with the "look and feel" of a managed landscape (Gustafson and Crow 1994).

Four treatments (small harvest units and low-intensity harvest, small har-vest units and high-intensity harvest, large harvest units and low-intensity harvest, and large harvest units and high-intensity harvest) were applied to the study area and their impacts on landscape structure compared in a 150-year simulation. The small harvest units averaged 2.5 pixels (s.d. = 0.6) or 0.225 ha in size, while the large harvest units averaged ten times as large (25 pixels [s.d. = 1.2] or 2.25 ha in size). Low-intensity harvesting was defined as 1% of the forest harvested per decade, compared with 7% harvested per decade for high intensity. The rotation length (or minimum age at which a stand could be harvested) was eighty years. Simulations begin with the assumption that all forest on the study site was greater than eighty years of age (figure 3.1a), with the result that HARVEST distributed cutting units randomly across the study site.

The coupling of HARVEST with GIS allows us to express the output from the model as a visual product (fig. 3.1b, c, d, and e). Figure 3.1b–e shows the amount of interior forest, edge, and nonforest following 150 years of harvest activity for each combination of cutting unit size and intensity of harvest. Harvested stands appear as openings until they exceed 20 years of age.

When harvest units were distributed across an entire management unit, substantial forest fragmentation and loss of interior forest conditions occurred even with large cutting units (size) and low harvest intensity (fig. 3.1b). Extreme decreases in forest interior and extreme increases in forest edge occurred with all other combinations of size and intensity (fig. 3.1c–e). Low-intensity harvesting and small cutting units still produced a landscape with little forest interior (fig. 3.1c) compared with the initial conditions. With high-intensity harvesting, forest interior was largely eliminated despite the use of fewer but larger cutting units (fig. 3.1d). Small size and high intensity completely eliminated interior conditions in the Pleasant Run planning unit (fig. 3.1e). Small nonforest openings, even when ephemeral, were scattered throughout the forest with the small size, high-intensity combination.

It is obvious from these results that maintaining forest interior will be difficult without establishing zones in which no harvesting is allowed. This is the strategy currently being followed by the HNF, and an experimental analysis using HARVEST confirms that zoning forest use is more important than either the distribution of harvest units (for example, scattered vs. aggregated), intensity of harvest (area harvested per unit of time), or size of harvest units in reducing forest fragmentation at the forest level (Gustafson and Crow 1996).

LANDSCAPE DESIGN

Nearly half of the earth's land surface is under some form of intensive management in which the dominant spatial features are created by human actions (Klopatek et al. 1979). The remaining lands exist in a natural or seminatural state. To comprehend the significance of this figure, we can compare some general characteristics of natural and human-dominated landscapes. In natural landscapes, the basic matrix is typically extensive in its distribution, and as a result, natural landscapes tend to have high connectiveness among matrix elements (Godron and Forman 1983). High variation in patch size is also the norm, with many small patches and a few large patches present. While sharp, abrupt edges do occur in natural landscapes, gradual or ''soft'' edges and low contrast among landscape elements are common.

In contrast, the matrix in human-dominated landscapes is likely to be fragmented into small patches, and the patches become regular, more geometric in shape when compared to the natural landscape (Godron and Forman 1983). Large habitat patches are lost (Mladenoff et al. 1993). Fragmentation results in low connectiveness in the human-dominated landscape, and high contrast among patches is a common feature.

What are the implications of these changes in landscape pattern? What are the inadvertent cumulative effects at the landscape, regional, and global scales of the multitude of human actions implemented at local levels? It is axiomatic that the composition of the landscape and the arrangement of these elements in time and space strongly affect the actual and potential values derived from the land. With this in mind, landscapes should be designed based on ecological principles to derive and sustain the full spectrum of human benefits from our biosphere. The historic approach to land-use planning is haphazard and piecemeal at best or nonexistent at worst. When we consider present land-use practices along with broad-scale natural processes and population growth, basic trends are obvious—losses of productive potential, reductions in biological diversity, degradations of landscapes. All these trends lead to human impoverishment. Sustainability—regardless if in ecological systems, resource management, or economic development—requires a time frame of human generations, and a focus on the interactions between ecological integrity and human aspirations.

Landscape ecology is a young science and so its theoretical structures and methodologies are not well developed. Broad conceptual areas, including spatial and temporal scales, spatial context, boundaries and spatial heterogeneity, pattern and process, are identified as emerging concepts that might serve as loci for developing theory.

While setting goals and defining objectives are still obvious initial requirements for resource management and planning, the next logical step is to define a composition and a structure for the landscape that will produce the desired benefits. To be effective, planning and management need to be spatially and temporally explicit. Only then can we evaluate the cumulative effects of our management actions, and only then can we determine whether we are indeed good stewards of the land.

A landscape perspective supplements the more traditional species and forest-stand approaches to resource management. But a species-by-species or stand-by-stand approach to wildlife or forest management is inadequate given the increased demands on natural resources resulting from a growing human population. Instead, a more comprehensive approach, an ecosystem approach, is required. A landscape perspective, with its emphasis on aggregates of interacting ecosystems, is a component of ecosystem management.

LITERATURE CITED

Addicott, J. F., J. M. Ahl, M. F. Antolin, D. K. Padilla, J. S. Richardson, and D. A. Soluk. 1987. Ecological neighborhoods: scaling environmental patterns. Oikos 49: 340–346.

Allen, T. F. H., and T. W. Hoekstra. 1986. The critical role of scaling in land modelling. Pages 9–13 *in* R. Gelinas, D. Bond, and B. Smit, editors. Perspectives on land modelling. Polyscience Publications, Montreal, Quebec.

Allen, T. F. H., and T. B. Starr. 1982. Hierarchy: perspectives for ecological complexity. University of Chicago Press, Chicago, Illinois.

Baskent, E. Z., and G. A. Jordan. 1995. Characterizing spatial structure of forest landscapes. Canadian Journal of Forest Research 25:1830–1849.

Brown, B. J., and T. F. Allen. 1989. The importance of scale in evaluating herbivory impacts. Oikos 54:189–194.

Buechner, M. 1987. Conservation in insular parks: simulation models of factors affecting the movement of animals across park boundaries. Biological Conservation 41:57–76.

Charbonneau, R., and G. M. Kondolf. 1993. Land use change in California, USA: nonpoint source water quality impacts. Environmental Management 17:453–460.

Chen, J., J. F. Franklin, and T. A. Spies. 1993. Contrasting microclimate among clearcut, edge, and interior of old-growth Douglas-fir forest. Agricultural and Forest Meteorology 3:219–237.

Cornell, H. V., and J. H. Lawton. 1992. Species interactions, local and regional processes, and limits to the richness of ecological communities: a theoretical perspective. Journal of Animal Ecology 61:1–12.

Crow, T. R. 1989. Landscape ecology: an eclectic science for the times. Pages 30–34 in G. Rink and C. A. Budelsky, editors. Proceedings 7th Central Hardwood Conference, USDA Forest Service, General Technical Report NC-132. North Central Forest Experiment Station, St. Paul, Minnesota.

———. 1990. Conservation biology and landscape ecology: new perspectives for resource managers. Pages 47–54 in Proceedings, National Convention Society of American Foresters. SAF, Washington, D.C.

———. 1991. Landscape ecology: the big picture approach to resource management. Pages 55–65 in D. J. Decker, M. E. Krasny, G. R. Goff, C. R. Smith, and D. W. Gross, editors. Challenges in the conservation of biological resources, a practitioner's guide. Westview Press, Boulder, Colorado.

Dillon, P. J., and W. B. Kirchner. 1975. The effects of geology and land use on the export of phosphorus from watersheds. Water Research 9:135–148.

Dunning, J. B., B. J. Danielson, and H. R. Pulliam. 1992. Ecological processes that affect populations in complex landscapes. Oikos 65:169–175.

Fahrig, L. 1992. Relative importance of spatial and temporal scales in a patchy environment. Theoretical Population Biology 41:300–314

Forman, R. T. T. 1987. The ethics of isolation, the spread of disturbance, and landscape ecology. Pages 213–229 in M. G. Turner, editor. Landscape heterogeneity and disturbance. Springer-Verlag, New York, New York.

Forman, R. T. T., and P. N. Moore. 1992. Theoretical foundations for understanding boundaries in landscape mosaics. Pages 236–258 in A. J. Hansen and F. di Castri, editors. Landscape boundaries, consequences for biotic diversity and ecological flows. Springer-Verlag, New York.

Godron, M., and R. T. T. Forman. 1983. Landscape modification and changing ecological characteristics. Pages 12–28 in H. A. Mooney and M. Godron, editors. Disturbance and ecosystems. Springer-Verlag, New York, New York.

Gustafson, E. J., and T. R. Crow. 1994. Modeling the effects of forest harvesting on landscape structure and the spatial distribution of cowbird brood parasitism. Landscape Ecology 9:237–248.

———. 1996. Simulating the effects of alternative forest management strategies on landscape structure. J. Environ. Management 46:77–94.

Harris, L. D. 1984. The fragmented forest. University of Chicago Press, Chicago, Illinois.

Hoekstra, T. W., T. F. H. Allen, and C. H. Flather. 1991. Implicit scaling in ecological research. BioScience 41:148–154.

Holling, C. S. 1992. Cross-scale morphology, geometry, and dynamics of ecosystems. Ecological Monographs 62:447–502.

Janzen, D. H. 1986. The eternal external threat. Pages 286–303 in M. E. Soulé, editor. Conservation biology: the science of scarcity and diversity. Sinauer Associates, Sunderland, Massachusetts.

Johnson, C. A. 1993. Material fluxes across wetland ecotones in northern landscapes. Ecological Applications 3:424–440.

Klopatek, J. M., R. J. Olson, C. J. Emerson, and J. L. Joness. 1979. Land-use conflicts with natural vegetation in the United States. Environmental Conservation 6:191–199.

Kolasa, J., and S. T. A. Pickett (editors). 1991. Ecological heterogeneity. Springer-Verlag, New York, New York.

Lillesand, T. M., and R. W. Keifer. 1987. Remote sensing and image interpretation. John Wiley and Sons, New York, New York.

Lord, J. M., and D. A. Norton. 1990. Scale and the spatial concept of fragmentation. Conservation Biology 4:197–202.

Luvall, J. C., and H. R. Holbo. 1991. Thermal remote sensing methods in landscape ecology. Pages 127–152 in M. G. Turner and R. H. Gardner, editors. Quantitative methods in landscape ecology. Springer-Verlag, New York, New York.

McLaughlin, J. F., and J. Roughgarden. 1992. Predation across spatial scales in heterogeneous environments. Theoretical Population Biology 41:277–299.

Mladenoff, D. J., M. A. White, J. Pastor, and T. R. Crow. 1993. Comparing spatial pattern in unaltered old-growth and disturbed forest landscapes. Ecological Applications 3:293–305.

Naveh, Z., and A. S. Lieberman. 1984. Landscape ecology. Springer-Verlag, New York, New York.

Noss, R. F., and L. D. Harris. 1986. Nodes, networks, and MUMs: preserving diversity at all scales. Environmental Management 10:299–309.

O'Neill, R. V., D. L. DeAngelis, J. B. Waide, and T. F. H. Allen. 1986. A hierarchical concept of ecosystems. Princeton University Press, Princeton, New Jersey.

O'Neill, R. V., J. R. Krummel, R. H. Gardner, G. Sugihara, B. Jackson, D. L. DeAngelis, B. T. Milne, M. G. Turner, B. Zygmunt, S. Christensen, V. H. Dale, and R. L. Graham. 1988. Indices of landscape pattern. Landscape Ecology 1:153–162.

O'Neill, R. V., S. J. Turner, V. I. Cullinan, D. P. Coffin, T. Cook, W. Conley, J. Brunt, J. M. Thomas, M. R. Conley, and J. Gosz. 1991. Multiple landscape scales: an intersite comparison. Landscape Ecology 5:137–144.

Pahl-Wostl, C. 1993. Food webs and ecological networks across temporal and spatial scales. Oikos 66:415–432.

Pattee, H. H. 1973. The physical basis and origin of hierarchical control. Pages 73–108 in H. H. Pattee, editor. Hierarchy theory: the challenge of complex systems. George Braziller, New York, New York. USA.

Patten, D. T. 1991. Human impacts in the Greater Yellowstone Ecosystem: evaluating sustainability goals and eco-redevelopment. Conservation Biology 5:405–411.

Probst, J. R., and T. R. Crow. 1991. Integrating biological diversity and resource management. Journal of Forestry 89:12–17.

Pulliam, H. R. 1988. Sources, sinks, and population regulation. American Naturalist 132:652–661.

Pyke, G. H. 1984. Animal movements: an optimal foraging approach. Pages 7–31 *in* I. R. Swingland and P. J. Greenwood, editors. The ecology of animal movement. Clarendon Press, Oxford, England.

Roland, J. 1993. Large-scale forest fragmentation increases the duration of tent caterpillar outbreak. Oecologia 93.25–30.

Schonewald-Cox, C. 1988. Boundaries in the protection of nature reserves. BioScience 38:480–486.

Schonewald-Cox, C., M. Buechner, R. Sauvajot, and B. A. Wilcox. 1992. Cross-boundary management between national parks and surrounding lands: a review and discussion. Environmental Management 16:273–282.

Sharpley, A. N., T. C. Daniel, and D. R. Edwards. 1993. Phosphorus movement in the landscape. Journal of Production Agriculture 6.492–500.

Simon, H. A. 1973. The organization of complex systems. Pages 3–27 *in* H. H. Pattee, editor. Hierarchy theory: the challenge of complex systems. George Braziller, New York, New York.

Solheim, S. L., W. S. Alverson, and D. M. Waller. 1987. Maintaining biotic diversity in national forests: the necessity for large blocks of mature forest. Endangered Species Technical Bulletin 4(8).

Turner, M. G. 1989. Landscape ecology: the effect of pattern on process. Annual Review of Ecology and Systematics 20:171–197.

———. 1990. Spatial and temporal analysis of landscape patterns. Landscape Ecology 4:21–30.

——— (editor). 1987. Landscape heterogeneity and disturbance. Springer-Verlag, New York, New York.

USDA Forest Service. 1991. Plan Amendment, Land and Resource Management Plan, Hoosier Nation Forest. USDA Forest Service, Eastern Region. Milwaukee, Wisconsin.

Watt, A. S. 1947. Pattern and process in the plant community. Journal of Ecology 35:1–22.

Wiens, J. A. 1989. Spatial scaling in ecology. Functional Ecology 3:385–397.

———. 1976. Population responses to patchy environments. Annual Review of Ecology and Systematics 7:81–120.

———. 1992. Ecological flows across landscape boundaries: a conceptual overview. Pages 217–235 *in* A. J. Hansen and F. di Castri, editors. Landscape boundaries. Springer-Verlag, New York, New York.

Wiens, J. A., C. S. Crawford, and J. R. Gosz. 1986. Boundary dynamics: a conceptual framework for studying landscape ecosystems. Oikos 45:421–427.

Wilcove, D. S., C. H. McClellan, and A. P. Dobson. 1986. Habitat fragmentation in the temperate zone. Pages 237–256 *in* M. E. Soulé, editor. Conservation biology: the science of scarcity and diversity. Sinauer Associates, Sunderland, Massachusetts.

Chapter 4 The Role of Keystone Ecosystems in Landscapes

Phillip deMaynadier and Malcolm Hunter, Jr.

One of the prerequisites of ecosystem management is learning to examine ecological phenomena at larger spatial and temporal scales than are typical of traditional natural resource management. In particular, forest ecosystem managers need to look beyond the borders of a stand and the length of a rotation. One step in this direction can involve defining ecosystems at large spatial scales. According to the classic image, an ecosystem (which we define as a group of interacting populations and their environment) usually operates at the scale of a lake or a field, something we would probably measure in hectares or acres. However, ecosystems can be defined as much smaller (for example, a fallen log or a tidepool) or larger (for example, the Greater Yellowstone Ecosystem [Grumbine 1990] or Gaia, the entire planet conceived as a single large ecosystem [Lovelock 1988]).

An alternative approach is to continue defining ecosystems at more conventional scales and then to examine large-scale phenomena from the perspective of a landscape, which we define as ''a group of interacting ecosystems.'' For example, a large forested valley with various forest stands forming a matrix, and with wetlands, streams, and meadows embedded in the matrix, could readily be considered a landscape. Energy, soil, water, air, nutrients, and organisms will routinely move among the ecosystems of this landscape. Landscape ecology is the science that deals with these movements and their consequences (Forman 1995).

Studies on the effects of ecosystem degradation, exploitation, and conversion have been important contributions for people who are trying to manage natural resources carefully (Bormann and Likens 1967, 1979, Ashmore 1985, Maltby 1988, and others). However, most of this work has focused on identifying the consequences of anthropocentric impacts on structural and functional properties of individual ecosystems. Relatively little attention has been given to the functional role that ecosystems have within the surrounding landscape, and the potential synergistic effects that may occur following degradation or loss of particular ecosystems. Interactions among ecosystems have important ramifications for ecological integrity, and thus it is important that practitioners of ecosystem management learn to think like landscape ecologists. In this chapter, we will attempt to highlight the importance of ecosystem function at a landscape scale by describing a new idea, keystone ecosystems. Before discussing keystone ecosystems we need to review an antecedent concept, that of keystone species.

KEYSTONE SPECIES

The term *keystone species* is prominent in the lexicon of conservation biology because conservation of species that play key roles in ecosystems is a critical task (deMaynadier and Hunter 1994). The term has its origins in a narrower concept, *keystone predators,* referring to predators that maintain the diversity of a community by preventing any one prey species from competitively excluding other species (Paine 1969). Subsequently, *keystone* has been used to describe species or guilds that fill a variety of roles. Examples include fruit-bearing tree species that provide food for frugivores at a season when fruit is scarce (Terborgh 1986, Lambert and Marshall 1991); elephants *(Loxodonta africana)* that are critical to the dispersal and germination of certain seeds (Lieberman et al. 1987); crabs (Smith et al. 1991) and rodents (Brown and Heske 1990) that affect plant communities by disturbing the soil; and beavers *(Castor* spp.) and alligators *(Alligator mississippiensis)* that create new environments for lentic communities by damming streams and creating wallows, respectively (Naiman et al. 1988, Finlayson and Moser 1991). Broader uses of the term have led some to question the concept's clarity and application (Mills et al. 1993). Recent consensus is emerging, however, around an operational definition that identifies keystone species as those that have a greater role in maintaining ecosystem structure or function than one would predict based on their abundance or biomass (Hunter 1990, Power and Mills

1995). Some writers have used the term for any species that plays a critical ecological role regardless of its abundance, for example, kelp in a kelp forest (Thorn-Miller and Catena 1991) or Douglas firs in a Douglas fir forest (Wilson 1992). We would refer to these as *dominant species* and reserve the term *keystone* for species whose role is larger than would be commensurate with their abundance. This narrower usage is consistent with Paine's (1969) architectural metaphor of a keystone in an arch and with the original concept of keystone predators.

KEYSTONE ECOSYSTEMS: DEFINITION AND EXAMPLES

This narrower keystone concept can be extended in scale from the role of species in ecosystems to the role of ecosystems in landscapes, which we define as a group of interacting ecosystems. A keystone ecosystem is one whose role in landscape structure and function is greater than one would predict from its area. There are two primary mechanisms by which keystone ecosystems influence landscapes: (1) by shaping landscape disturbance regimes or (2) by providing a resource that is limiting within the landscape.

Linear ecosystems may often serve as barriers that impede the spread of disturbance. For example, in forest or grassland landscapes that experience regular fires, aquatic ecosystems, especially large rivers, can play a keystone role by allowing unburned areas to persist beyond a natural firebreak (Bergeron 1991). During a widespread fire these areas may serve as refugia, later providing a postdisturbance source for recolonizations. Barrier reef ecosystems play a keystone role in maintaining associated mangrove forest and seagrass ecosystems by providing a self-maintaining breakwater, buffering the coastline from damaging wave action (Johannes and Hatcher 1986). Barrier beach ecosystems often function in a similar fashion in temperate zones by providing the low-energy conditions necessary for coastal saltmarsh ecosystems.

Conversely, keystone ecosystems can operate by facilitating the spread of a natural disturbance. For example, in the northeastern United States many remnant pitch pine *(Pinus rigida)*–scrub oak *(Quercus ilicifolia)* forests exist in isolated patches too small to sustain the natural disturbance regime of lightning-started fires and consequently are threatened with hardwood succession from surrounding communities. However, where these patches are connected by intervening fire-permeable communities such as ridge-top chestnut oak *(Quercus montana)*–ericaceous woodlands in Pennsylvania, they are more likely to persist in the landscape without management intervention (personal observation). Riverine ecosystems may profoundly influence the structure and composition of adjacent ecosystems (for example, floodplain forests, sand-bar thickets, herbaceous meadows, vernal pools, and swamps) through cyclical flooding and ice-scour events. Shankman (1993) suggests that periodic flooding and channel migration is the most important natural disturbance mechanism affecting surrounding spatial heterogeneity in bottomlands of the southeastern coastal plain of the United States. He specifically predicts the loss of black willow *(Salix nigra)*, cottonwood *(Populus deltoides)*, and silver maple *(Acer saccharinum)* forest communities in less than one hundred years and a decline in bald cypress *(Taxodium distichum)*–dominated stands in those areas where riverine disturbance regimes have been prevented through dam and channelization projects.

There are also many examples of keystone ecosystems that provide a limiting resource in their particular landscape context, including *Spartina* salt-marsh ecosystems embedded in an estuary landscape/seascape offering nursery habitat and primary production (Pomeroy and Wiegert 1981), desert springs or oases in desert landscapes providing food, water, and shade, especially during droughts (Noy-Meir and Goodall 1985), and pelagic upwellings or polynyas in seascapes where nutrients are otherwise scarce or unavailable (France and Sharp 1992). An ecosystem whose location facilitates the movement of dominant or keystone species across the landscape, such as a forested corridor connecting two isolated patches of forest, could be considered a keystone element if habitat for movement is a limited resource. A residual patch of forest in an agricultural matrix can affect the whole landscape by providing breeding and resting cover for animals that forage over a larger area. Conversely, disturbances such as fires and treefalls can create early-successional ecosystems in landscapes dominated by mature forests, and these may have keystone roles. For example, Levey (1990) suggests that treefall gaps, by allowing light to the forest floor and understory, provide a regeneration niche for many fruiting species. Furthermore, he presents data for fruit production in Costa Rican treefall gaps which suggest that treefall gaps may operate as shifting, small-scale keystone ecosystems by providing a large and reliable source of fruit during dry seasons, a limiting period for frugivore populations.

A particular ecosystem may provide a limiting resource for a species that has a keystone role in the ecosystems that dominate the landscape. For example, large herbivores such as elephants may play keystone roles across a whole landscape, moving among ecosystems and altering successional dynamics by consuming large quantities of vegetation and spreading seeds (Lieberman et al. 1987). An aquatic ecosystem that provides water for these keystone herbivores during droughts could thus operate as a keystone ecosystem for the entire landscape. Similarly, at temperate and boreal latitudes, an ecosystem that provides winter habitat for populations of keystone herbivores might constitute a keystone ecosystem in the overall landscape. Conifer forests that provide critical winter yards for white-tailed deer *(Odocoileus virginianus)* are often targeted for special management and might be an example of this phenomenon.

Finally, hedgerow networks in agricultural landscapes are a keystone ecosystem that both provide a limiting resource and influence the local disturbance regime. Although covering relatively little area, hedgerows contribute an important structural element capable of influencing patterns of movement (Wegner and Merriam 1979), species composition, and successional dynamics across an entire agricultural landscape (Forman and Baudry 1984). Hedgerows also help dampen the intensity of common agricultural disturbances, including wind- and water-driven soil erosion (Forman 1995) and insect pest population fluctuations that might be moderated in fields that have adequate cover habitat for insect predators (Pollard 1968, 1971).

Admittedly, most of the examples outlined above provide only anecdotal evidence of the keystone ecosystem concept. Empirical evidence of the keystone concept operating at a landscape scale will be hard to collect. The critical experiment would be to "remove" the proposed keystone ecosystem in a particular setting and evaluate the landscape's response to the loss. Difficulties of replication and control aside, one would need to use care in specifying the landscape responses to be measured: loss of adjacent ecosystems, altered disturbance regimes, new patch dynamic equilibria, loss of keystone or dominant species, among others. Perhaps the most likely tests will be conducted without experimental design in landscapes undergoing ecosystem loss and fragmentation, where the fate of a previously intact region is monitored after the loss of a proposed keystone element.

OTHER USES OF THE KEYSTONE CONCEPT

It is notable that the keystone concept has been applied only to species (and, in this chapter, ecosystems) that have positive roles as perceived by most conservationists. For example, exotic species are not called keystone species although their influence can be profound. Similarly, introduced landscape elements—for example, a powerline right-of-way or transportation corridor bisecting a forest—are unlikely to be called keystone ecosystems.

We would suggest that this usage continue, and that an alternative adjective like *disrupter* be used for species or ecosystems whose role is perceived as negative. Of course, it will not always be easy to decide whether the role of a species is positive or negative, and there will even be cases, depending on context, in which the same species can have both a keystone and disrupter role in the same type of ecosystem. For example, one hundred years ago, when the white-tailed deer was nearly extinct in most of the eastern United States, conservation biologists might have spoken of its keystone role as a herbivore. Today, with deer populations in some areas at unnaturally high levels, some conservation biologists, concerned with overgrazing of understory plants and productivity of ground-nesting birds, speak of white-tailed deer as a disrupter species (Alverson et al. 1988, McShea and Rappole 1992).

The keystone concept has already been extended to other biological entities of interest to conservation biologists, such as genes that produce a cascading effect on other genes (regulatory genes) and elements that are important to biota at very low concentrations (micronutrients). By expanding the idea to ecosystem's roles at the landscape scale, this chapter proposes a logical extension of the concept.

CONSERVATION IMPLICATIONS

The fact that individual species losses may have cascading effects on a surrounding web of interacting species is well recognized by ecologists and is often used as a rallying point for the protection of biodiversity. We are suggesting an analogous notion—that few natural ecosystems exist within a landscape vacuum, and that some ecosystems may play a keystone role in their influence upon landscape structure and function. The conservation implications of this idea are important because the loss or degradation of keystone ecosystems may result in the functional disruption of an entire landscape. Obviously, if a reserve boundary fails to encompass an adjoining keystone ecosystem as we have defined it, the reserve risks losing what it was originally designed to protect. In order to maintain the integrity of a particular ecosystem, we need to ask what other surrounding ecosystems are critical to its persistence. Many interdependent "functional mosaics" can be recognized (Noss 1987), but practicality and cost dictate a limit to the number of ecosystems that can effectively be protected in a given landscape, thus highlighting the utility of the keystone ecosystem concept.

Traditionally, the assignment of conservation value to ecosystems has retained a species-level approach in which those systems that are particularly species rich or harbor a large number of rarities are the priority for protection efforts. Recently, this effort has been complemented by a higher, ecosystem-level approach that assigns conservation value to rare or regionally representative ecosystem examples (Noss 1987). Recognition of keystone ecosystems suggests a landscape-level approach that assigns conservation value to ecosystems, based not only on the level of diversity that they themselves contain but also on the functional significance a given ecosystem may have in maintaining ecological processes at a landscape scale. Admittedly, our understanding of these large-scale ecological processes is rather limited, and thus we need to be conservative about applying the keystone ecosystem concept. In other words, if an ecosystem is known to have a keystone role we should strive to protect it, but in the absence of a known keystone role, we should not assume an ecosystem has little value.

Finally, the purpose of this essay is not just to introduce a fashionable term to the vocabulary of ecosystem managers, but rather to highlight the importance of ecosystem *function,* particularly in those special cases where a relatively small area is playing a large role in landscape dynamics.

We would like to thank Tony Davis, Michael Jennings, Reed Noss, and Raymond O'Connor for reviewing the manuscript. Maine Agricultural and Forest Experiment Station Paper No. 2010.

LITERATURE CITED

Alverson, W. S., D. M. Waller, and S. L. Solheim. 1988. Forests too deer: edge effects in northern Wisconsin. Conservation Biology 2:348–358.

Ashmore, M., N. Bell, and J. Rutter. 1985. The role of forest damage in West Germany. Ambio 14:81–87.

Bergeron, Y. 1991. The influence of island and mainland lakeshore landscapes on boreal forest fire regimes. Ecology 72:1980–1992.

Bormann, F. H., and G. E. Likens. 1967. Nutrient cycling. Science 155:424–429.

———. 1979. Catastrophic disturbance and the steady state in the northern hardwood forest. American Scientist 67:660–669.

Brown, J. H., and E. J. Heske. 1990. Control of a desert-grassland transition by a keystone rodent guild. Science 250:1705–1707.

deMaynadier, P., and M. L. Hunter, Jr. 1994. Keystone support. BioScience 44:2.

Finlayson, M., and M. Moser (editors). 1991. Wetlands. International Waterfowl and Wetlands Research Bureau, Oxford, England.

Forman, R. T. T. 1995. Land mosaics: the ecology of landscapes and regions. Cambridge University Press, Cambridge, England.

Forman, R. T. T., and J. Baudry. 1984. Hedgerows and hedgerow networks in landscape ecology. Environmental Management 8:495–510.

France, R., and M. Sharp. 1992. Polynyas as centers of organization for structuring the integrity of arctic marine communities. Conservation Biology 6:442–446.

Grumbine, E. 1990. Protecting biological diversity through the greater ecosystem concept. Natural Areas Journal 10(3):114–120.

Hunter, M. L. 1990. Wildlife, forests, and forestry. Prentice Hall, Englewood Cliffs, New Jersey.

Johannes, R. E., and B. G. Hatcher. 1986. Shallow tropical marine environments. Pages 371–382 in M. E. Soulé, editor. Conservation biology: the science of scarcity and diversity. Sinauer Associates, Sunderland, Massachusetts.

Lambert, F. R., and A. G. Marshall. 1991. Keystone characteristics of bird-dispersed Ficus in a Malaysian lowland rain forest. Journal of Ecology 79:793–809.

Levey, D. J. 1990. Habitat-dependent fruiting behavior of an understory tree, Miconia centrodesma, and tropical treefall gaps as keystone habitats for frugivores in Costa Rica. Journal of Tropical Ecology 6:409–420.

Lieberman, D., M. Lieberman, and C. Martin. 1987. Notes on seeds in elephant dung from Bio National Park, Ghana. Biotropica 19:365–369.

Lovelock, J. 1988. The ages of gaia. Norton, New York, New York.

Maltby, E. 1988. Wetland resources and future prospects: an international perspective. Pages 3–14 in J. Zelazny and J. S. Feierabend, editors. Wetlands: increasing our wetland resources. National Wildlife Federation, Washington, D.C.

McShea, W. J., and J. H. Rappole. 1992. White-tailed deer as keystone species within forest habitats of Virginia. Virginia Journal of Science 43:177.

Mills, L. S., M. E. Soulé, and D. F. Doak. 1993. The keystone-species concept in ecology and conservation. BioScience 43:219–224.

Naiman, R. J., C. A. Johnston, and J. C. Kelley. 1988. Alteration of North American streams by beaver. BioScience 38:753–762.

Noss, R. F. 1987. From plant communities to landscapes in conservative inventories: a look at the Nature Conservancy (USA). Biological Conservation 41:11–37.

Noy-Mier, I. 1985. Desert ecosystem structure and function. Pages 93–103 in M. I. Evenardi, I. Noy-Mier, and D. W. Goodall, editors. Hot deserts and arid shrublands. Elsevier Science, New York, New York.

Paine, R. T. 1969. A note on trophic complexity and community structure. American Naturalist 103:91–93.

Pollard, E. 1971. Hedges VI. Habitat diversity and crop pests: a study of Brevicorgne brassicae and its symphid predators. Journal of Applied Ecology 8:751–780.

Pomeroy, L. R., and R. G. Wiegert (editors). 1981. The ecology of a salt marsh. Springer-Verlag, New York, New York.

Power, M. E., and L. S. Mills. 1995. The keystone cops meet in Hilo. Trends in Ecology and Evolution 10:182–184.

Shankman, D. 1993. Channel migration and vegetation patterns in the southeastern coastal plain. Conservation Biology 7:176–183.

Smith, T. J., III, K. G. Boto, S. D. Frusher, and R. L. Giddins. 1991. Keystone species and mangrove forest dynamics: the influence of burrowing by crabs on soil nutrient status and forest productivity. Estuarine, Coastal and Shelf Science 33:419–432.

Terborgh, J. 1986. Keystone plant resources in the tropical forest. Pages 330–344 in

M. E. Soulé, editor. Conservation biology. Sinauer Associates, Sunderland, Massachusetts.

Thorne-Miller, B., and J. G. Catena. 1991. The living ocean: understanding and protecting marine biodiversity. Island Press, Washington, D.C.

Wegner, J. F., and G. Merriam. 1979. Movements by birds and small mammals between a wood and adjoining farmland habitats. Journal of Applied Ecology 16: 349–357.

Wilson, E. O. 1992. The diversity of life. Harvard University Press, Cambridge, Massachusetts.

Chapter 5 **Maintaining the Integrity of Managed Ecosystems: The Challenges of Preserving Rare Species**

Stanley A. Temple

Why include a chapter on preserving rare species in a discussion of ecosystem management? After all, proponents of ecosystem-scale conservation have often criticized attempts to save individual rare species, especially endangered species, as being major distractions from efforts to keep entire communities of species safe by providing habitat in managed ecosystems (Hutto et al. 1987, Scott et al. 1988). Although many authors have addressed the challenges of rescuing critically endangered species (Temple 1978, Scott et al. 1994), only recently have conservation biologists begun to focus similar attention on keeping populations of rare species—that are not yet endangered—from declining to critically small size (Soulé 1987). Because of the impracticality of managing these rare species one by one, as is typical of endangered species, this challenge will fall largely to ecosystem managers. They will be expected to maintain managed ecosystems in conditions that will keep even naturally rare species from becoming endangered. Retaining these naturally rare species in managed ecosystems is part of maintaining ecosystem integrity (Mitchell et al. 1990). An ecosystem that fails to accommodate the needs of its rarest elements cannot be considered complete and reflects inadequate planning and management.

RARE SPECIES AS IMPORTANT COMPONENTS OF
ECOSYSTEM INTEGRITY

Ecologists have struggled with the challenge of identifying appropriate indicators of ecosystem health (National Academy of Sciences 1986), and various measures of ecosystem integrity have been proposed. Of these, the
presence and persistence of rare species within the biotic community are now
viewed as among the more important indicators of ecosystem integrity (Karr
1990, 1991, 1993). This idea is hardly new, however. The importance of
managing ecosystems to accommodate all species, even the rare ones, was
eloquently promoted almost fifty years ago by Aldo Leopold (1949): "To
keep every cog and wheel is the first precaution of intelligent tinkering."
Leopold (1939) also proposed a "Golden Rule," to which all human actions
in nature—including ecosystem management—should conform: "A thing is
right when it tends to preserve the integrity, stability and beauty of the biotic
community. It is wrong when it tends otherwise." Ecosystem integrity is an
appropriate goal for ecosystem management, and rare species represent one
important component of that integrity.

The logic of this ethic of ecosystem integrity is compelling (Callicott
1991), and it has become widely accepted (Woodley et al. 1993). But what
evidence do we have that rare species are so important in the functioning of
natural ecosystems that we ought to make special efforts to accommodate
them as part of the goal of maintaining integrity? Certainly skeptics can rally
contrary evidence (Main 1982). Many ecosystems have lost abundant members of their biotic community without catastrophic consequences. Our eastern deciduous forest, for example, lost a dominant plant, the American
chestnut *(Castanea dentata),* and a superabundant animal, the passenger pigeon *(Ectopistes migratorius),* over a span of a few decades without becoming obviously dysfunctional (Bormann and Likens 1979).

Although it can be difficult to justify the value of individual rare species,
the importance of rare species derives in part from their prevalence in natural
communities. In most communities, relatively rare species account for the
majority of species present; only a few species are relatively abundant. This
pattern is illustrated clearly in "dominance-diversity" curves for plant communities (Whittaker 1965) and "rank-abundance" curves for animal communities (MacArthur 1960, May 1975). An example from the bird
community of the forest ecosystem in the Apostle Islands of Lake Superior
(Temple 1990) illustrates a typical pattern (fig. 5.1). Eighty-six (79%) of the
community's 109 breeding bird species account for less than 1% each of the
total individuals present in the ecosystem. In short, rare species must be

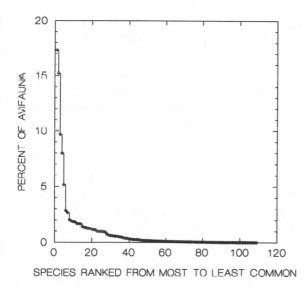

Fig. 5.1. Rank-abundance curve for the land birds of the Apostle Islands, Lake Superior (Temple 1990).

considered important components of integrity because there are so many of them in most natural ecosystems and because the cumulative impacts of their diverse ecological niches must be great.

Furthermore, some individual rare species have interactive roles in an ecosystem that are much more important than their low abundance and biomass suggest. These "keystone species" (Paine 1969, Terborgh 1986, 1988), which can be rare, have widespread impacts on the rest of the biotic community and the entire ecosystem. Loss of a keystone species typically produces "cascading effects" throughout the ecosystem that can result in the loss of additional species (Pimm 1984). These keystone species may be top predators, primary producers that are the base of a food chain, symbiotic partners, or players in other key interactions. Their significance is often obscure until they are missing from an ecosystem (Temple 1977, Owen-Smith 1987). Because of the sheer numbers and potential importance of rare species, ecosystem managers ought to understand them better than they do.

PATTERNS OF RARITY IN NATURE

Because we have only a single adjective, *rare,* to describe species that are relatively scarce, there is a tendency to imagine that they are a uniform group presenting similar challenges. Nothing could be further from the truth. The deceptively simple term belies a complex group of species requiring different management strategies.

Several attempts to categorize rarity have established the twin concepts of prevalence and intensity as basic to understanding patterns of rarity. Prevalence is a measure of the distribution of a species, whereas intensity is a measure of its density. Figure 5.2 illustrates how the concepts of prevalence and intensity can be used to define a range of conditions that we lump together as rarity. Building on these two concepts, Rabinowitz (1981) proposed seven classes of rarity based on the local density of a species (ranging from dense to sparse), its habitat specificity (ranging from broad to narrow), and its geographic range (ranging from extensive to restricted). This scheme can be viewed as an eight-celled matrix (table 5.1) in which one cell (the upper left-hand one) describes abundant species, while the remaining seven cells describe various types of rarity.

For example, consider the following four rare species: The peregrine falcon *(Falco peregrinus)* has a cosmopolitan geographic range and occurs in most of the world's ecosystems, yet it is considered rare because it always exists at relatively low densities. The red mangrove *(Rhizophora mangle)* exists in high-density, largely monotypic stands throughout the tropics, but only where a very specific set of habitat conditions occur. The Aldabra tortoise *(Geochelone gigantea)* exists at high densities throughout the range of habitats on Aldabra, a tiny island in the Indian Ocean. The Philippine eagle *(Pithocophagus philippinensis)* is an extreme example of rarity. It is a sparse top predator found only in mature tropical forests of two islands in the Philippines. In each of these examples, the species fits into a cell of the rarity matrix and is considered to be rare, but there are clearly major dissimilarities in the nature of their rarity. Different types of rarity require very different approaches from conservation biologists intent on maintaining these species in managed ecosystems.

CAUSES OF RARITY

The type of rarity doesn't necessarily tell us much about underlying causes, yet these causes must also be understood to formulate appropriate strategies of ecosystem management. Causes of rarity fall into several broad categories, including environmental stress, resource limitations and interspecific interactions. Environmental stresses are often imposed by physical conditions. For example, species often become rare at the periphery of their range because the environment is too extreme (for example, too hot or cold, wet or dry,

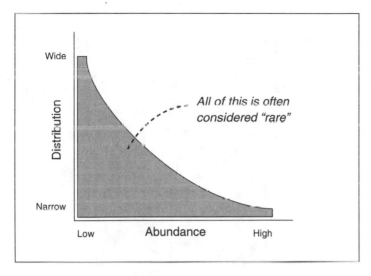

Fig. 5.2. Rarity as a function of prevalence (distribution) and intensity (density) (after Fiedler and Ahouse 1992).

concentrated or dilute). Resource limitations can be imposed by scarcity of a key requirement, such as availability of space (the cause of island endemics' rarity), food (the cause of top predators' rarity), essential symbionts (the cause of obligate mutualists' rarity), or ephemeral habitats (the cause of seral species' rarity). Some species are rarer than the limits imposed by the carrying capacity of their habitat because of interactions with other species. In these cases, predators, competitors, pathogens, parasites, or human beings reduce the abundance or range of a species below the limits set by availability of essential resources.

ACCOMMODATING RARE SPECIES IN MANAGED ECOSYSTEMS

How can details regarding a species' rarity lead to strategies for ensuring its persistence in a managed ecosystem? It is perhaps easiest to frame the process in terms of four major strategic aspects of managing ecosystems: the location of the managed ecosystem, its context, its size, and its internal patchiness or heterogeneity.

Location of a Managed Ecosystem

Decisions about where to locate a managed ecosystem on the landscape are usually made on the basis of factors other than the needs of rare species.

Table 5.1 Classification of rarity based on geographic range, habitat specificity, and local density

Population density	Extensive range, broad specificity	Extensive range, narrow specificity	Restricted range, broad specificity	Restricted range, narrow specificity
Dense	Locally abundant over a large range in several habitats	Locally abundant over a large range in a specific habitat	Locally abundant in several habitats but restricted geographically	Locally abundant in a specific habitat but restricted geographically
Sparse	Constantly sparse over a large range and in several habitats	Constantly sparse in a specific habitat but over a large range	Constantly sparse in a specific habitat but over a large range	Constantly sparse and geographically restricted in a specific habitat

Source: Rabinowitz 1981.

Often the decision comes late enough in the development of a landscape that the position of remnant natural ecosystems is predetermined by previously established, competing land uses. But in cases where there are still options, the needs of rare species might be considered.

Recently, there have been a number of proposals for siting ecosystem reserves on the basis of biodiversity. One approach, exemplified by gap analysis (Scott et al. 1987, 1991a, 1991b), is to seek out areas where the geographic ranges and habitats of a large number of species overlap. Ecosystem reserves sited in these locations might be expected to accommodate a high species richness and make optimal use of managed ecosystems for preserving biological diversity. Another approach, exemplified by the International Council for Bird Preservation's program of identifying priority areas for conservation (Bibby et al. 1992), promotes locating ecosystem reserves in areas with high concentrations of range-restricted, local endemics (Pimm and Gittleman 1992). This approach attempts to accommodate the needs of species that are rare because of their small ranges.

Although each approach has merits, the needs of rare species are not addressed in the same ways. Locating reserves on the basis of species richness alone has a potential for overlooking the habitats of range-restricted, local endemics that are frequently confined to small areas that have unique species but unremarkable species richness. Focusing on areas of high endemism alone could fail to accommodate the needs of species whose rarity derives from their low density, not their restricted distribution.

A compromise that skirts this dilemma and uses the best of both approaches is the "coarse filter/fine filter" strategy adopted by The Nature Conservancy (Noss 1987a, Brown 1991). Under this strategy, most species (perhaps 85–90%) are accommodated in managed ecosystems sited to capture high levels of species richness (the coarse filter), while the needs of unprotected species found in other locations are addressed with reserves specifically sited on the basis of the restricted distributions of rare species (the fine filter).

Context of a Managed Ecosystem

Remnant natural ecosystems are increasingly embedded in landscape matrices dominated by human development. As most landscapes are steadily modified, it seems inevitable that many managed natural ecosystems will eventually become isolated patches in the landscape mosaic (Urban et al. 1987). When that happens, the spatial relations between isolated natural patches and between natural patches and the developed matrix will determine the fate of many rare species (Noss 1983). Two issues, isolation and edge effects, have been identified as threats to some rare species in isolated natural ecosystems (Temple 1994).

Isolation is a serious threat to certain rare species (especially those with low densities and narrow habitat specificity) that become confined to a remnant ecosystem and lose their genetic and demographic linkages to populations in other remnants (Wilcox and Murphy 1985, Pulliam 1988). Once dispersal movements are prevented by the excessive distance between remnants or unsuitable habitat in the matrix, populations of rare species become closed, and the small size of the closed populations of rare species threatens their continued existence (Ehrlich and Murphy 1987, Temple 1991).

Prescriptions for keeping these isolation-sensitive, rare species as viable members of the biotic community have taken three forms: keeping remnant ecosystems close enough together to permit dispersal, linking remnant ecosystems with dispersal corridors, or translocating individuals artificially to recreate dispersal. Each isolation-sensitive, rare species will have its own effective dispersal distance and tolerance for the matrix habitat. Ecosystem managers need to take the life history and ecology of these species into account in planning the spatial relations between remnant natural ecosystems (Temple 1994).

Dispersal corridors can in some situations help ensure the persistence of isolation-sensitive rare species (Simberloff and Cox 1987, Hudson 1991, Saunders and Hobbs 1991, Simberloff et al. 1992). The species of concern must be shown to use the corridors for dispersal movements, and the risks associated with the corridors must be assessed before these types of landscape linkages are created (Noss 1987b, Soulé 1991, Harrison 1992).

If isolated natural ecosystems cannot be connected or located in close proximity, populations of isolation-sensitive species can still be linked through translocations that mimic the scale of natural dispersal (Allendorf 1983, Lacy 1987, Griffith et al. 1989). Such translocations can serve to reduce inbreeding rates, to provide population "rescue effects" (Brown and Kodric-Brown 1977), or to reestablish a population of a rare species that has been extirpated.

The landscape surrounding a managed ecosystem affects rare species through edge effects as well as isolation. Wherever dissimilar landscape elements are abruptly juxtaposed, boundary or edge effects are inevitable (Harris 1988). Both the physical and biotic environments near edges are distinctly different from environments deep in the interior of an ecosystem remnant. A variety of edge-related phenomena (changes in rates of predation, browsing, competition, and other biotic interactions, as well as changes in microclimate) cause problems for a group of rare species with narrow habitat requirements that do not exist near alien habitats along edges (Janzen 1986, Schonewald-

Cox 1988). These edge-sensitive species can become increasingly rare in remnant ecosystems that are dominated by edges if the managed area is small in size or has a shape that precludes the existence of a core area far from edges (Temple 1986, Temple and Cary 1988). Ensuring the viability of edge-sensitive, rare species involves planning the size and shape of a managed ecosystem to minimize edges where external threats occur (Harris 1984, Hunter 1990, Shafer 1990). Once edge-sensitive rare species have been identified, estimation of the distances from edges that ensure their fitness can guide the design process.

The spatial scale of concern to ecosystem managers can be extensive for some rare species, especially seasonal migrants that depend on widely separated habitat patches. In some cases, a migratory species may be rare in one seasonally occupied ecosystem because its population size is limited elsewhere (either along the migratory pathway or in the other ecosystem it occupies). In the complexity of such cases, an ecosystem manager must carefully evaluate when and where the population's size is being limited in order to direct management activities effectively (Temple 1988).

Size of a Managed Ecosystem

Perhaps no other aspect of managed ecosystems has received as much attention as size or area (Shafer 1990). Guided initially by the patterns discovered by island biogeographers (Diamond 1975, Simberloff and Abele 1982), ecosystem managers recognize that the smaller a remnant ecosystem is the fewer low-density, rare species it will accommodate. The classic species-area curve demonstrates the rapid loss of rare species from ecosystems below a threshold area. An example from birds in the Apostle Islands is provided in figure 5.3. It has been suggested that this pattern reveals a "minimum critical area" for an ecosystem that will accommodate an intact biological community with viable populations of nearly all its species (Lovejoy and Oren 1981, Usher 1986). Conner (1988) cautions that a population must typically be greater than its minimum viable size to be ecologically functional within its ecosystem. A "minimum dynamic area" is large enough to permit dynamic fluctuations while accommodating most species associated with the ecosystem's biotic community (Pickett and Thompson 1978, White 1987).

Area-related losses fall primarily on members of the community that normally exist at low densities because of their life history and ecology. The lower their natural density the smaller the size of the population in a remnant. If the largest population that could exist in a managed ecosystem is below viable size, its extirpation becomes increasingly likely. The technique of population viability analysis (Shaffer 1981, Soulé 1987) allows managers to ap-

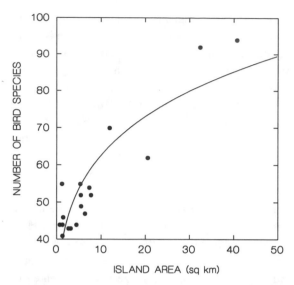

Fig. 5.3. A species-area curve for land birds of the Apostle
Islands, Lake Superior (Temple 1990).

proximate the population size that provides an acceptable level of security
for the population. Multiplying the viable population size by the species'
density yields a rough estimate of the area required to support a viable population (Temple 1992).

Many area-sensitive species are important though rare members of an ecosystem's biotic community. They include top predators, large-sized organisms, and ecological specialists. Failure to provide for the spatial needs of
these naturally rare species leads to their extirpation.

Internal Heterogeneity of Managed Ecosystems

Most natural ecosystems exist as a patchwork of ecological conditions created
by disturbance and irregularities in the environment (Pickett and White 1985).
Disturbances typically cause a mosaic of patches in various successional
stages (Harris 1984), while natural irregularities impose variety in the climax
conditions on a particular site. This internal heterogeneity accommodates a
variety of rare organisms that are narrow habitat specialists. Some are associated only with ephemeral seral conditions, while others are associated only
with unique edaphic or microclimatic conditions. In all cases, these rare species depend on the existence of a network of patches where their unique
habitat conditions exist (Pickett and Thompson 1978, White 1987).

A managed ecosystem will retain these rare habitat specialists only if it provides a spatially and temporally appropriate mosaic of conditions within a minimum dynamic area. These patterns must reflect the needs of dynamic metapopulations. Most of these rare species exist as a network of subpopulations linked to one another by dispersal movements and governed by local extinction and recolonization (that is, a metapopulation). In the case of disturbance-dependent species, as one subpopulation declines and disappears when succession overtakes its ephemeral habitat, new subpopulations are forming as dispersing individuals colonize new patches of habitat created by disturbance (Gilpin 1987, Harrison and Quinn 1989, Hanski 1989). The continued existence of such rare species depends on a dynamic, shifting mosaic of patches in various stages of succession (Menges 1990, Murphy et al. 1990). Most ecosystems have characteristic disturbance regimes that keep predictable proportions of the ecosystem in various seral stages (Pickett and Thompson 1978). In some ecosystems, mature (old-growth) conditions dominate, while in others earlier seral stages are more prevalent. Managers must be aware of these patterns created by natural disturbance regimes and try to mimic them in managed ecosystems lest rare species associated with a particular stage be discriminated against.

Rare, narrow-habitat specialists associated with unique local conditions can be especially challenging to ecosystem managers. Their metapopulations are frequently composed of such widely separated subpopulations that any individual managed ecosystem may contain few (Bleich et al. 1990). In such cases, attention to subpopulations outside the managed area may be important to maintain subpopulations within (Noss and Harris 1986).

RARE SPECIES AS CORNERSTONES OF MANAGED ECOSYSTEMS

In some cases, managers plan for the needs of rare species after many features of a managed ecosystem have already been established. But there are also cases in which the needs of rare species are paramount, and the entire managed ecosystem revolves to some extent around them (Miller and Bratton 1987). This situation has become more prevalent since specific rare species (endangered species) have been singled out for special conservation attention. The Endangered Species Act of 1973, for example, mandates that the needs of listed species receive high priority. In response to the special status of endangered species, conservationists may design some managed ecosystems largely to accommodate one or a few species.

This approach is not necessarily at odds with the more holistic goals of managing an ecosystem to preserve its entire biotic community (Noss 1990). When the featured rare species are particularly demanding ones, meeting their needs can coincidentally accommodate the needs of many other less-demanding species (Eisenberg 1980, Myers 1987, Eisenberg and Harris 1989). These "umbrella" species, which are often rare, have been proposed as surrogates for accommodating the needs of an entire community. In the case of some rare top predators, it may be correct that if their populations are viable and functional, the ecosystem is likely to maintain its integrity, although not all species will be accommodated equally (Murphy and Wilcox 1986).

Caution is necessary, however, to avoid the pitfall of engaging in narrow, species-specific management that addresses only the needs of a featured species, especially if it is at the expense of other nonfeatured species (Landres et al. 1988). Umbrella species are most useful to ecosystem planning if they are managed in the context of ecosystem integrity. A similar situation exists for "indicator species," some of which are rare, chosen to be surrogates for their ecosystem. Managing an ecosystem primarily to accommodate these species has many well-recognized risks (Landres et al. 1988).

Rare species of all types are important components of managed ecosystems. They represent some of the most fragile elements of ecosystem integrity, and their continued existence in a managed ecosystem is a good indicator of successful planning and management. Meeting their needs requires careful attention to features—such as the location, size, context, and structure of the ecosystem—that are also important for maintaining biological diversity. Guidelines to help managers accommodate rare species and biodiversity are continually being developed and refined by conservation biologists. In this rapidly changing arena, ecosystem managers must stay aware of emerging concepts and shifting paradigms (Murphy 1989). Nonetheless, the guiding principles and enabling technologies for accommodating rare species within managed ecosystems must be developed rapidly to counterbalance the ever-expanding threats to their continued existence.

LITERATURE CITED

Allendorf, F. 1983. Isolation, gene flow, and genetic differentiation among populations. Pages 51–65 *in* C. Schonewald-Cox et al., editors. Genetics and conservation: a reference for managing wild animal and plant populations. Benjamin/Cummings, Menlo Park, California.

Bibby, C., N. Collar, M. Crosby, M. Heath, C. Imboden, T. Johnson, A. Long, A.

Sattersfield, and S. Thirgood. 1992. Putting biodiversity on the map: priority areas for global conservation. International Council for Bird Preservation, Cambridge, England.

Bleich, V., J. Wehausen, and S. Holl. 1990. Desert-dwelling mountain sheep: conservation implications of a naturally fragmented distribution. Conservation Biology 4:383–389.

Bormann, F. H., and G. E. Likens. 1979. Pattern and process in a forested ecosystem. Springer-Verlag, New York, New York.

Brown, B. 1991. Landscape protection and The Nature Conservancy. Pages 66–71 in W. Hudson, editor. Landscape linkages and biodiversity. Island Press, Washington, D.C.

Brown, J. H., and A. Kodric-Brown. 1977. Turnover rates in insular biogeography: effects of immigration on extinction. Ecology 58:445–449.

Callicott, J. B. 1991. Conservation ethics and fishery management. Fisheries 16:22–28.

Conner, R. N. 1988. Wildlife populations: minimally viable or ecologically functional. Wildlife Society Bulletin 16:80–84.

Diamond, J. M. 1975. The island dilemma: lessons of modern biogeographic studies for the design of natural reserves. Biological conservation 7:129–146.

Ehrlich, P., and D. Murphy. 1987. Conservation lessons from long-term studies of checkerspot butterflies. Conservation Biology 1:122–131.

Eisenberg, J. F. 1980. The density and biomass of tropical mammals. Pages 35–55 in M. Soulé and B. Wilcox, editors. Conservation Biology: an evolutionary-ecological perspective. Sinauer, Sunderland, Massachusetts.

Eisenberg, J. F., and L. Harris. 1989. Conservation: a consideration of evolution, population and life history. Pages 99–108 in D. Western and M. Pearl, editors. Conservation for the twenty-first century. Oxford University Press, New York, New York.

Fiedler, P., and J. Ahouse. 1992. Hierarchies of cause: toward an understanding of rarity in vascular plant species. Pages 23–47 in P. Fiedler and S. Jain, editors. Conservation biology: the theory and practice of nature conservation, preservation and management. Chapman and Hall, New York, New York.

Gilpin, M. E. 1987. Spatial structure and population vulnerability. Pages 125–139 in M. Soulé, editor. Viable populations for conservation. Cambridge University Press, Cambridge, England.

Griffith, B., J. M. Scott, J. W. Carpenter, and C. Reed. 1989. Translocation as a species conservation tool: status and strategy. Science 245:477–480.

Hanski, I. 1989. Metapopulation dynamics: does it help to have more of the same? Trends in Ecology and Evolution 4:113–114.

Harris, L. 1984. The fragmented forest: island biogeographic theory and the preservation of biological diversity. University of Chicago Press, Chicago, Illinois.

———. 1988. Edge effects and conservation of biological diversity. Conservation Biology 2:330–332.

Harrison, R. 1992. Toward a theory of inter-refuge corridor design. Conservation Biology 6:293–295.

Harrison, S., and J. Quinn. 1989. Correlated environments and the persistence of metapopulations. Oikos 56:293–298.

Hudson, W. E., editor. 1991. Landscape linkages and biodiversity. Island Press, Washington, D.C.

Hunter, M. L. 1990. Wildlife, forests, and forestry: principles of managing forests for biological diversity. Prentice-Hall, Englewood Cliffs, New Jersey.

Hutto, R. L., S. Reel, and P. B. Landres. 1987. A critical evaluation of the species approach to biological conservation. Endangered Species Update 4:1–4.

Janzen, D. 1986. The eternal external threat. Pages 286–303 in M. Soulé, editor. Conservation biology: the science of scarcity and diversity. Sinauer, Sunderland, Massachusetts.

Karr, J. R. 1990. Biological integrity and the goal of environmental legislation: lessons for conservation biology. Conservation Biology 4:244–250.

———. 1991. Biological integrity: a long-neglected aspect of water resource management. Ecological Applications 1:66–84.

———. 1993. Measuring biological integrity: lessons from streams. Pages 83–104 in Ecological integrity and the management of ecosystems. St. Lucie Press, Delray Beach, Florida.

Lacy, R. 1987. Loss of genetic diversity from managed populations: interacting effects of drift, mutation, immigration, selection and population subdivision. Conservation Biology 1:143–158.

Landres, P. B., J. Verner, J. W. Thomas. 1988. Ecological uses of vertebrate indicator species: a critique. Conservation Biology 2:316–328.

Leopold, A. 1939. A biotic view of the land. Journal of Forestry 37:727–730.

———. 1949. A Sand County almanac. Oxford University Press, Oxford, England.

Lovejoy, T., and D. Oren. 1981. The minimum critical size of ecosystems. Pages 7–12 in R. Burgess and D. Sharpe, editors. Forest island dynamics in man-dominated landscapes. Springer-Verlag, New York, New York.

MacArthur, R. 1960. On the relative abundance of species. American Naturalist 94:25–36.

Main, A. R. 1982. Rare species: precious or dross? Pages 163–179 in R. H. Groves and W. Ride, editors. Species at risk: research in Australia. Australian Academy of Science, Canberra.

May, R. 1975. Patterns of species abundance and diversity. Pages 81–120 in M. Cody and J. Diamond, editors. Ecology and evolution of communities. Belknap Press, Cambridge, Massachusetts.

Menges, E. 1990. Population viability analysis for an endangered plant. Conservation Biology 4:52–62.

Miller, R. I., and S. P. Bratton. 1987. A regional strategy for reserve design and placement based on an analysis of rare and endangered species' distribution patterns. Biological Conservation 39:255–268.

Mitchell, R., C. Sheviak, and D. Leopold (editors). 1990. Ecosystem management: rare species and significant habitats. New York State Museum Bulletin No. 471.

Murphy, D. 1989. Conservation and confusion: wrong species, wrong scale, wrong conclusions. Conservation Biology 4:203–204.

Murphy, D., K. Freas, and S. Weiss. 1990. An environment metapopulation approach to population viability analysis for a threatened invertebrate. Conservation Biology 4:41–51.

Murphy, D., and D. Wilcox. 1986. Butterfly diversity in natural habitat fragments: a test of the validity of vertebrate-based management. Pages 287–292 in J. Verner, M. Morrison, and C. J. Ralph, editors. Wildlife 2000: modelling habitat relationships of terrestrial vertebrates. University of Wisconsin Press, Madison, Wisconsin.

Myers, N. 1987. The extinction spasm impending: synergisms at work. Conservation Biology 1:14–21.

National Academy of Sciences. 1986. Ecological knowledge for environmental problem solving: concepts and case studies. National Academy Press, Washington, D.C.

Noss, R. 1983. A regional landscape approach to maintain diversity. BioScience 33: 700–706.

———. 1987a. From plant communities to landscapes in conservation inventories: a look at The Nature Conservancy (USA). Biological Conservation 41:11–37.

———. 1987b. Corridors in real landscapes: a reply to Simberloff and Cox. Conservation Biology 1:159–164.

———. 1990. Indicators for monitoring biodiversity: a hierarchical approach. Conservation Biology 4:355–364.

Noss, R., and L. Harris. 1986. Nodes, networks and MUMS: preserving diversity at all scales. Environmental Management 10:299–309.

Owen-Smith, N. 1987. Pleistocene extinctions: the pivotal role of mega-herbivores. Paleobiology 13:351.

Paine, R. T. 1969. A note on trophic complexity and community stability. American Naturalist 103:91–93.

Pickett, S., and J. Thompson. 1978. Patch dynamics and the design of nature reserves. Biological Conservation 13:27–37.

Pickett, S., and D. White (editors). 1985. The ecology of natural disturbance and patch dynamics. Academic Press, New York, New York.

Pimm, S. L. 1984. The complexity and stability of ecosystems. Nature 307:321–326.

Pimm, S. L., and J. L. Gittleman. 1992. Biological diversity: where is it? Science 255:940.

Pulliam, H. R. 1988. Sources, sinks, and population regulation. American Naturalist 132:652–661.

Rabinowitz, D. 1981. Seven forms of rarity. Pages 205–217 in H. Synge, editor. The biological aspects of rare plant conservation. J. Wiley & Sons, New York, New York.

Saunders, D. A., and R. J. Hobbs (editors). 1991. The role of corridors in nature conservation. Surrey Beatty & Sons, Sydney, Australia.

Schonewald-Cox, C. M. 1988. Boundaries in the protection of nature reserves. BioScience 38:480–486.

Scott, J. M., B. Csuti, and S. Caicco. 1991a. Gap analysis: assessing protection needs. Pages 15–26 in W. Hudson, editor. Landscape linkages and biodiversity. Island Press, Washington, D.C.

Scott, J. M., B. Csuti, and F. Davis. 1991b. Gap analysis: an application of geographic information systems for wildlife species. Pages 167–179 in D. Decker, M. Krasny, G. Goff, C. Smith, and D. Gross, editors. Challenges in the conservation of biological resources: a practitioner's guide. Westview Press, Boulder, Colorado.

Scott, J. M., B. Csuti, J. D. Jacobi, and J. E. Estes. 1987. Species richness: a geographic approach to protecting future biological diversity. BioScience 37:782–788.

Scott, J. M., B. Csuti, K. Smith, J. E. Estes, and S. Caicco. 1988. Beyond endangered species: an integrated conservation strategy for the preservation of biological diversity. Endangered Species Update 5:43–48.

Scott, J. M., S. A. Temple, D. Harlow, and M. Shaffer. 1994. Restoration and management of endangered species. Pages 531–539 in T. A. Bookhout, editor. Research and management techniques for wildlife and habitats. The Wildlife Society, Bethesda, Maryland.

Shafer, C. L. 1990. Nature reserves: island theory and conservation practice. Smithsonian Institution Press, Washington, D.C.

Shaffer, M. 1981. Minimum population sizes for species conservation. BioScience 31:131–134.

Simberloff, D., and L. Abele. 1982. Refuge design and island biogeographic theory: effects of fragmentation. American Naturalist 120:41–50.

Simberloff, D., and J. Cox. 1987. Consequences and costs of conservation corridors. Conservation Biology 1:63–71.

Simberloff, D., J. A. Farr, J. Cox, and D. Mehlman. 1992. Movement corridors: conservation bargains or poor investments. Conservation Biology 6:493–505.

Soulé, M. 1991. Theory and strategy. Pages 91–104 in W. Hudson, editor. Landscape linkages and biodiversity. Island Press, Washington, D.C.

——— (editor). 1987. Viable populations for conservation. Cambridge University Press, Cambridge, England.

Temple, S. A. 1977. Plant-animal mutualism: coevolution with dodo leads to near extinction of plant. Science 197:885–886.

———. 1986. Predicting impacts of fragmentation on forest birds: a comparison of two models. Pages 301–304 in J. Verner, M. Morrison, and C. J. Ralph, editors. Wildlife 2000: modelling habitat relationships of terrestrial vertebrates. University of Wisconsin Press, Madison, Wisconsin.

———. 1988. What's behind declines in bird populations? Passenger Pigeon 50:133–138.

———. 1990. Patterns of abundance and diversity of birds on the Apostle Islands. Passenger Pigeon 52:219–224.

———. 1991. The role of dispersal in the maintenance of bird populations in a fragmented landscape. Acta Congressus Internationalis Ornithologici 20:2298–2305.

———. 1992. Population viability analysis of a sharp-tailed grouse metapopulation in Wisconsin. Pages 750–758 in D. R. McCollough and R. Barrett, editors. Wildlife 2001: populations. Elsevier Applied Science, London.

———. 1994. A natural history of fragmented landscapes. Miscellaneous publication of State University of New York College of Environmental Science and Forestry, Syracuse, New York.

——— (editor). 1978. Endangered birds: management techniques for preserving threatened species. University of Wisconsin Press, Madison, Wisconsin.

Temple, S. A., and J. R. Cary. 1988. Modelling the dynamics of forest-interior bird populations in a fragmented landscape. Conservation Biology 2:341–347.

Terborgh, J. 1986. Keystone plant species in the tropical forest. Pages 330–340 in M. Soulé, editor. Conservation Biology. Sinauer, Sunderland, Massachusetts.

———. 1988. The big things that run the world: a sequel to E. O. Wilson. Conservation Biology 2:402–403.

Urban, D., R. O'Neill, and H. Shugart. 1987. Landscape ecology. BioScience 37:119–127.

Usher, M. B. 1986. Wildlife conservation evaluation: attributes, criteria and values. Pages 3–44 in M. B. Usher, editor. Wildlife conservation evaluation. Chapman and Hall, London.

Whittaker, R. H. 1965. Dominance and diversity in land plant communities. Science 147:250–260.

White, P. S. 1987. Natural disturbance, patch dynamics and landscape patterns in natural areas. Natural Areas Journal 7:14–22.

Wilcox, B. A., and D. D. Murphy. 1985. Conservation strategy: the effects of fragmentation on extinction. American Naturalist 125:879–887.

Woodley, S., J. Kay, and G. Francis. 1993. Ecological integrity and the management of ecosystems. St. Lucie Press, Delray Beach, Florida.

Chapter 6 **Managing the Invisible: Ecosystem Management and Macronutrient Cycling**
Clive A. David

Forest management practices frequently alter nutrient cycling processes and budgets, and thus potentially affect ecosystem productivity. Yet little attention has been paid to the implications of altered nutrient cycles in alternative management systems, even though they are fundamental to sustainability. In this chapter, I examine ecosystem management techniques in relation to macronutrient cycling in forested ecosystems. This chapter is not intended to be an extensive review of the literature on nutrient cycles, nor is it a detailed discussion of the related scientific issues. I have aimed to provide useful information to forest managers, administrators, and field personnel who have had only an introduction to nutrient cycling concepts and issues. Specifically, my objectives are to (i) review important concepts of macronutrient cycling, (ii) discuss potential impacts of proposed ecosystem management techniques on macronutrient cycling, and (iii) suggest actions for managers. The emphasis is on temperate forest ecosystems of North America.

Macronutrients are elements used in relatively large quantities by plants. By definition, they attain concentrations greater than 500 mg kg⁻¹ in tissues of mature plants. Nitrogen (N), phosphorus (P), potassium (K), calcium (Ca), magnesium (Mg), and sulfur (S) are included in this group; carbon (C), hydrogen (H), and oxygen (O) are traditionally excluded (Soil Science Society of America 1987). Many processes and pathways are involved in the cycling and availability of macronutrients in terrestrial ecosystems (fig. 6.1). The sources and forms of these nutrients (table 6.1) affect both their rates of supply to plants and the potential for their losses from systems. Forested ecosystems represent a wide variety of abiotic and biotic conditions across the continent, with concomitant ranges of variation in the magnitudes, rates, and patterns of nutrient cycling. Each nutrient has its own relations and circulation patterns within and among ecosystems (Stone 1975b, 1979, Cole and Rapp 1981). More detailed discussions of nutrient cycling specific to forested systems and their management can be found elsewhere (e.g., Bormann and Likens 1979, Clark and Rosswall 1981, Stewart and Rosswall 1982, Stone 1984, Gessel et al. 1990, Harvey and Neuenschwander 1991).

ECOSYSTEM MANAGEMENT TECHNIQUES

How is ecosystem management different from traditional forest management? Although the concept of ecosystem management is still evolving, a few principles emerge. In the context of this chapter, perhaps the most important of these are *sustainability* and *ecological integrity*. Sustainability derives from our expectation that, with proper management, ecosystems will continually provide goods and services in perpetuity (Franklin this volume). Some argue that this can be achieved only when natural ecological processes are maintained. They advocate (i) setting goals for production of goods and services within the ecological constraints of the ecosystem, and (ii) maintaining ecosystem structural complexity, function, and processes over a broad range of spatial scales (Gosz 1992, Grumbine 1994, Salwasser and Pfister 1994, Franklin this volume). What constitutes ecological integrity is not well defined. Some view the fully functional ecosystem as representing ecological integrity; others consider the following to be necessary to achieve integrity (Grumbine 1994):

(i) Viable populations of native species;
(ii) Maintenance of natural disturbance regimes;
(iii) Reintroduction of native extirpated species; and
(iv) Representation of the natural range of variation of ecosystems.

We do not know to what extent these components of ecosystem integrity are necessary to achieve sustainability.

Despite our uncertainty surrounding "correct" ecosystem management, several forest management practices are thought to contribute to perpetuation of integrity and sustainability (table 6.2). Suggested practices reflect concerns

Table 6.1 Summary of the principal sources and ionic forms of macronutrients in forest soils

Element	Solid phase sources (unavailable)	Principal ionic forms	Available forms in acid forest soils	References
Nitrogen	Proteins; other organic compounds	NH_4^+, NO_3^-, NO_2^-	NH_4^+, NO_3^-	Carlyle (1986), Pritchett and Fisher (1987), Troeh and Thompson (1993)
Phosphorus	Apatite; Fe-, Al-, and Ca-phosphates; organic compounds	$H_2PO_4^-, HPO_4^{2-}$	$H_2PO_4^-, HPO_4^{2-}$	Brady (1990), Krause (1991), Troeh and Thompson (1993)
Potassium	Micas; feldspars; other minerals	K^+	K^+	Brady (1990), Krause (1991), Troeh and Thompson (1993)
Sulfur	Proteins; other organic compounds; pyrites, gypsum	S^{2-}, SO_4^{2-}	SO_4^{2-}	Brady (1990), Krause (1991), Troeh and Thompson (1993)
Calcium	Calcite; gypsum; feldspars; other minerals	Ca^{2+}	Ca^{2+}	Pritchett and Fisher (1987), Krause (1991), Troeh and Thompson (1993)
Magnesium	Ferromagnesian minerals; dolomite; other minerals	Mg^{2+}	Mg^{2+}	Pritchett and Fisher (1987), Krause (1991), Troeh and Thompson (1993)

about loss of biological diversity, which has been a principal force behind the move toward ecosystem management (Probst and Crow 1991, Grumbine 1994). However, many actions and treatments intended to protect or enhance diversity are based on hypotheses that have not been fully tested (Norcross 1991, DeBell and Curtis 1993, Franklin this volume). For this reason, an adaptive management approach to ecosystem management is recommended (Franklin this volume). Ecosystem management does not require manipulation of the entire landscape; neither does it demand its application on every management unit (Seymour and Hunter 1992, Franklin this volume). Managers can modify their practices on one or more units while using traditional practice on others for comparison.

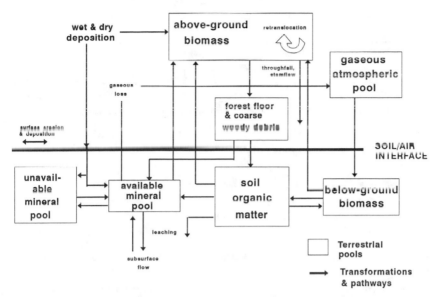

Fig. 6.1. Generalized representation of the principal macronutrient pools, transformations, and pathways involved in nutrient cycling under natural conditions in forested ecosystems. Definitions or descriptions of the terms and processes involved can be found in Swift et al. 1979, Carlyle 1986, Pritchett and Fisher 1987, Brady 1990, Bockheim and Leide 1990, and Aber and Melillo 1991.

Silvicultural Systems, Spatial Patterns, and Biological Legacies

Modified silvicultural systems have been developed for the Pacific Northwest and the Northeast (table 6.3). At first glance, the western systems appear to differ little from traditional systems (see Smith 1986, e.g.) in the degree of stand removals (USDA Forest Service 1990). Seymour and Hunter (1992) viewed two-aged systems as mimicking the natural disturbance patterns of spruce-fir (*Picea* spp.-*Abies* spp.) forests in Maine. For these forests, they proposed a "two-cohort, irregular shelterwood system." Two-aged silvicultural systems have been recognized only recently as a separate, legitimate set of silvicultural methods (Society of American Foresters 1993). The two-cohort system (Seymour and Hunter 1992) does not seem to depart markedly from the basic concept of irregular shelterwood systems. However, irregular shelterwood systems have been used much less in North America than their even-aged and uneven-aged counterparts (Smith 1986, Seymour 1992, Seymour and Hunter 1992).

Table 6.2 Summary of recommended forest management practices for achievement of ecosystem management objectives

Practice	Spatial scale	Importance	References
Creation/retention of biological legacies (large green trees, snags, coarse woody debris)	Site	Wildlife, invertebrate habitat; soil fertility; microorganisms; structural complexity; regeneration	Franklin (1989, 1993), Hopwood (1991), Gosz (1992), Seymour and Hunter (1992)
Enhancement of biological diversity (genetics, species, ecosystem)	Site; landscape	Habitat as above; system complexity, resilience, integrity	Salwasser (1990), Hopwood (1991), Probst and Crow (1991), Gosz (1992)
Use of spatial patterns, spatial scales, and time scales appropriate to ecosystem features	Site; landscape	Complexity, diversity as above; ecosystem recovery	Franklin (1989, 1993), Kimmins (1989), Probst and Crow (1991), Hopwood (1991), Brooks and Grant (1992), Seymour and Hunter (1992)
Reintroduction of preintervention (extirpated) species	Site; landscape	Complexity, diversity as above	Probst and Crow (1991), Seymour and Hunter (1992)
Control of cumulative effects of operations on watersheds and landscape	Landscape	Aquatic habitat, riparian zones, channel stability	Hopwood (1991), Probst and Crow (1991)

Because of the overall similarity between the recommended and traditional silvicultural systems, it is unlikely that great differences in nutrient cycling would emerge. Much would depend on what debris and vegetation materials are considered to be part of a "stand" (table 6.3), and on how residues from these materials are distributed. The timber extraction system, species composition, and site quality also will influence nutrient cycling responses. For example, whole-tree harvesting can increase nutrient removal by three to six times (Stone 1979). This harvesting system also removes much of the woody debris that ecosystem management recommends leaving in the forest. Whole-tree harvesting of deciduous species

Table 6.3 Examples of proposed silvicultural systems for ecosystem management

Modified system	% stand retained	Root traditional system	References
Even-aged			
Light conifer retention	10–30	Patch clearcut	USDA Forest Service (1990), Smith (1986)
Moderate conifer retention	20–50	Seed-tree	USDA Forest Service (1990), Smith (1986)
Heavy conifer retention	20–70	Uniform shelterwood	USDA Forest Service (1990), Smith (1986)
Group removal	70–90	Group shelterwood	USDA Forest Service (1990), Smith (1986)
Two-aged			
Two-cohort irregular shelterwood	(not stated)	Irregular shelterwood	Seymour and Hunter (1992)
Uneven-aged			
Group removal	70–90/entry	Group selection	USDA Forest Service (1990)

may result in much higher nutrient loss than similar harvesting of coniferous species.

While silvicultural systems under ecosystem management (table 6.3) retain much of the character of the traditional silvicultural systems, they differ in principle by including and emphasizing one or more ecosystem management practices (table 6.2) (USDA Forest Service 1990). The priority in ecosystem management is maintaining ecosystem integrity rather than commodity production. Retaining substantial biological legacies—in particular, dead and down material and structure—is perhaps the most significant departure from traditional practice. Removal of dead and down material by conventional site preparation can greatly reduce the biological diversity of a site (Hunter 1990). Retention of at least five to ten snags per hectare is desirable. Franklin (1989) recommended that twenty to thirty-seven green trees per hectare should be left during the entire second rotation. Hopwood (1991) presented a vivid illustration of the marked difference in character between harvested sites with biological legacies retained and those logged using traditional practices.

Biological Diversity

Biological diversity enhances the functioning, productivity, and stability of ecosystems (Baskin 1994, Tilman and Downing 1994). Effects of increasing species richness on function and stability appear to be greatest where relatively low numbers of species are involved (Tilman and Downing 1994), a common situation in many intensely managed temperate forest ecosystems.

Plant species mixtures have been advocated as a means of increasing soil fertility and site productivity in a variety of forest ecosystems characterized by few species. Mixtures may enhance ecosystem function and stability in such systems (Binkley 1986, Aber and Melillo 1991, Kelty 1992). Fisher (1990), however, argues that much of the evidence supporting the belief that the presence of tree species ameliorates soils is anecdotal and that more rigorous investigation is needed. Clearly, different species have different effects on nutrient cycles. Therefore, a subtle distinction must be made between mixtures that truly increase soil fertility and those that reallocate portions of a relatively fixed nutrient reservoir among ecosystem pools and compartments. For example, starting from similar conditions on old agricultural fields in the Lake States, aspen *(Populus tremuloides)* and white spruce *(Picea glauca)* plantation sites had more than 40% of their Ca pool in the forest floor or aboveground vegetation after forty years. By contrast, two corresponding pine *(Pinus* spp.) plantation sites held more than 75% of their Ca pool in the mineral soil. The total mass of Ca was the same for all four ecosystems. Compared to the pine sites, the aspen and spruce sites exhibited less acid forest floor layers but more acid A horizons (Aber and Melillo 1991). The aspen and spruce acted as "cation pumps" (Pritchett and Fisher 1987, Aber and Melillo 1991) to increase Ca allocations to aboveground compartments and alter forest floor and mineral soil chemical characteristics. By definition, enhanced fertility potentially derives from (i) increased abundance (total amounts) of nutrients in the soil, (ii) increases in the relative proportions of plant-available nutrient pools in the soil, or (iii) increased rates of supply of plant-available forms of nutrients (Soil Science Society of America 1987). Species that act as cation pumps or that otherwise affect nutrient allocations do not materially enhance *fertility* yet may enhance nutrient availability and cycling. Stone (1975a) considered that species with nitrogen-fixing symbionts are the only ones that increase total amounts of a nutrient (N).

The principle of maintaining or increasing biological diversity also applies to soil microbial and faunal populations, but this has received little attention in forest management. It is well-known that bacteria, fungi and other microflora influence nutrient availability and system productivity by mediating such processes as decomposition, mineralization, and nutrient transformations and by forming important symbiotic relationships (for example, for dinitrogen fixation,* mycorrhizae) with higher plants (Alexander 1977, Swift et al. 1979, Carlyle 1986, Pritchett and Fisher 1987). However, soil fauna of all sizes also exert major influences on nutrient cycling processes. For example, faunal communities affect decomposition rates both directly (for example, by assimilation and excretion) and indirectly (for example, by comminution of materials and microbial grazing) (Swift et al. 1979, Setala and Huhta 1990). In temperate deciduous forests, earthworms can increase litter decomposition rates greatly over similar forests lacking such populations (Staaf 1987, Alban and Berry 1994). For example, earthworms introduced to a Minnesota forest site caused a mor humus form to disappear and be incorporated with the mineral soil within fourteen years (Alban and Berry 1994). The chemical composition of woody materials makes them relatively resistant to decay, and their decomposition affects nutrient availability and retention (Ausmus 1977, Swift et al. 1979). Four major stages in wood decomposition—colonization, exploitation, invasion, and postinvasion—each involve a different suite of organisms. Arthropod and annelid populations play a major role in regulating processes in the colonization and invasion stages (Ausmus 1977, Swift et al. 1979). Soil megafauna also can affect ecosystem productivity; for example, burrowing by crabs contributed to increased productivity and reproductive activity of mangrove (*Rhizophora* spp.) forests (Smith et al. 1991). Such burrowing and digging animals as moles, skunks, and marmots help maintain a porous soil and incorporate organic matter. All of these organisms should respond positively to retention of woody debris, organic matter, and structure in the managed forest.

Spatial and Temporal Scales

Ecosystem management should be considered in the context of both spatial and temporal scales (Grumbine 1994). Woodmansee (1990) described a useful hierarchy of ecological units for considering spatial scales. The smallest biological units are the cell, organ, and organism; their physical analogs are the soil aggregate, horizon, and pedon. The next level is the site (landscape element or ecotope), and beyond that lie the site cluster, landscape, and so on. The site is a fundamental unit, composed of a biotic community existing on a soil polypedon. It suggests a spatial scale ranging from approximately 100 m² to perhaps one square kilometer (Woodmansee 1990, Zonneveld

---

* Conversion of molecular nitrogen to ammonia and subsequently to organic combinations (Soil Science Society of America 1987).

1990). Such a range encompasses the scales of most macronutrient cycling processes. The following discussion focuses upon activities and processes at the site level.

Under ecosystem management, one of the goals that contributes to sustaining ecological integrity is that of using appropriate time scales. Management periods that maintain the evolutionary potential of species and ecosystem should be employed (Grumbine 1994). Kimmins (1989) calls for managers to use "ecological rotation," defined as the time taken by an ecosystem to recover its predisturbance (or some desired) condition following disturbance. In considering landscapes, Turner et al. (1993) question whether the concept of return to a predisturbance state or equilibrium (homeostatic stability) should be applied to communities that are constantly changing and adapting to their environment. Instead, they argue that return to the normal dynamics (termed homeorhetic stability) of a system is more applicable. In either case, estimated recovery rates will vary depending on the attributes used as indexes of recovery. Some processes may require nearly 200 years (Bormann and Likens 1979). In Oregon, rotations of 130–250 years have been planned for stands that would otherwise have been managed on a 90–100-year rotation (Hopwood 1991). Thus, ecosystem management may involve rotation lengths of as much as two times those commonly used for timber production in many northern forests.

MACRONUTRIENT CYCLES AND AVAILABILITY

Nutrient pools and pathways vary according to the specific element and site conditions involved. Reliable estimates of such fluxes as weathering and plant uptake have been difficult to obtain (Pritchett and Fisher 1987, Bockheim and Leide 1990). The relative importance of different pathways may also change with time. For example, the mineral uptake pathway is crucial in young stands, whereas in older stands retranslocation* becomes more important. The time of canopy closure is pivotal. Generally, rates of elemental uptake increase rapidly up to that point, then decrease gradually over time. After closure, there is relatively little increase in elemental accumulation in foliage. During the preclosure period, nutrient accumulation shifts progressively from understory vegetation to canopy trees. This trend is gradually reversed as succession progresses and the canopy opens up (Cole and Rapp 1981, Carlyle 1986, Pritchett and Fisher 1987, Aber and Melillo 1991). This

* Internal cycling or redistribution within plants (Bockheim and Leide 1990; van den Driessche 1991).

suggests that nutrients are more vulnerable to loss from the ecosystem when the canopy is removed and that silvicultural practices that minimize the time of open canopy will conserve nutrients.

In the discussion that follows, greater emphasis is placed on N cycling because N is more often limiting in temperate forest ecosystems. For macronutrients that occur as cations, the same ions make up both the principal forms in solution and the plant-available forms. This is not true for the anion forms (table 6.1). Generally, appreciable quantities of sulfide (S^{2-}) and nitrite (NO_2^-) do not accumulate in well-drained or well-aerated soils (Tisdale and Nelson 1975, Carlyle 1986, Troeh and Thompson 1993). For reasons outlined later, concentrations of nitrate (NO_3^-) and the divalent orthophosphate ion (HPO_4^{2-}) are usually low in acid soils (Pritchett and Fisher 1987, Brady 1990, Krause 1991).

Nitrogen

Nitrogen is generally accepted to be the principal growth-limiting nutrient in most temperate forest systems, where more than 90% of total soil N may be in organic matter and thus unavailable to plants (Vitousek 1982, Carlyle 1986, Pritchett and Fisher 1987). In any system, it is the nutrient required in the largest amounts; however, the requirement values vary widely across systems. Availability and cycling of N are strongly dependent on decomposition and other microbially mediated processes. Those that are particularly important in forested ecosystems include mineralization-immobilization, nitrification, dinitrogen fixation (biological N fixation), denitrification, and retranslocation.

Mineralization-immobilization, nitrification. Mineralization is the microbial conversion of an element from an organic form into an inorganic state. This occurs during decomposition and depends upon both organic matter and microbial populations that work in conjunction with soil fauna. Immobilization is the conversion of an element from an inorganic form to the organic form in microbial or plant tissues (Soil Science Society of America 1987). Uptake into tissues of higher plants relies on the mutualistic relations between roots and mycorrhizal fungi. The availability of N depends on the extent to which mineralization exceeds immobilization (Swift et al. 1979). Mineralization, or decomposition, is the key process governing N availability in temperate forest ecosystems. Controlling factors include the physical and chemical environment (for example, temperature, moisture, soil acidity), the types and numbers of decomposer organisms present, and the quantity and quality of the organic substrate (physical nature, chemical composition, C/N ratios, and so on). These factors are highly interactive (Swift et al. 1979,

Gosz 1981). It seems clear that organic matter, favorable temperature and moisture, and well-aerated soil will favor N availability. Because of the emphasis on leaving woody debris, and the corresponding effect on diversity, ecosystem management practices tend to support these more favorable conditions relative to traditional practices.

Annual decomposition rates vary tremendously over the range of temperate forest ecosystems, and they vary within a given system both with the substrate material and time (Swift et al. 1979, Rustad 1994). For example, loss rates approximating 0.1% by weight were reported for litter under conifers in the Pacific Northwest. Rates approximating 1.0% by weight were reported for litter under oak (*Quercus* spp.) and pine forests in the South (Kimmins et al. 1985, Krause 1991). The mean turnover time (Cole and Rapp 1981) of organic matter and nutrients varies with latitude and ecosystem type. Boreal forests exhibit the longest turnover time, with coniferous forests retaining N more than forty times as long as temperate deciduous forests (Cole and Rapp 1981). Estimates of annual N mineralization rates also vary considerably over the range of temperate forested ecosystems (Gosz 1981, Carlyle 1986, Pritchett and Fisher 1987). In a 55-year-old northern hardwood ecosystem in the Northeast, only 1.5% (approximately 70 kg N ha^{-1}) of the total N content of the forest floor and mineral soil was mineralized annually (Bormann and Likens 1979, Swift et al. 1979). Short growing seasons, low mean growing season and annual temperatures, fungal domination of decomposition processes, mycorrhizal suppression of decomposition, and unfavorable chemical composition of organic inputs all contribute to limiting N availability and cycling rates (Gadgil and Gadgil 1975, Alexander 1977, Meentemeyer 1978, Swift et al. 1979, Berg and Lindberg 1980, Gosz 1981, Van Cleve et al. 1981, Alexander 1983, Krause 1991).

Nitrification* rates are thought to be low in most temperate forested ecosystems. As soil acidity increases, bacterial populations and activities are correspondingly inhibited. This includes both the heterotrophic bacterial decomposers and the autotrophic nitrifying bacteria (Alexander 1977, Swift et al. 1979). When mineralization rates are low, ectomycorrhizae can outcompete the nitrifiers for the limited pool of ammonium (NH_4^+) nitrogen (Alexander 1983). Under relatively undisturbed forest conditions, the NH_4^+ ion is the most common form of available N. Small amounts of soluble organic N may be taken up directly by trees (Heal et al. 1982); however, the bulk of N uptake under most temperate forest conditions is in the NH_4^+ form (table

---

* Biological oxidation of ammonium (NH_4^+) to nitrite and nitrate (Soil Science Society of America 1987).

6.1). Therefore, such factors as pH, temperature of the soil, soil moisture and gas exchange, and organic matter in a form available to microorganisms involved in the mineralization process will influence availability of N. Mixtures of trees that include those capable of cation pumping can return cations to the surface in the form of fine litter, consequently influencing the availability of N. So, too, will the maintenance of canopy closure levels that help to maintain more favorable temperature and moisture conditions in the surface soil.

Such forest ecosystem characteristics as species composition, forest floor characteristics, and faunal activity combine with the environmental controls to influence N cycling and availability. The relatively high C/N ratio and low decomposability of litter from northern coniferous forests can interact with the conditions of low temperature, acid soils, fungal decomposition, and low levels of faunal activity to slow decomposition and tie up N in the system (Tamm 1950, Tamm 1982, Krause 1991). The characteristics of forest floors and humus forms (for example, mor, moder, or mull) reflect these influences (Green et al. 1993). For example, in a late-successional coniferous forest in the Sierra Nevada Mountains, fungal activity may have accounted for translocation of as much as 9 kg N ha^{-1} from the mineral soil into the forest floor. This magnitude was similar to those estimated for other fluxes (for example, uptake) within the system (Hart and Firestone 1991). Faunal influences on decomposition also extend to specific effects on the macronutrients. For example, the presence of soil fauna can increase releases of N and other nutrients from branchwood, forest floor, and mineral soil materials; fauna can also decrease fungal immobilization of N through microbial grazing (Swift 1977, Anderson and Ineson 1983, Setala et al. 1990).

Dinitrogen fixation. Dinitrogen fixation can contribute substantial quantities of N to forested ecosystems (Cole and Rapp 1981). The process involves anaerobic reduction and has a relatively high demand for energy, P, Mg, and molybdenum (Postgate 1982, Fortin et al. 1984). Nonsymbiotic fixation (by heterotrophic microorganisms and cyanobacteria or blue-green algae) can occur in the canopy, decaying logs, and soil; in most temperate forests, annual contributions are less than 5 kg N ha^{-1} yr^{-1}, but rates of 10–50 kg N ha^{-1} yr^{-1} have been reported (Fortin et al. 1984, Carlyle 1986, Pritchett and Fisher 1987). Leguminous trees and other nitrogen-fixing species occur in many natural temperate forest ecosystems in North America, but estimates of their contributions are not well established; some legumes do not nodulate or fix N (Fortin et al. 1984, Pritchett and Fisher 1987). Rates for leguminous woody species are thought to be much lower than those of herbaceous plants (Carlyle 1986). Fixation associated with such plants as alder (*Alnus* spp.), *Ceanothus* spp., and *Comptonia peregrina* involves the *Frankia* actinomycete. Fixation rates of 12–300 kg ha^{-1} yr^{-1} and greater have been reported for such actinorhizal plants. In red alder *(Alnus rubra)* ecosystems on the Pacific coast, net accumulation and fixation rates of 35–130 kg ha^{-1} yr^{-1} were noted (Cole and Rapp 1981, Fortin et al. 1984, Binkley 1986, Van Miegroet et al. 1990). Dinitrogen fixation can contribute large amounts of N to forested ecosystems, but the N does not immediately enter the available pool (fig. 6.1). However, the relatively N-rich litter, root exudates, and other organic compounds that are ultimately released promote rapid decomposition, mineralization, and cycling rates (Fortin et al. 1984, Binkley 1986, Pritchett and Fisher 1987, Van Miegroet et al. 1990).

Dinitrogen fixation is strongly affected by environmental factors. Because of the carbohydrate energy requirement, symbiotic fixation is also strongly influenced by factors affecting photosynthesis and growth of the host plant (Postgate 1982, Fortin et al. 1984, Carlyle 1986, Pritchett and Fisher 1987). Fixation with actinorhizal plants appears to be less sensitive to several factors than rhizobial fixation; however, much less is known about actinorhizal fixation than about rhizobial fixation. The organisms of both types can be adversely affected by water stress, low levels of P and nutrient cations, and high levels of available soil N. Nodule formation by rhizobia is largely unsuccessful below pH 4.5–5.1, whereas actinorhizal associations are common at and below this range. Particularly on less fertile sites, fostering conditions favorable to good growth of trees should also favor dinitrogen fixation. In turn, this fixation will tend to increase the pool of N for cycling in the ecosystem.

Denitrification and leaching losses. Denitrification processes result in loss of N from the ecosystem, but they have not been widely studied. Annual losses from undisturbed upland forests are usually less than 1.0 kg^{-1} ha^{-1} yr^{-1}. Losses can be much higher after disturbance or at particular times of the year (Goodroad and Keeney 1984, Robertson and Tiedje 1984, Davidson et al. 1990). Nitrification levels in undisturbed temperate forested ecosystems are thought to be negligible, with correspondingly low leaching losses (Cole and Rapp 1981, Carlyle 1986, Pritchett and Fisher 1987). Major disturbance is usually followed by a period of increased nutrient availability in terrestrial systems, which may also lead to increased leaching or loss through surface water runoff (Bormann and Likens 1979). The phenomenon has been termed the "assart effect" (Tamm 1979, Kimmins 1989), or "ash-bed effect" where fire is involved (Humphreys and Lambert 1965). Contributing factors include additions of fresh organic matter to decomposers, increased temperature, decreased vegetative uptake, relaxation of mycorrhizal suppression of decomposition, and increased runoff. This varies considerably from one site to another. The factors and influences involved were reviewed in detail by Vitousek and colleagues (Vitousek et al. 1982, Vitousek 1983), who observed that lower-quality sites and those with low initial N availability levels exhibited low or delayed nitrate losses. The most important mechanisms preventing or delaying nitrate losses were low net N mineralization rates and lags in the onset of nitrification.

Phosphorus

Organic forms account for some 10–50% of total soil P, with the remainder occurring in mineral complexes (table 6.1). Availability and cycling are strongly influenced by both biotic and abiotic processes. The climatic and biological controls operating on N mineralization affect P similarly. However, chemical reactions in the soil exert strong control on P availability. The geological origin and age of the landscape and parent materials, weathering intensity, soil reaction, and availability of Ca are particularly important (Brady 1990, Krause 1991, Troch and Thompson 1993).

Phosphorus contents of soils in the United States tend to be highest in the Northwest and lowest in the Southeast. This general pattern derives from an interaction involving the climate, the nature of the parent materials, and the duration and intensity of weathering (Krause 1991, Troeh and Thompson 1993). Under acid conditions, iron and aluminum form insoluble phosphate complexes, while Ca does the same under alkaline conditions. Therefore, inorganic P forms are at a theoretical peak of availability at pH values between 5.5 and 7. Atmospheric deposition of P is usually negligible. Organic matter is a major source on forested sites; it is often the principal source in many instances (Pritchett and Fisher 1987, Aber and Melillo 1991). In the United States, low P availability occurs principally in coastal plain soils of the Southeast, but also on organic and volcanic ash soils (Pritchett and Fisher 1987, Krause 1991). However, P limitations can occur under other conditions as well. In Alaska, organic matter formed the primary reservoir of P in a white spruce ecosystem on floodplain soils. In that system, periodic siltation caused P limitations on growth (Van Cleve and Harrison 1985). Declining stands of sugar maple *(Acer saccharum)* growing on Inceptisols and Spodosols in the Appalachian highlands of Quebec exhibited acute P deficiencies. It is unclear whether this limitation was a factor contributing to the decline (Bernier and Brazeau 1988). Although the importance may vary from one forest ecosystem to another, it is clear that management favoring retention and availability of P will maintain or perhaps enhance productivity.

Under natural forested conditions, it is generally accepted that leaching losses of P are usually negligible. However, leaching of P to deeper horizons can occur in soils with very low levels of iron and aluminum, such as some coastal and organic soils (Pritchett and Fisher 1987, Aber and Melillo 1991). Elsewhere, organic forms of P can move to deeper horizons and be lost from forest sites through leaching (Frossard et al. 1989, Xiao et al. 1991). Losses of P from the ecosystem, or to deeper, less available horizons, are usually least in undisturbed forests. Therefore, compared to traditional practices of forest management, ecosystem management practices should minimize losses from disturbance.

Sulfur

With the exception of biological fixation, pools and transformation processes for S are similar to those of N. In forested ecosystems, up to 90% of the total soil pool of S can be held in organic forms. Availability and cycling are therefore heavily influenced by microbially mediated transformations (for example, mineralization-immobilization, biological oxidation, and reduction), and by the abiotic controls affecting such processes. However, because the available pool (fig. 6.1) for S is supplied by both the weathering and mineralization pathways, S is more readily available than N and P (Pritchett and Fisher 1987, Brady 1990, Krause 1991). As we shall see, there are also human-caused deposits to many ecosystems. Sulfur limitations have been reported in forested ecosystems of western United States and Canada (Pritchett and Fisher 1987, Krause 1991).

Potassium, Calcium, Magnesium

Available supplies of Ca and Mg are usually ample in most forested soils, with K less so. For both coniferous and deciduous tree species, annual uptake of Ca and Mg is usually in excess of minimal requirements (Cole and Rapp 1981). Coarse-textured soils or soils that lack colloidal clay have been associated with K and Mg deficiencies (Pritchett and Fisher 1987, Krause 1991). Results of fertilization trials in young pine stands on wet Spodosols in the Southeast also suggest that K limitations might exist (Neary et al. 1990). Bernier and Brazeau (1988) concluded that K deficiencies were contributing to decline of sugar maple, and perhaps threatening the integrity of maple ecosystems on Inceptisols and Spodosols in the Quebec Appalachians. The soils were derived from mafic rocks and were generally rich in Mg and poor in K (Bernier and Brazeau 1988). Management that favors retention or accumulation of soil organic matter will favor retention of K, Ca, and Mg in the ecosystem.

Inputs from Acid Deposition

Questions arise concerning the effects of acid deposition and associated inputs of N and S to terrestrial systems (fig. 6.1) (Foster 1989). In North America elevated inputs of H^+, NO_3^-, and SO_4^{2-} occur principally in eastern and northeastern regions; annual input levels are 8–20 kg S ha^{-1} and 2–6 kg N ha^{-1}. There are north-south and east-west gradients, with the highest deposition occurring south of the Great Lakes basin (Binkley et al. 1989, Foster 1989, Rennie 1990, Aber and Mclillo 1991). Increasing inputs have been observed in pristine areas of western Canada (British Columbia Ministry of Forests 1987).

Several potential influences of acid deposition on forest ecosystems have been described, and it has been hypothesized that acid deposition is responsible for observed forest declines in eastern North America and Europe. Impacts include soil acidification, cation leaching, aluminum toxicity, and reduced growth or death of vegetation. Potential direct effects on vegetation include damage to leaf and stem cuticle, disturbance of metabolism and growth, interference with reproductive processes, and synergistic interactions with other environmental stress factors. Potential indirect effects include accelerated leaching of substances from foliar organs, increased susceptibility to drought, frost, and other environmental stress factors, alteration of symbiotic associations, and alteration of host/parasite interactions. Since both N and S are plant nutrients, acid deposition may also improve tree nutrition (Cole and Rapp 1981, Morrison 1984, Foster 1989). Clear supporting evidence exists for only some of these impacts. For example, acid deposition has caused increased acidity in the soils of many European forest ecosystems (Tamm and Hallbäcken 1988, van Breemen 1990). However, the relative contributions to pH decline from natural acidity and acidic deposition is largely unknown (Foster 1989). Excess inputs of N have been linked to frost damage, but recent studies have cast considerable doubt on this hypothesis (Foster 1989).

Atmospheric inputs of N are thought to cause nutrient imbalances in trees as well as increased leaching of NO_3^- from forest ecosystems (Foster 1989). Although undisturbed forest systems tend not to lose much N, it has been hypothesized that long-term depositional inputs can stress them sufficiently to lead to their decline (Ågren and Bosatta 1988, Aber et al. 1989). The hypothesis and related considerations hold serious implications for management activities and long-term productivity of systems. For example, simulation studies by Aber et al. (1991) suggest that the interaction of N deposition, harvesting, and N transformation processes could hasten the onset of N losses and ecosystem decline. Forest canopies can attenuate such inputs; however, basic cations are released and the potential for leaching losses can be increased (Johnson and Reuss 1984, Jepson and Bockheim 1985). The exchange and leaching effects vary with the tree species involved. In New Brunswick forests, with the exception of white birch *(Betula papyrifera)*, the canopies of broad-leaved species had a greater neutralizing effect on rain acidity than conifer canopies (Mahendrappa 1990). Because of potential deleterious impacts, ecosystem management practices are probably especially important in areas subject to acid deposition.

Effects of Large Mammal Populations

Animal influences are generally incorporated into the pools and pathways associated with aboveground biomass, forest floor and coarse woody debris, and soil organic matter (fig. 6.1). Large mammalian herbivores can facilitate nutrient cycling and ecosystem productivity. They affect both the timing and the rates of nutrient return to a site and alter nutrient flux patterns across the landscape (Frank and McNaughton 1992). Herbivores can also change successional patterns through preferential browsing and influence the frequencies of fire disturbances (Baskin 1994, Irwin et al. 1994). The presence of significant numbers of large mammalian herbivores signals that special attention should be paid to their influences on nutrient cycling and ecosystem composition.

Carbon Sequestration

While C is not typically considered a macronutrient, its loss from terrestrial systems and the associated implications for the global C cycle and climate change are serious environmental concerns. Harvesting and other forest management activities can have major impacts on the C cycle, both locally and globally. Carbon storage (sequestration) also has financial and economic implications related to timber production. While detailed consideration of this issue is beyond the scope of this essay, it is worthwhile to illustrate the influence that forest management decisions can have on C storage dynamics.

Carbon storage in U.S. forest ecosystems has increased steadily over the past forty years, and this trend is expected to continue. However, most of this increase has occurred in the eastern regions, while the West has actually shown declines of approximately 10% (Birdsey et al. 1993). Harmon et al. (1990) argue that conversion of late-successional ("old-growth") forests in the Pacific Northwest has contributed significantly to the atmospheric C pool. They estimate that more than 57% of the C removed in harvesting from a late-successional forest may ultimately be lost to the atmosphere. Further, simulation studies suggested that the intensively managed plantations that usually replace such forests would not match the original storage rates for at least two hundred years; nor would the plantations decrease atmospheric C pools in general (Harmon et al. 1990). By contrast, aspen grown on forty-year rotations in the Lake States may sequester more than three times as much C as late-successional aspen ecosystems, with no loss of soil C from harvesting (Alban and Perala 1992). Alban and Perala (1992) considered this aspen management regime to be most appropriate for mitigating global atmospheric C increases, while one with a higher proportion of shade-tolerant species maximized local C storage.

Species and site differences among late-successional forest ecosystems contribute greatly to differences in C storage dynamics, and consequently to the impacts of harvesting practices. For example, in the Lake Superior region of Ontario, the total C reserves of a 300-year-old sugar maple ecosystem were more than twice those of a 62-year-old jack pine *(Pinus banksiana)* ecosystem (348,200 and 161,000 kg ha^{-1}, respectively). Net annual C storage in the maple trees was 1.5 times that of the pines. However, the pines allocated 56% of this C to woody tissues, compared with only 21% for the maples. The mass of the C stored annually in woody tissues of the pines was 1.5 times that of the maples. For these reasons, conventional shortwood or tree-length harvesting of the pines would remove about the same proportion of total ecosystem C (33%) as whole-tree harvesting of the maples (Morrison et al. 1993).

On the economic side, C sequestration rates, fluxes, and losses in the timber production process can have a significant influence on net present values and timber cash flows. Management prescriptions that take C storage dynamics into account can provide greater long-term flexibility in an uncertain future (Hoen and Solberg 1994). Therefore, from the economic, ecological, and environmental viewpoints, the C sequestration issue is important to managers. Such decisions as those concerning harvesting systems, species choices for reforestation, intensity of site preparation, amounts of coarse woody debris retained, intensity of management, and rotation ages can all affect C storage dynamics. In general, ecosystem management would seem to favor C sequestering.

POTENTIAL IMPACTS OF ECOSYSTEM MANAGEMENT TECHNIQUES

I shall now examine impacts of proposed ecosystem management practices on macronutrient cycling and availability. To provide a perspective in the discussion, the focus is on two major sources of variation (Stone 1975b):

(i) The disturbance of the forest canopy and soil surface; and

(ii) The nature of posttreatment practices and the speed and composition of vegetative regrowth.

Explicit considerations of the recommendations involving time scales and cumulative effects (table 6.2) are excluded. They are difficult to place within the context of source (i) above, and aspects related to source (ii) are covered under biological legacies. Readers are reminded of the generalized nature of the discussion in the face of the broad range of forested ecosystems and conditions.

Macronutrient cycling in forest ecosystems involves a wide range of pathways and processes that are not immediately evident to managers. Yet these invisible dynamics are easily affected by forest management activities. The effects of harvesting practices are variable across temperate forested ecosystems. In the Hubbard Brook forest, harvesting increased stream water concentrations of most ions measured. Losses after harvesting and restriction of revegetation ranged from 6 times as much for Mg to 160 times as much for N (Bormann and Likens 1979). The treatments that yielded these results were more severe than most forest management practices. Nevertheless, studies in the same and other regions have confirmed that such exports do occur after harvesting, but in smaller amounts (Stone 1975b). In Douglas fir ecosystems of the Pacific Northwest, however, losses from undisturbed ecosystems were comparatively low, and clearcutting did not increase them. Clearcutting and burning led to a 10–30% reduction in cation reserves, but 70–90% of the cations mobilized remained within the rooting zone (Cole and Johnson 1979). In three different forest types in the northeastern United States, estimates of combined losses in harvested products and increased leaching (annual basis) ranged up to 379 kg ha^{-1} for N, 54 kg ha^{-1} for P, 253 kg ha^{-1} for K, 558 kg ha^{-1} for Ca, and 50–65 kg ha^{-1} for Mg (Hornbeck et al. 1990). Increased availability and mobility of anions increases leaching losses of both cations and anions (Vitousek 1983).

Creation and Retention of Biological Legacies

Most silvicultural treatments (table 6.3) involve some degree of canopy removal. Changes in microclimatic conditions at the soil surface will increase as the amount of canopy removal increases. Since even partial cuttings such as thinning can increase soil temperature and decomposition rates (Piene 1978), a period of increased nutrient availability (assart effect) and increased stream exports of macronutrients may occur. Factors affecting its magnitude and duration will include the degree of canopy removal, the degree of forest floor and mineral soil disturbance, the distribution and retention of standing dead trees and coarse woody debris, the nature and intensity of any site preparation operations, and the species involved in regrowth. Compared to harvesting alone, site preparation operations can cause marked increases in rates of nutrient transformations, the amounts lost, and the duration of release under widely different forest conditions (Vitousek et al. 1992, Johansson 1994).

If standing dead trees and woody debris are retained, the mass of forest floor materials will be increased; this should have an effect opposite to the litter removal practices that led to severe soil and diversity depletion in western Europe a century ago (Stone 1979, Pritchett and Fisher 1987). The overall pool of nutrients retained on the site should be greater, as would be the variety of microsites and habitats that are beneficial to microbial and faunal communities. The numbers and diversity of such populations and activities should increase, thus increasing the overall biological diversity of the site. Inputs from nonsymbiotic dinitrogen fixation may also increase as a result. Therefore, compared to traditional harvesting, the overall effect of retaining biological legacies on nutrient cycling should be to protect and diversify its pathways and processes.

Enhancement of Biological Diversity, Species Reintroductions

Enhancing biological diversity should affect nutrient retention and cycles only to the extent that vegetative cover or the soil surface is disturbed in the process. For example, thinning, small openings created by group selection or small clearcuts, or single-tree selection might be recommended to increase diversity of native species in a forest. Creation of species mixtures in the canopy or species reintroductions may have beneficial effects on nutrient retention and availability. For example, deciduous tree species introduced into coniferous ecosystems can increase cycling rates and availability because of their higher rates of uptake and more rapid decomposition of deciduous litter. Increases in the absolute pool of nutrients will probably occur only if species with nitrogen-fixing associations are encouraged. Otherwise, only rates of cycling among compartments of the existing pool may be increased.

Soil acidity may be increased with proliferation of coniferous species. Although Stone (1975a) argued that the scientific evidence was not strong, more recent reports from long-term monitoring support the generalization. For example, the presence of conifers caused increases in surface soil acidity under stands in Sweden over a sixty-year period (Tamm and Hallbäcken 1988).

Increases were also noted over a thirty-year period in an old-field pine ecosystem on an Ultisol in the southern United States. In that case, losses of basic cations at rates that exceeded resupply were also noted (Richter et al. 1994).

Introductions involving soil fauna (for example, earthworms) may increase rates of macronutrient cycling and availability. Increasing large-mammal populations has the potential for major influences, including more rapid nutrient cycling and species shifts as a result of selective browsing.

SUGGESTED COURSES OF ACTION FOR MANAGERS

Virtually any management action has an impact on nutrient cycling processes. Some may cause temporary increases in rates of cycling or availability, or loss of nutrients from the site. Interactions occur among abiotic influences, the stage of development of the system, its vegetative composition, microbial and faunal activities, the numbers and types of plant propagules, the nature and intensities of both initial and subsequent management treatments, and human-caused external influences. Farnum (1994) identified several considerations of soil and productivity that he considered important for forest practices. They can be extended to ecosystem management practices and include:

(i) Action by all landowners to protect the soil and maintain its productivity;

(ii) Diagnosing and maintaining the proper level and balance of all nutrients for each soil type;

(iii) Preserving the amount and function of soil organic matter; and

(iv) Developing both methods of assessment and (where necessary) ameliorative treatments for the health of soil organisms.

Achieving these goals would be a monumental task in any forest management system but would probably be impossible without an ecosystem management approach. So far, I have discussed ecosystem management actions and how they can contribute to sustaining macronutrient cycling processes and productivity at the site level. To the extent that diversity is important for maintaining soil fertility, practices should also protect diversity within management units. Practices include increasing structural complexity, extending rotation lengths, minimizing fragmentation, protection of riparian corridors, and enhancing connectivity between units. These practices tend to be contrary to intensive management.

Stone (1979) posed several questions regarding impacts of intensive management. I have modified them to address successful implementation of eco-system management:

(i) What levels of nutrient removal can each ecosystem sustain with no decrease in productive capacity?

(ii) What elements will become limiting first in a given system, and how do systems differ in their responses?

(iii) What unplanned secondary changes (for example, species composition, habitat diversity, pest problems) are likely as a result of altered cycling patterns?

(iv) How can we objectively predict the nature and magnitude of possible decreases in productivity?

(v) What measures can be devised to avoid decreases in productivity, or even to increase it?

We should also ask questions specifically focused on macronutrient cycling. For example, what baseline levels apply in specific soils and ecosystems? What are appropriate measurements? What is the role of natural disturbance? What critical ranges of function values exist? How do we develop standards and techniques that take natural rates of change in ecosystem processes into account? How are temporal and spatial variability to be separated and quantified? (Gosz 1992, Farnum 1994).

Much of the information that managers need to address these questions is lacking. However, available knowledge can be used together with adaptive management (Gosz 1992, Salwasser and Pfister 1994, Franklin this volume). This approach is based on the tenet that management involves continual learning that cannot be divided into separate functions, although full knowledge and optimum productivity probably is never achieved (Walters 1986). Designing adaptive management strategies involves four measures (Walters 1986):

(i) Identifying explicit and hidden objectives, constraints, and factors;

(ii) Representation of explicit models of dynamic behavior, with clearly stated assumptions and predictions;

(iii) Representation of uncertainty and how it might evolve in relation to management actions; and

(iv) Design of balanced policies that provide continued resource production and understanding.

Therefore, the adaptive management approach to ecosystem management requires as full an incorporation of nutrient cycling dynamics as knowledge allows. Franklin (this volume) indicated that ecosystem management incorporates the philosophy rather than the formality of adaptive management. However, if ecosystem management is to succeed, the complexity of nutrient cycling in temperate forested ecosystems demands a formal approach, particularly including the first two measures listed above. This approach has been termed "active adaptive" management (Walters and Holling 1990).

Successful active adaptive management in forest ecosystems requires knowledge of nutrient dynamics and productivity. Data based on short-term studies may give little or no indication of long-term trends in nutrient-cycling processes or ecosystem productivity (Rustad 1994). The components and processes of the systems are strongly interconnected; comparatively small changes, from the manager's viewpoint, in the characteristics of initial treatments can affect both the path and nature of the response. Examples of this from forestry include a change from tree-length to whole-tree extraction or in the type of equipment used in mechanical site preparation. Differences in short-term responses (for example, in species composition) may affect subsequent paths and outcomes. External influences (for example, changes in atmospheric conditions) or unexpected events (for example, insects and disease, long-delayed seed dormancy and germination) also influence responses. These characteristics of forest ecosystems indicate that they may be chaotic; that is, they may exhibit turbulence, nonlinearity, and feedback (Briggs and Peat 1989, Allaby 1994). To the extent that forest ecosystems are chaotic, ecosystem-specific management models will have lower predictive capabilities.

Adaptive management requires well-defined objectives. The importance of nutrient cycling processes is frequently considered in general terms only. Application of the adaptive management approach to a specific forest ecosystem should include goals that are as specific as possible for each macronutrient cycle. It would be unrealistic to assume that we can achieve quantifiable or sustainable management of the complex of nutrient cycling processes within a short period of time.

Long-Term Actions

The knowledge base concerning macronutrient dynamics should be increased considerably. This will involve rigorous collection and synthesis of data from existing studies as well as additional research. Monitoring must be a crucial part of adaptive management. Because of ecosystem-specific differences, the scope of needed research parallels that of the IBP studies (see Reichle 1981). Such an effort will be expensive. High costs have been an inhibiting factor to the establishment of credible monitoring programs in the past (Franklin this volume). Ecosystem management requires both an interdisciplinary approach and interagency cooperation (Grumbinc 1994, Franklin this volume). Cooperative research on nutrient cycling and monitoring might lessen the cost for a single management unit or agency. One example would be the formation of research cooperatives that involve various public agencies and private ownerships.

Within the context of the foregoing discussion, a sequence of actions is recommended for administrators and managers of extensive land holdings (for example, a national, state, or county forest; lands of a large private industrial concern). An interdisciplinary team (for example, comprising representatives of several agencies, or a task force of specialists drawn together by a private company) should spearhead the effort. This team should:

(i) Decide what constitutes the set of "ecosystems" or "ecosystem management units" and clearly define the components and boundaries of each unit using ecological criteria;

(ii) Make a determined effort to identify, gather, and synthesize all available relevant data for each type of ecosystem management unit (EMU);

(iii) Identify which nutrients are likely to be limiting, systematically identify the weaknesses in their knowledge base for each type of EMU, and implement investigations to strengthen these areas;

(iv) Set up a permanent monitoring network aimed at determining existing pool sizes, fluxes, rates of change, and other factors under natural conditions in each type of EMU;

(v) Use existing and emerging data to decide on the components of integrity for each type of EMU and develop provisional answers (where possible) to the questions of Stone (1979) and Farnum (1994) outlined earlier;

(vi) Define specific objectives for each type of EMU and design theoretical ecosystem management systems (similar to silvicultural systems) that may achieve these;

(vii) Define clearly how "success" will be measured in terms of productivity, sustainability, or the specific dynamics and processes of each type of EMU; and

(viii) Implement the systems at (vi), using active adaptive management in appropriate areas of each type of EMU and integrating activities and units across the landscape, for example, by applying the "triad approach" of Seymour and Hunter (1992) to allocating land use across the ownership.

Managers of smaller ownerships could establish links with the projects of larger agencies or concerns. Because ecosystem management efforts need integration across the landscape, such incorporations would benefit both larger and smaller ownerships. Each type of EMU may require a separate team of specialists to implement steps (ii) to (viii).

Several examples suggest approaches, techniques, and information sources that managers can use in implementing the steps listed above. The IBP studies (Reichle 1981) demonstrate a useful mode of organization, and the resulting data sets are important sources of baseline information. Kimmins et al. (1985) summarized the literature on biomass and nutrient dynamics for forest ecosystems ranging from the boreal to the subtropics. Gregersen et al. (1990) presented detailed guidelines for managers in planning and management of forestry research. Powers and Van Cleve (1991) and Gordon et al. (1993) are two examples of the types of protocols and methods that are needed for manipulative studies in nutrient cycling and productivity research. The Total Ecosystem Management Strategies project (Ticknor 1993) demonstrates collaboration among private concerns in covering a relatively modest area (slightly more than 10,000 ha). By contrast, Canada's "model forests" program exemplifies efforts involving multifaceted partnerships (for example, including industry, nongovernmental organizations, and communities) and large tracts of land (in one example, 367,000 ha). This project also indicates the magnitude of the costs associated with such efforts. For a five-year period, the Canadian Government's contribution alone will be $800,000 (U.S.) per model forest, with additional funds and in-kind support coming from the partner agencies (Brand and LeClaire 1994).

Immediate Actions

Despite limited knowledge of nutrient dynamics and other aspects of forested ecosystems, field managers can take action to achieve ecosystem management goals. The first involves self- and continuing education. Increased understanding of the concepts and processes will help managers understand recommendations and adapt them to their specific ecosystems. Understanding will also facilitate interagency or interdisciplinary communication. Moreover, it will also enhance managers' professional comfort levels and confidence in decision making during a period of changing management paradigms. The same arguments make it generally desirable for personnel at all levels to strive to attain at least a working understanding of nutrient cycling and related processes, and more generally of ecological processes in forested ecosystems. Where feasible, they should develop familiarity with the essentials of analytical and modeling methods that may be applied in adaptive management.

Franklin (this volume) has stressed the crucial role of dead and down materials. A tree's influence extends well beyond the time of its death—some two hundred or more years (Hunter 1990, Franklin this volume). If ages of organic matter fractions are an indication, this influence could well last five hundred to one thousand years (Swift et al. 1979). The temporal pattern of the importance of a tree to the structure and processes of an eastern temperate forest ecosystem can be conceptualized (fig. 6.2). Managers can modify practices to increase the amount of woody debris and snags in the forest. Examples include extended rotations, increased numbers of snag trees, girdling or injecting low value trees, or delimbing trees when possible where they are felled.

In addition to organic matter, soil porosity can be a critical regulator of key soil processes influencing ecosystem productivity (Powers et al. 1990, Powers and Van Cleve 1991). Porosity certainly is important to many macronutrient cycles, aeration status influences the functioning of several of the transformation pathways discussed earlier. Thus, periodic measurement of such variables as bulk density and infiltration rates on a given site should provide an additional indicator of ecosystem integrity. Increased litter cover, mixing deciduous trees into conifer cultures, and avoiding running heavy equipment over wet soil can protect soil porosity. Elevated concentrations of soil NO_3^- might be useful as an alarm signal. Periodic measurement of NO_3^- concentrations in the soil or associated stream water after management manipulations may serve to highlight potential trouble spots.

Considering the importance of macronutrient cycles to ecosystem productivity, one thing is certain: if ecosystem sustainability is a principal objective of ecosystem management, many forest management practices must be modified. Long-term effects of the ecosystem management practices and techniques recommended thus far (tables 6.2 and 6.3) are unknown. Tenner (1991) warned about "revenge effects," the unintended consequences of human ingenuity. In a similar vein, the cautionary words of DeBell and Curtis (1993:26) are peculiarly apt where the impacts of ecosystem management on nutrient cycling processes are concerned: "Past mistakes should teach skepticism about widespread adoption of untested practices." However, none of the recommended ecosystem management practices would be expected to have negative effects on nutrient retention or availability, in either the long term or the short term.

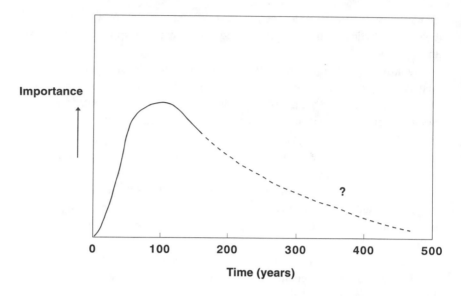

Fig. 6.2. Conceptual representation of the pattern of influence over time of a tree in a forested ecosystem.

I thank M. Boyce, J. Cook, R. D. Hammer, A. Haney, and R. Rogers for valuable comments on earlier versions of this chapter.

LITERATURE CITED

Aber J. D., and J. M. Melillo. 1991. Terrestrial ecosystems. Saunders College Publishing, Philadelphia, Pennsylvania.

Aber, J. D., J. M. Melillo, K. J. Nadelhoffer, J. Pastor, and R. D. Boone. 1991. Factors controlling nitrogen cycling and nitrogen saturation in northern temperate forest ecosystems. Ecological Applications 1:303–315.

Aber, J. D., K. J. Nadelhoffer, P. Steudler, and J. M. Melillo. 1989. Nitrogen saturation of northern forest ecosystems. BioScience 39:378–386.

Ågren, G. I., and E. Bosatta. 1988. Nitrogen saturation of terrestrial ecosystems. Environmental Pollution 54:185–197.

Alban, D. H., and E. C. Berry. 1994. Effects of earthworm invasion on morphology, carbon, and nitrogen of a forest soil. Applied Soil Ecology 1:243–249.

Alban, D. H., and D. A. Perala. 1992. Carbon storage in Lake States aspen ecosystems. Canadian Journal of Forest Research 22:1107–1110.

Alexander, I. J. 1983. The significance of ectomycorrhizas in the nitrogen cycle. Pages 69–93 in A. J. Lee, S. McNeill, and I. H. Rorison, editors. Nitrogen as an ecological factor. The 22d Symposium of the British Ecological Society, Blackwell Scientific Publications, London.

Alexander, M. 1977. Introduction to soil microbiology. 2d edition. John Wiley and Sons, New York, New York.

Allaby, M. (editor). 1994. The concise Oxford dictionary of ecology. Oxford University Press, Oxford, England.

Anderson, J. M., and P. Ineson. 1983. Interactions between soil arthropods and microorganisms in carbon, nitrogen, and mineral element fluxes from decomposing leaf litter. Pages 413–432 in A. J. Lee, S. McNeill, and I. H. Rorison, editors. Nitrogen as an ecological factor. The 22d Symposium of the British Ecological Society, Blackwell Scientific Publications, London.

Ausmus, B. S. 1977. Regulation of wood decomposition rates by arthropod and annelid populations. Pages 180–192 in U. Lohm and T. Persson, editors. Soil organisms as components of ecosystems. Ecological Bulletins (Stockholm) 25, Swedish Natural Science Research Council.

Baskin, Y. 1994. Ecologists dare to ask: how much does diversity matter? Science 264:202–203.

Berg, B., and T. Lindberg. 1980. Is litter decomposition retarded in the presence of mycorrhizal roots in forest soil? Internal Report 95, Swedish Coniferous Forest Project, Institute of Ecology and Environmental Research, Swedish University of Agricultural Sciences, Uppsala.

Bernier, B., and M. Brazeau. 1988. Foliar nutrient status in relation to sugar maple dieback and decline in the Quebec Appalachians. Canadian Journal of Forest Research 18:754–761.

Binkley, D. 1986. Forest nutrition management. John Wiley and Sons, New York, New York.

Binkley, D., C. T. Driscoll, H. L. Allen, P. Schoeneberger, and D. McAvoy. 1989. Acidic deposition and forest soils. Ecological Studies Series 72. Springer-Verlag, New York, New York.

Birdsey, R. A., A. J. Plantinga, and L. S. Heath. 1993. Past and prospective carbon storage in United States forests. Forest Ecology and Management 58:33 40.

Bockheim, J. G., and J. E. Leide. 1990. Estimating nutrient uptake in forest ecosystems. Pages 155–177 in S. P. Gessel, D. S. Lacate, G. F. Weetman, and R. F. Powers, editors. Sustained productivity of forest soils. Proceedings, Seventh North American Forest Soils Conference, Faculty of Forestry, University of British Columbia, Vancouver.

Bormann, F. H., and G. E. Likens. 1979. Pattern and process in a forested ecosystem. Springer-Verlag, New York, New York.

Brady, N. C. 1990. The nature and properties of soils. 10th edition. Macmillan Publishing Company, New York, New York.

Brand, D. G., and A. M. LeClaire. 1994. The model forests programme: international cooperation to define sustainable management. Unasylva 176, 45:51–58.

Briggs, J., and F. D. Peat. 1989. Turbulent mirror. Harper and Row, New York, New York.

British Columbia Ministry of Forests. 1987. Acid rain—no threat to B.C., yet! Information Forestry 14(1):3.

Brooks, D. J., and G. E. Grant. 1992. New approaches to forest management, part one. Journal of Forestry 90(1):25–28.

Carlyle, J. C. 1986. Nitrogen cycling in forested ecosystems. Forestry Abstracts 47: 307–336.

Clark, F. E., and T. Rosswall (editors). 1981. Terrestrial nitrogen cycles. Ecological Bulletins (Stockholm) 33, Swedish Natural Science Research Council.

Cole, D. W., and D. Johnson. 1979. The cycling of elements within forests. Pages 185–198 in P. E. Heilman, H. W. Anderson, and D. M. Baumgartner, editors. Forest soils of the Douglas-fir region. Cooperative Extension Service, Washington State University, Pullman.

Cole, D. W., and M. Rapp. 1981. Elemental cycling in forest ecosystems. Pages 341–409 in D. E. Reichle, editor. Dynamic properties of forest ecosystems. International Biological Programme 23, Cambridge University Press, Cambridge, England.

Davidson, E. A., D. D. Myrold, and P. M. Groffman. 1990. Denitrification in temperate forest ecosystems. Pages 196–220 in S. P. Gessel, D. S. Lacate, G. F. Weetman, and R. F. Powers, editors. Sustained productivity of forest soils. Proceedings, Seventh North American Forest Soils Conference, Faculty of Forestry, University of British Columbia, Vancouver.

DeBell, D. S., and R. O. Curtis. 1993. Silviculture and new forestry in the Pacific Northwest. Journal of Forestry 91(12):25–30.

deMaynadier, P., and M. Hunter, Jr. This volume. The role of keystone ecosystems in landscapes.

Farnum, P. 1994. What will we want from tomorrow's forest? Pages 20–25 in Society of American Foresters. Foresters together: meeting tomorrow's challenges. Proceedings, 1993 Society of American Foresters National Convention, Nov. 7–10, Indianapolis, Indiana. Society of American Foresters, Bethesda, Maryland.

Fisher, R. F. 1990. Amelioration of soils by trees. Pages 290–300 in S. P. Gessel, D. S. Lacate, G. F. Weetman, and R. F. Powers, editors. Sustained productivity of forest soils. Proceedings, Seventh North American Forest Soils Conference, Faculty of Forestry, University of British Columbia, Vancouver.

Fortin, J. A., L. Chatarpaul, and A. Carlisle. 1984. The role of nitrogen fixation in intensive forestry in Canada. Part 1, Principles, practice, and potential. Petawawa National Forestry Institute, Canadian Forestry Service, Information Report PI-X-28.

Foster, N. W. 1989. Acidic deposition: what is fact, what is speculation, what is needed? Water, Air, and Soil Pollution 48:299–336.

Frank, D. A., and S. J. McNaughton. 1992. The ecology of plants, large mammalian herbivores, and drought in Yellowstone National Park. Ecology 73:2043–2058.

Franklin, J. F. 1989. Toward a new forestry. American Forests 95(11/12):37–44.

———. 1993. Lessons from old-growth: fueling controversy and providing direction. Journal of Forestry 91(12):10–13.

Frossard, E., J. W. B. Stewart, and R. J. St. Arnaud. 1989. Distribution and mobility of phosphorus in grassland and forest soils of Saskatchewan. Canadian Journal of Soil Science 69:401–416.

Gadgil, R. L., and P. D. Gadgil. 1975. Suppression of litter decomposition by mycorrhizal roots of Pinus radiata. New Zealand Journal of Forest Science 5:33–41.

Gessel, S. P., D. S. Lacate, G. F. Weetman, and R. F. Powers (editors). 1990. Sustained productivity of forest soils. Proceedings, Seventh North American Forest Soils Conference, Faculty of Forestry, University of British Columbia, Vancouver.

Goodroad, L. L., and D. R. Keeney. 1984. Nitrous oxide emissions from soils during thawing. Canadian Journal of Soil Science 64:187–194.

Gordon, A. G., D. M. Morris, and N. Balakrishnan. 1993. Impacts of various levels of biomass removals on the structure, function, and productivity of black spruce ecosystems: research protocols. Ontario Forest Research Institute, Ontario Ministry of Natural Resources, Forest Research Report no. 109.

Gosz, J. R. 1981. Nitrogen cycling in coniferous ecosystems. Pages 405–426 in F. E. Clark, and T. Rosswall, editors. Terrestrial nitrogen cycles. Ecological Bulletins (Stockholm) 33, Swedish Natural Science Research Council.

———. 1992. Sustainable forest ecosystem management: interpretations from the Sustainable Biosphere Initiative. William P. Thompson Memorial Lecture XVI, School of Forestry, Northern Arizona University, Flagstaff.

Green, R. N., R. L. Trowbridge, and K. Klinka. 1993. Towards a taxonomic classification of humus forms. Forest Science Monograph 29, Society of American Foresters, Bethesda, Maryland.

Gregersen, H. M., A. L. Lundgren, and D. N. Bengston. 1990. Planning and managing forestry research: guidelines for managers. Food and Agriculture Organization of the United Nations, Rome. Forestry Paper 96.

Grumbine, R. E. 1994. What is ecosystem management? Conservation Biology 8:27–38.

Harmon, M. E., W. K. Ferrell, and J. F. Franklin. 1990. Effects on carbon storage of conversion of old-growth forests to young forests. Science 247:699–702.

Hart, S. C., and M. K. Firestone. 1991. Forest floor–mineral soil interactions in the internal nitrogen cycle of an old-growth forest. Biogeochemistry 12:103–127.

Harvey, A. E., and L. F. Neuenschwander (editors). 1991. Proceedings: management and productivity of western-montane forest soils. USDA Forest Service, Intermountain Research Station, General Technical Report INT-280.

Heal, O. W., M. J. Swift, and J. M. Anderson. 1982. Nitrogen cycling in United Kingdom forests: the relevance of basic ecological research. Philosophical Transactions of the Royal Society of London, B. Biological Sciences 296:427–444.

Hoen, H. F., and B. Solberg. 1994. Potential and economic efficiency of carbon sequestration in forest biomass through silvicultural management. Forest Science 40:429–451.

Hopwood, D. 1991. Principles and practices of new forestry. British Columbia Ministry of Forests (Canada), Land Management Report no. 71.

Hornbeck, J. W., C. T. Smith, Q. W. Martin, L. M. Tritton, and R. S. Pierce. 1990. Effects of intensive harvesting on nutrient capitals of three forest types in New England. Forest Ecology and Management 30:55–64.

Humphreys, F. R., and M. J. Lambert. 1965. An examination of a forest site which has exhibited the ash-bed effect. Australian Journal of Soil Research 3(1):81–94.

Hunter, M. L. 1990. Wildlife, forests, and forestry. Prentice Hall, Englewood Cliffs, New Jersey.

Irwin, L. L., J. G. Cook, R. A. Riggs, and J. M. Skovlin. 1994. Effects of long-term grazing by big game and livestock in the Blue Mountains forest ecosystems. USDA Forest Service, Pacific Northwest Research Station, General Technical Report PNW-GTR-325.

Jepson, E. A., and J. G. Bockheim. 1985. Acidic deposition influences on biogeochemistry of four forest ecosystems in northwestern Wisconsin. Pages 509–521 in D. E. Caldwell, J. A. Brierley, and C. L. Brierley, editors. Planetary ecology. Van Nostrand Reinhold Company, New York, New York.

Johansson, M.-B. 1994. The influence of soil scarification on the turn-over rate of slash needles and nutrient release. Scandinavian Journal of Forest Research 9:170–179.

Johnson, D. W., and J. O. Reuss. 1984. Soil-mediated effects of atmospherically deposited sulphur and nitrogen. Philosophical Transactions of the Royal Society of London, B. Biological Sciences 305:383–392.

Kelty, M. J. 1992. Comparative productivity of monocultures and mixed-species stands. Pages 125–141 in M. J. Kelty, B. C. Larson, and C. D. Oliver, editors. The ecology and silviculture of mixed-species forests. Kluwer Academic Publishers, Boston, Massachusetts.

Kimmins, J. P. 1989. Ecological implications of successional manipulation. Pages 9–16 in B. A. Scrivener and J. A. MacKinnon, editors. Learning from the past, looking to the future. Proceedings, Northern Silviculture Committee's Winter Workshop, Prince George, February 2–3, 1988. Canadian Forestry Service/British Columbia Ministry of Forests, FRDA Report 030.

Kimmins, J. P., D. Binkley, L. Chatarpaul, and J. de Catanzaro. 1985. Biogeochemistry of temperate forest ecosystems: literature on inventories and dynamics of biomass and nutrients. Canadian Forestry Service, Petawawa National Forestry Institute, Information Report PI-X-47E/F.

Krause, H. H. 1991. Nutrient form and availability in the root environment. Pages 1–24 in R. van den Driessche, editor. Mineral nutrition of conifer seedlings. CRC Press, Boca Raton, Florida.

Mahendrappa, M. K. 1990. Partitioning of rainwater and chemicals into throughfall and stemflow in different forest stands. Forest Ecology and Management 30:65–72.

Meentemeyer, V. 1978. Macroclimatic and lignin control of litter decomposition rates. Ecology 59:465–472.

Morrison, I. K. 1984. Acid rain: a review of literature on acid deposition effects in forest ecosystems. Forestry Abstracts 45:483–506.

Morrison, I. K., N. W. Foster, and P. W. Hazlett. 1993. Carbon reserves, carbon cycling, and harvesting effects in three mature forest types in Canada. New Zealand Journal of Forestry Science 23:403–412.

Neary, D. G., E. J. Jokela, N. B. Comerford, S. R. Colbert, and T. E. Cooksey. 1990. Understanding competition for soil nutrients: the key to site productivity on southeastern Coastal Plain Spodosols. Pages 432–450 in S. P. Gessel, D. S. Lacate, G. F. Weetman, and R. F. Powers, editors. Sustained productivity of forest soils. Proceedings, Seventh North American Forest Soils Conference, Faculty of Forestry, University of British Columbia, Vancouver.

Norcross, E. J. 1991. Ecosystem management in a dynamic society. Pages 146–148 in D. C. Le Master, and G. R. Parker, editors. Ecosystem management in a dynamic society. Proceedings of a conference, Nov. 19–21, West Lafayette, Indiana. Department of Forestry and Natural Resources, Purdue University.

Piene, H. 1978. Effects of increased spacing on carbon mineralization rates and temperature in a stand of young balsam fir. Canadian Journal of Forestry Research 8: 398–406.

Postgate, J. R. 1982. Biological nitrogen fixation: fundamentals. Philosophical Transactions of the Royal Society of London, B. Biological Sciences 296:375 385.

Powers, R. F., D. H. Alban, R. E. Miller, A. E. Tiarks, C. G. Wells, P. E. Avers, R. G. Cline, R. O. Fitzgerald, and N. S. Loftus. 1990. Sustaining site productivity in North American forests: Problems and prospects. Pages 49–79 in S. P. Gessel, D. S. Lacate, G. F. Weetman, and R. F. Powers, editors. Sustained productivity of forest soils. Proceedings, Seventh North American Forest Soils Conference, Faculty of Forestry, University of British Columbia, Vancouver.

Powers, R. F., and K. Van Cleve. 1991. Long-term ecological research in temperate and boreal forest ecosystems. Agronomy Journal 83:11–24.

Pritchett, W. L., and R. F. Fisher. 1987. Properties and management of forest soils. 2d edition. John Wiley and Sons, New York, New York.

Probst, J. R. and T. R. Crow. 1991. Integrating biological diversity and resource management. Journal of Forestry 89(2):12–17.

Reichle, D. E. (editor). 1981. Dynamic properties of forest ecosystems. International Biological Programme 23, Cambridge University Press, Cambridge, England.

Rennie, P. J. 1990. Some threats to sustaining forest yields in North America: research challenges. Pages 6–22 in S. P. Gessel, D. S. Lacate, G. F. Weetman, and R. F. Powers, editors. Sustained productivity of forest soils. Proceedings, Seventh North American Forest Soils Conference, Faculty of Forestry, University of British Columbia, Vancouver.

Richter, D. D., D. Markewitz, C. G. Wells, H. L. Allen, R. April, P. R. Heine, and B. Urrego. 1994. Soil chemical change during three decades in an old-field loblolly pine (*Pinus taeda* L.) ecosystem. Ecology 75:1463–1473.

Robertson, G. P., and J. M. Tiedje. 1984. Denitrification and nitrous oxide production in successional and old-growth Michigan forests. Soil Science Society of America Journal 48:383–389.

Rustad, L. E. 1994. Element dynamics along a decay continuum in a red spruce ecosystem in Maine, USA. Ecology 75:867–879.

Salwasser, H. 1990. Conserving biological diversity: a perspective on scope and approaches. Forest Ecology and Management 35:79–90.

Salwasser, H., and R. D. Pfister. 1994. Ecosystem management: from theory to practice. Pages 150–161 *in* W. W. Covington and L. F. DeBano, technical coordinators. Sustainable ecological systems: implementing an ecological approach to land management. USDA Forest Service, Rocky Mountain Forest and Range Experiment Station, General Technical Report RM-247.

Schaedle, M. 1991. Nutrient uptake. Pages 25–59 *in* R. van den Driessche, editor. Mineral nutrition of conifer seedlings. CRC Press, Boca Raton, Florida.

Setala, H., and V. Huhta. 1990. Evaluation of the soil fauna impact on decomposition in a simulated coniferous forest soil. Biology and Fertility of Soils 10:163–169.

Setala, H., E. Martikainen, M. Tyynismaa, and V. Huhta. 1990. Effects of soil fauna on leaching of nitrogen and phosphorus from experimental systems simulating coniferous forest floor. Biology and Fertility of Soils 10:170–177.

Seymour, R. S. 1992. The red spruce-balsam fir forest of Maine: evolution of silvicultural practice in response to stand development patterns and disturbances. Pages 217–244 *in* M. J. Kelty, B. C. Larson, and C. D. Oliver, editors. The ecology and silviculture of mixed-species forests. Kluwer Academic Publishers, Boston, Massachusetts.

Seymour, R. S., and M. L. Hunter. 1992. New Forestry in eastern spruce–fir forests: principles and applications to Maine. Maine Agricultural Experiment Station, University of Maine, Miscellaneous Publication 716.

Smith, D. M. 1986. The practice of silviculture. 8th edition. John Wiley and Sons, New York, New York.

Smith, T. J., K. G. Boto, S. D. Frusher, and R. L. Giddins. 1991. Keystone species and mangrove forest dynamics: the influence of burrowing by crabs on soil nutrient status and forest productivity. Estuarine, Coastal and Shelf Science 33:419–432.

Society of American Foresters. 1993. SAF Silviculture Working Group Newsletter, October, 1993. Bethesda, Maryland.

Soil Science Society of America. 1987. Glossary of soil science terms. Madison, Wisconsin.

Staaf, H. 1987. Foliage litter turnover and earthworm populations in three beech forests of contrasting soil and vegetation types. Oecologia (Berlin) 72:58–64.

Stewart, W. D., and T. Rosswall (organizers). 1982. The nitrogen cycle. Philosophical

Transactions of the Royal Society of London, B. Biological Sciences 296:299–576.

Stone, E. L. 1975a. Effects of species on nutrient cycles and soil change. Pages 15–28 *in* N. B. Comerford and D. G. Neary, compilers. 1985. Forestry and soils: the contributions of Dr. Earl L. Stone to forest soil science. Soil Science Department, Institute of Food and Agricultural Sciences, University of Florida, Gainesville.

―――. 1975b. Nutrient release through forest harvest: a perspective. Pages 155–174 *in* N. B. Comerford and D. G. Neary, compilers. 1985. Forestry and soils: the contributions of Dr. Earl L. Stone to forest soil science. Soil Science Department, Institute of Food and Agricultural Sciences, University of Florida, Gainesville.

―――. 1979. Nutrient removals by intensive harvest: some research gaps and opportunities. Pages 179–199 *in* N. B. Comerford and D. G. Neary, compilers. 1985. Forestry and soils: the contributions of Dr. Earl L. Stone to forest soil science. Soil Science Department, Institute of Food and Agricultural Sciences, University of Florida, Gainesville.

―――（editor）. 1984. Forest soils and treatment impacts. Proceedings, Sixth North American Forest Soils Conference, Department of Forestry, Fisheries and Wildlife, University of Tennessee, Knoxville.

Swift, M. J. 1977. The role of fungi and animals in the immobilisation and release of nutrient elements from decomposing branchwood. Pages 193–202 *in* U. Lohm and T. Persson, editors. Soils organisms as components of ecosystems. Ecological Bulletins (Stockholm) 25, Swedish Natural Science Research Council.

Swift, M. J., O. W. Heal, and J. M. Anderson. 1979. Decomposition in terrestrial ecosystems. Studies in Ecology 5. University of California Press, Berkeley.

Tamm, C. O. 1979. Productivity of Scandinavian forests in relation to changes in management and environment. Irish Forestry 36:111–120.

―――. 1982. Nitrogen cycling in undisturbed and manipulated boreal forest. Philosophical Transactions of the Royal Society of London, B. Biological Sciences 296:419–425.

Tamm, C. O., and L. Hallbäcken. 1988. Changes in soil acidity in two forest areas with different acid deposition: 1920s to 1980s. Ambio 17:56–61.

Tamm, O. 1950. Northern coniferous forest soils. Translated from the Swedish by M. L. Anderson. Scrivener Press, Oxford, England.

Tenner, E. 1991. Revenge theory. Harvard Magazine, March–April, pp. 27–30.

Ticknor, W. D. 1993. The TEMS approach. American Forests 99(7/8):34–36.

Tilman, D., and J. A. Downing. 1994. Biodiversity and stability in grasslands. Nature 367:363–365.

Tisdale, S. L., and W. L. Nelson. 1975. Soil fertility and fertilizers. 3d edition. Macmillan Publishing Company, New York, New York.

Troeh, F. R., and L. M. Thompson. 1993. Soils and soil fertility. 5th edition. Oxford University Press, New York, New York.

Turner, M. G., W. H. Romme, R. H. Gardner, R. V. O'Neill, and T. K. Kratz. 1993.

A revised concept of landscape equilibrium: disturbance and stability on scaled landscapes. Landscape Ecology 8:213–227.

USDA Forest Service. 1990. Draft environmental impact statement: Shasta Costa timber sales and integrated resource projects, Siskiyou National Forest. USDA Forest Service, Pacific Northwest Region.

van Breemen, N. 1990. Deterioration of forest land as a result of atmospheric deposition in Europe: a review. Pages 40–48 *in* S. P. Gessel, D. S. Lacate, G. F. Weetman, and R. F. Powers, editors. Sustained productivity of forest soils. Proceedings, Seventh North American Forest Soils Conference, Faculty of Forestry, University of British Columbia, Vancouver.

Van Cleve, K., R. Barney, and R. Schlenter. 1981. Evidence of temperature control of production and nutrient cycling in two interior Alaska black spruce ecosystems. Canadian Journal of Forest Research 11:258–273.

Van Cleve, K., and A. F. Harrison. 1985. Bioassay of forest floor phosphorus supply for growth. Canadian Journal of Forest Research 15:156–162.

van den Driessche, R. 1991. Effects of nutrients on stock performance in the forest. Pages 229–260 *in* R. van den Driessche, editor. Mineral nutrition of conifer seedlings. CRC Press, Boca Raton, Florida.

Van Miegroet, H., D. W. Cole, and P. S. Homann. 1990. The effect of alder forest cover on site fertility and productivity. Pages 333–354 *in* S. P. Gessel, D. S. Lacate, G. F. Weetman, and R. F. Powers, editors. Sustained productivity of forest soils. Proceedings, Seventh North American Forest Soils Conference, Faculty of Forestry, University of British Columbia, Vancouver.

Vitousek, P. M. 1982. Nutrient cycling and nutrient use efficiency. American Naturalist 119:553–572.

———. 1983. Mechanisms of ion leaching in natural and managed ecosystems. Pages 129–144 *in* H. A. Mooney, and M. Godron, editors. Disturbance and ecosystems. Springer-Verlag, New York, New York.

Vitousek, P. M., S. W. Andariese, P. A. Matson, L. Morris, and R. L. Sanford. 1992. Effects of harvest intensity, site preparation, and herbicide use on soil nitrogen transformations in a young loblolly pine plantation. Forest Ecology and Management 49:277–292.

Vitousek, P. M., J. R. Gosz, C. C. Grier, J. M. Melillo, and W. A. Reiners. 1982. A comparative analysis of potential nitrification and nitrate mobility in forest ecosystems. Ecological Monographs 52:155–177.

Walters, C. 1986. Adaptive management of renewable resources. Macmillan Publishing Company, New York, New York.

Walters, C., and C. S. Holling. 1990. Large-scale management experiments and learning by doing. Ecology 71:2060–2068.

Woodmansee, R. G. 1990. Biogeochemical cycles and ecological hierarchies. Pages 57–71 *in* I. S. Zonneveld, and R. T. Forman, editors. Changing landscapes: an ecological perspective. Springer-Verlag, New York, New York.

Xiao, X. J., D. W. Anderson, and J. R. Bettany. 1991. The effect of pedogenetic processes on the distribution of phosphorus, calcium, and magnesium in Gray Luvisols. Canadian Journal of Soil Science 71:397–410.

Zonneveld, I. S. 1990. Scope and concepts of landscape ecology. Pages 3–20 *in* I. S. Zonneveld, and R. T. Forman, editors. Changing landscapes: an ecological perspective. Springer-Verlag, New York, New York.

DISTURBANCE

Management activities should conserve or restore natural ecosystem disturbance patterns.
—Merrill Kaufmann and colleagues

Natural processes will be relied on to control populations of native species to the greatest extent possible.
—National Park Service Management Policies

Overleaf: The endangered Karner blue butterfly *(Lycaeides melissa samuelis)* at Fort McCoy military reservation in central Wisconsin (courtesy of A. Bidwell). This species occurs on oak savannahs that are maintained by fire. Because of fire suppression and agricultural development throughout the midwestern United States, oak savannah is listed among the twenty-one rarest ecosystem types in North America (Noss and Peters 1995).

Chapter 7 **Applied Disequilibriums:
Riparian Habitat Management
for Wildlife**
Mark S. Boyce and Neil F. Payne

In this chapter we review the role of riparian zone management in the context of ecosystem management, with particular reference to wildlife species. We conclude that management to maintain or restore disturbance regimes is fundamental to ecosystem management in riparian areas.

Ecosystem management promises fundamental changes in land management practices, especially on public lands. New priorities contesting traditional economic considerations include biodiversity and protection of threatened and endangered species, ecological-process management, establishment of ecological baseline preserves, and aesthetics. Riparian areas often receive special attention in ecosystem management because they provide corridors, often contain many species, and might contain unique assemblages of species. Indeed, good reason exists for conserving riparian zones because more than 70% of original riparian areas in North America have been substantially altered by removal of beaver (*Castor* spp.) and by human developments, including agriculture, channelization, dams, logging, and road construction (Megahan and King 1985).

The dynamic nature of riparian ecosystems—frequently perturbed and perpetually changing (Botkin 1990)—acts to structure ecosystems in time and space (Sousa 1984). Human development usually attempts to stabilize riparian areas: to reduce the frequency and extent of natural disturbances, for example, flooding, wildfires, and herbivory.

Riparian zones tend to be narrow, restricted by the extent that water influences adjacent vegetation. Often they are small, and therefore overlooked in such large-scale assessments as GAP analysis (Scott et al. 1993). Yet riparian areas are important because of their high productivity and diversity. Although they occupy only 1% of North America's landscapes, riparian areas host over 80% of our threatened and endangered species (Kie et al. 1994).

DEFINITIONS

While reviewing the literature, we found that most references to riparian zones were for areas next to streams and rivers. Yet the dictionary definition of *riparian* includes areas surrounding lakes and ponds as well as stream areas. Kie et al. (1994:671) defined riparian areas as "the sum of the terrestrial and aquatic components characterized (1) by the presence of permanent or ephemeral surface or subsurface water, (2) by water flowing through channels defined by the local physiography, and (3) by the presence of obligate, occasionally facultative, plants requiring readily available water and rooted in aquatic soils derived from alluvium." Our discussion will concentrate on riparian zones associated with flowing waters.

Riparian zone and riparian vegetation are not synonymous. The riparian zone can extend into upland areas and consist of upland plants as well as riparian plants, depending on how wide the riparian zone is defined. Larger (higher-order) streams tend to have wider zones of riparian vegetation. A riparian area associated with a low-gradient stream is much wider than a riparian zone associated with steep-gradient headwater streams (Elmore and Beschta 1987).

ATTRIBUTES

Biodiversity

Most attention in biological diversity conservation has been on mammals, birds, and flowering plants, largely because these taxa engender most interest. Yet most biological diversity in ecosystems exists among smaller organisms, including insects, nematodes, and fungi. For the most part we do not know how to manage the smaller, often obscure, organisms, leading Noss and Scott (this volume) to argue that the only effective way to preserve biodiversity is ecosystem protection.

Diversity of species tends to be high in riparian zones. Riparian areas are ecotones between aquatic and terrestrial communities and contain species from both aquatic and terrestrial communities, plus a few that are unique to riparian habitats. Some taxa are strongly associated with riparian areas, including most amphibians and reptiles (Jones 1988), willow flycatchers *(Empidonax traillii)*, muskrats *(Ondatra zibethicus)*, mink *(Mustela vison)*, and beaver.

Riparian areas also are diverse because of geomorphological processes, for example, bank erosion, sedimentation, channel movements, and oxbows. In most regions of the world riparian areas are seasonally flooded. The influx of nutrients and water permits development of plant communities that are more diverse and more productive than surrounding habitats (Kie et al. 1994).

Where riparian areas are surrounded by low vegetation, such as grassland, agricultural land, shrub-steppe, or desert, bird diversity seems to be influenced by the structural complexity of the riparian vegetation (Anderson and Olmart 1977, Bull and Skovlin 1982). Where riparian areas are surrounded by forest, with consequent less edge, bird diversity seems more influenced by composition of vegetation than floristic structure (Knopf and Samson 1996).

Edge Effects and Corridors

By their very nature, riparian areas have a large amount of edge, which benefits taxa that require different kinds of vegetation (Leopold 1933). Although forest interior species are lacking in most riparian zones, many bird species that are considered forest species in the uplands are found in riparian forests of the Upper Mississippi River (Knutson 1995). Because some predators concentrate in edge habitats, predation is high in riparian zones for some prey species (Wilcove 1985, 1987). Conversely, productive riparian zones can host source populations of animals that disperse to surrounding uplands, as Andersen (1994) observed for *Peromyscus maniculatus* in Arizona.

Riparian areas as corridors are not always beneficial (Noss 1987, Hobbs 1992). In grasslands and agricultural areas, riparian forests afford opportunity for dispersal of forest birds and mammals. Genetic exchange via corridors might be essential to maintain viable populations of certain species occupying small habitat fragments. However, corridors also can facilitate dispersal of such brood parasites as brown-headed cowbirds *(Molothrus ater)* and other edge species and exotics; corridors can encourage spread of disease and fire, as well as providing travel lanes for predators (Simberloff et al. 1992). Riparian corridors that have developed along the Platte River after agricultural development have resulted in the invasion of eastern woodland bird species into western states (Knopf 1992).

Ecosystem Functions

As well as providing important habitat for wildlife, riparian vegetation can reduce nonpoint pollution (Megahan and King 1985, Dickson 1989) and greatly reduce velocity of flood waters (Malanson 1993). Features of riparian environments valuable for wildlife habitat include (1) structural variation and woody debris in plant communities, (2) surface water and soil moisture, (3) spatial heterogeneity of habitats, and (4) corridors for migration and dispersal (Melanson 1993). Riparian features that often distinguish wildlife habitat include (1) vegetation type and size, for example, nesting and perching sites, mast; (2) size and shape of the site, which influences interior and edge species; (3) flooding, which affects food resources and nesting sites; and (4) elevation, which affects climate and topography (Melanson 1993).

DISTURBANCE

Community composition is often greatly influenced by the history of disturbances that have occurred on a site. Many species have adaptations to cope with disturbance, and their perpetuation can even require disturbance.

Flooding

Riparian zone specialists often have adapted to a flooding regimen, and altering the regimen can have severe consequences. For example, black stilts (*Himantopus novaezealandiae*) in New Zealand nest on gravel bars immediately following the spring floods created by snow melt in the mountains. Construction of dams for agriculture altered the flooding schedule so that nests of the black stilt are often flooded, contributing to the severely endangered status of this wading bird (Reed et al. 1993). Similarly, flooding renews sandbar habitats for least terns *(Sterna albifrons)* and piping plovers *(Charádrius melódus)* in the midwest United States (Sidle et al. 1992).

Plant composition and structure in riparian habitats can be altered substantially by the flooding regime. For example, reduced spring flooding along the Platte River, attributable to flood-control dams, has resulted in perennial vegetation colonizing stream banks. Cottonwood communities in some areas became decadent because no spring floods created seedling germination sites (Knopf and Scott 1990). Stream diversions can cause selective mortality of

Fig. 7.1. The wood turtle is listed as either threatened or endangered in Iowa, Minnesota, and Wisconsin. Habitat for nesting consists of eroded banks.

juvenile plants that can have long-term consequences for riparian vegetation (Smith et al. 1991).

Erosion caused by flowing waters can create essential habitats. For example, eroded sandy banks afford essential nesting habitat for wood turtles (*Clemmys insculpta*), which are locally threatened in the midwestern United States (Buech 1992; fig. 7.1). Gravel, which provides important habitat for certain invertebrates that fish eat and is required for trout and salmon spawning, occurs in streams as a consequence of erosion (Baltz and Moyle 1984). Flooding shapes the distribution of such sediments as accumulating silt, which provides substrate for the establishment of various woody plants, for example, willow (*Salix* spp.), birch (*Betula* spp.), poplars (*Populus* spp.), alder (*Alnus* spp.), and herbaceous species.

Waterfowl and other wildlife benefit when rivers meandering across flat terrain cut new channels, leaving isolated islands. In northern regions ice floes during breakup can be a major force shaping stream channels each spring and can affect wildlife. For example, beaver lodges built along a stream channel often will be destroyed by ice floes that scour the banks. Yet ice floes create stream course changes that can benefit beaver by affording them access to newly available forage (Boyce 1981). Deposition of eroded upstream sediments is essential for the development of beaver habitat in many streams (Gill 1972).

Fire

Fire can affect riparian areas because of the importance of downed woody debris in structuring stream channels and aquatic habitats. Postfire erosion and nutrient leaching also can cause influx of sediment and nutrients to streams, as occurred after one of the largest forest fires in recorded history of the United States, the Peshtigo fire, which burned 5,180 km² in 1871 (Wells 1968).

The range of consequences of fire was dramatically illustrated by the fires of 1988 in Yellowstone National Park. Much of the park burned lightly and recovered quickly so that the effects of the fires were a minor perturbation. But in areas of steep slopes that were intensively burned, massive mudflows altered stream courses. Intense fires in the Cache Creek drainage in northeastern Yellowstone Park resulted in such dramatic geomorphological changes to streams that communities will require decades if not centuries to recover from the fires (Minshall and Robinson 1992). Even more dramatic stream channel alterations were caused by the 1980 eruption of Mount St. Helens in Washington (Keller 1986).

Wind

Trees in riparian areas tend to have shallow root systems because of the high water table and thus are susceptible to windthrow, which can result in recruitment of large organic debris to the riparian area. Windthrow can be aggravated if clearcuts occur to the edge of the riparian zone, exposing the riparian forest to strong wind.

Biotic mechanisms

Grazing, browsing, pest outbreaks, and beaver activity are among biotic processes that affect riparian areas. Heavy use by elk *(Cervus elaphus)* along streams in Arizona resulted in reduced tall-willow communities that are important nesting and foraging areas for such songbirds as the willow flycatcher (Knopf et al. 1988). Heavy ungulate use of riparian willow in Yellowstone National Park has reduced the capacity of some areas to sustain white-tailed deer *(Odocoileus virginianus)* and beaver (Chadde 1989, Despain 1989,

Chadde and Kay 1991). Similarly, alligators *(Alligator mississippiensis)* in Florida and rhinos *(Rhinoceros unicornis)* in Nepal have been shown to alter riparian plant succession (Robinson and Bolen 1989, Dinerstein 1991). Beaver commonly alter the composition and structure of riparian vegetation (Parker et al. 1985, Barnes and Dibble 1986, Nolet et al. 1994).

High densities of hippos *(Hippopotamus amphibius)* in Africa have been shown to affect riparian vegetation (Laws 1981). Hippos typically spend the day in the water but emerge at night to graze nearby. Trampling on banks can result in sites that are highly susceptible to erosion. Activity of hippos in water increases turbidity. By defecating in the water, hippos stimulate the growth of algae and other aquatic plants. High nutrient loads in streams occupied by hippos have resulted in highly productive fisheries in Lake George, Uganda (Laws 1968).

Humans often believe that elk, beaver, and hippos must be culled to mitigate their influence on vegetation. Yet such decisions are value judgments for maintaining vegetation communities as we like them rather than decisions giving priority to natural ecological processes.

HUMAN-CAUSED DISTURBANCES

Economic loss and/or other difficulties for humans can result from natural processes that affect riparian areas. For example, fires and floods can destroy homes, crops, and timber. Wild ungulates can compete with domestic livestock for forage or effect changes in vegetation communities that are not desired by humans (see, for example, Chadde and Kay 1991; Kie et al. 1994). Likewise, beaver cut down trees and construct dams that flood roads, crops, and trees. Humans attempt to control these natural processes with fire suppression, dams, levees, and channelization to prevent flooding, as well as culling to decrease wild animals.

Attempts to control watercourses have had numerous negative consequences. Dams on the Columbia River have blocked migration of anadromous salmonids, contributing to recent enforcement of the Endangered Species Act that has closed important fisheries for these species. The levee system on the Mississippi River contributed to widespread flooding in 1993 (Sparks 1995). Dams for irrigation have reduced the flow of the Colorado River. Water diversion to cropland in Florida has altered vegetation in Everglades National Park and Florida Bay.

Although soil erosion is a natural process, levels at which it occurs in most of North America far exceed normal (Pimentel et al. 1995). Road construction, agriculture, mining, and urbanization have led to extensive erosion that has silted major watercourses, reduced water quality, and devastated fisheries. Logging has had far-reaching consequences for riparian environments. Although clearcutting might mimic wildfire by reducing ground cover, causing sloughing of slopes, erosion, and altered stream channels, the extent of clearcutting in many areas has dwarfed that of wildfires. Erosion associated with clearcutting also has contributed to salmonid declines in the western United States and has pitted fishermen against loggers in the debate over the management of Pacific Northwest forests (FEMAT 1993).

Livestock grazing in the western United States has caused bank erosion and siltation that affects salmonid spawning areas (Fleischner 1994). Again, although livestock grazing might be seen to mimic effects of natural ungulates grazing in riparian areas (Dodd 1992), the extent and manner of livestock grazing, as well as its seasonal distribution, is different from that of native ungulates (Julander and Jeffrey 1964).

MANAGING BY REDUCING HUMAN INTERFERENCE

By manipulating disturbance regimes, humans alter the natural composition and structure of plant and animal communities in riparian areas. These changes can result in the loss of some species and increased opportunities for the spread of exotic species (Hobbs and Huenneke 1992), or can change ecosystem functions that are desirable. Humans have not been notably successful at predicting consequences of their actions or fully understanding the ramifications of altering ecosystems.

Ecosystem management might seek to restore or sustain the natural disturbance regimen. Such ecological-process management could manage water release at existing dams to create spring flushing flows and to flood riparian areas periodically (Sparks 1995); lightning-caused fires could be allowed to burn in areas where they would not threaten human lives or property; and logjams could be allowed to choke streams in places, improving habitat for fish. Sometimes clearcutting, either extensive or patchcut, might be used to mimic effects of windthrow or fire in areas where allowing wildfires to burn is impractical. In many instances, reconstructing the natural disturbance regime will be difficult, but minimizing the level of human interference is likely to be an improvement (Sinclair 1983).

Erosion control is usually, but not always, beneficial. Because of agricultural development, much of North America has suffered extensive erosion of soils that have choked watercourses downstream. We have a perspective that all erosion is bad; indeed, management practices should attempt to control and reduce erosion because so much exists. But prevention of erosion in riparian areas that are not degraded by humans might not be justified. For example, programs initiated by Trout Unlimited, the U.S. Forest Service, and state agencies have so aggressively controlled erosion along certain trout streams that they have eliminated mud banks essential for nesting by wood turtles (Buech 1992).

Protection of riparian areas from natural disturbances can have complex ramifications. Small streams are affected to a great degree by allochthonous material (Sedell et al. 1989), and in general productivity and diversity of sites is higher when allochthonous material is dominated by deciduous tree leaves than by conifer needles. For example, the aquatic insects and fishes that feed on them are more abundant and diverse in stream segments receiving deciduous allochthonous material. Coniferous debris persists longer in streams than does hardwood debris (Anderson et al. 1978, Swanson and Lienkaemper 1978). But protected areas in many regions, for example, in the Pacific Northwest, will proceed through succession to conifer forest types with concomitant effects on stream productivity and diversity. Therefore, maintaining a natural disturbance regime on the landscape that maintains seral stages dominated by deciduous trees can be an important consideration for managing riparian habitats.

Similarly, fishes often depend upon spawning gravel that derives from alluvium that is eroded into the stream. Such gravel will be maintained only if the stream is allowed occasionally to erode banks and to flush out finer sediments; otherwise, gravel eventually will be buried by finer sediments. Silt accumulation and mud bank formation is essential for establishing some riparian plants, for example, willows and alders. Unnaturally excessive silt loads can degrade streams, but some erosion is needed to sustain riparian plant communities.

Managing most riparian areas mainly involves protection from human disturbance (Swift 1984, Oakley et al. 1985): (1) road construction, (2) logging, (3) excessive grazing, (4) agriculture, (5) stream channel modification, (6) reservoir development, (7) recreational development, and (8) urbanization. Groundwater pumping for industrial, municipal, and agricultural uses might become the most serious threat to riparian systems in regions of low rainfall (Ohmart and Anderson 1986). Tree cutting near a stream might have substantial negative effects on the stream ecosystem, but buffer strips of riparian and adjacent upland forests can largely mitigate the negative effects (Hunter 1990). Payne (1992) and Payne and Bryant (1994) described physical, chemical, and biological techniques to improve wetlands and adjacent uplands for wildlife, including planting and controlled grazing in riparian areas.

Advocating riparian protection schemes that prevent all bank erosion would be unwise. At the same time, it would be unwise to continue the severe stream degradation that occurs in many areas heavily grazed by livestock or farmed intensively. What is needed is a diversity of management along riparian zones, just as we need a diversity of treatments on upland sites, to maintain diversity (Hunter 1990).

Biodiversity and ecosystem function require maintenance of riparian disturbances and associated erosion and sedimentation. Protection (Hafner and Brittingham 1993) is not enough—we should attempt to maintain natural disturbance regimes of riparian areas. But we will be hard-pressed to know how much disturbance is enough or too much.

Adaptive management is an approach that allows us to cope with managing complex systems that we do not sufficiently understand. The process entails modeling the system first, then perturbing it through experimental management, then monitoring the response of the system to management, and finally returning to the model to revise it or update with new data (Walters 1986). Without such a rigorous interface between science and management, we will not learn from what we do.

D. W. Coble critically reviewed this manuscript and J. Varley provided helpful discussion. D. DonnerWright let us use the wood turtle photograph.

LITERATURE CITED

Andersen, D. C. 1994. Demographics of small mammals using anthropogenic desert riparian habitat in Arizona. Journal of Wildlife Management 58:445–454.

Anderson, B. W., and R. D. Ohmart. 1977. Vegetation structure and bird use in the lower Colorado River valley. Pages 23–34 in R. R. Johnson and D. A. Jones, editors. Importance, preservation, and management of riparian habitat. U.S. Forest Service General Technical Report RM-43.

Anderson, N. H., J. R. Sedell, L. M. Roberts, and F. J. Triska. 1978. The role of aquatic invertebrates in processing wood debris from coniferous forest streams. American Midland Naturalist 100:64–82.

Baltz, D. M., and P. B. Moyle. 1984. The influence of riparian vegetation on stream fish communities of California. Pages 183–187 in R. E. Warner and K. M. Hendrix, editors. California riparian systems. University of California Press, Berkeley.

Barnes, W. J., and E. Dibble. 1986. The effects of beaver in river-bank forest succession. Canadian Journal of Botany 66:40–44.

Botkin, D. B. 1990. Discordant harmonies: a new ecology for the twenty-first century. Oxford University Press, Oxford, England.

Boyce, M. S. 1981. Habitat ecology of an unexploited population of beavers in interior

Alaska. Pages 155–186 *in* J. A. Chapman and D. Pursley, editors. Proceedings of the Worldwide Furbearer Conference 2. University of Maryland, Frostburg.

Buech, R. R. 1992. Streambank stabilization can impact wood turtle nesting areas. Abstract. Proceedings of the Midwest Fish and Wildlife Conference 54:260.

Bull, E. L., and J. M. Skovlin. 1982. Relationships between avifauna and streamside vegetation. Transactions of the North American Wildlife and Natural Resources Conference 47:496–506.

Chadde, S. W. 1989. Willows and wildlife of the northern range, Yellowstone National Park. Pages 168–169 *in* R. E. Gresswell, B. A. Barton, and J. L. Kershner, editors. Riparian resource management. U.S. Bureau of Land Management, Billings, Montana.

Chadde, S. W., and C. E. Kay. 1991. Tall-willow communities on Yellowstone's northern range: a test of the "natural-regulation" paradigm. Pages 231–262 *in* R. B. Keiter and M. S. Boyce, editors. The Greater Yellowstone Ecosystem: redefining America's wilderness heritage. Yale University Press, New Haven, Connecticut.

Despain, D. G. 1989. Interpretation of exclosures in riparian vegetation. Page 188 *in* R. E. Gresswell, B. A. Barton, and J. L. Kershner, editors. Riparian resource management. U.S. Bureau of Land Management, Billings, Montana.

Dickson, J. G. 1989. Streamside zones and wildlife in southern U.S. forests. Pages 131–133 *in* R. E. Gresswell, B. A. Barton, and J. L. Kershner, editors. Riparian resource management. U.S. Bureau of Land Management, Billings, Montana.

Dinerstein, E. 1991. Effects of *Rhinoceros unicornis* on riverine forest structure in lowland Nepal. Ecology 73: 701–704.

Dodd, J. L. 1992. Viewpoint: an appeal for riparian zone standards to be based on real world models. Rangelands 14:332.

Elmore, W., and R. L. Beschta. 1987. Riparian areas: perceptions in management. Rangelands 9:260–265.

FEMAT. 1993. Forest ecosystem management: an ecological, economic, and social assessment. Forest Ecosystem Management Assessment Team, Government Printing Office, Washington, D.C.

Fleischner, T. L. 1994. Ecological costs of livestock grazing in western North America. Conservation Biology 8:629–644.

Gill, D. 1972. The evolution of a discrete beaver habitat in the Mackenzie River delta, Northwest Territories. Canadian Field-Naturalist 86:233–239.

Hafner, C. L., and M. C. Brittingham. 1993. Evaluation of a stream-bank fencing program in Pennsylvania. Wildlife Society Bulletin 21:307–315.

Hobbs, R. J. 1992. The role of corridors in conservation: solution or bandwagon? Trends in Ecology and Evolution 7:389–392.

Hobbs, R. J., and L. F. Huenneke. 1992. Disturbance, diversity, and invasion: implications for conservation. Conservation Biology 6:324–337.

Hunter, M. L., Jr. 1990. Wildlife, forests, and forestry: principles of managing forests for biological diversity. Prentice Hall, Englewood Cliffs, New Jersey.

Jones, K. B. 1988. Comparison of herpetofaunas of a natural and altered riparian ecosystem. Pages 222–227 in R. C. Szaro, K. E. Severson, and D. R. Patton, editors. Management of amphibians, reptiles, and small mammals in North America. U. S. Forest Service General Technical Report, RM-166.

Julander, O., and D. E. Jeffrey. 1964. Deer, elk, and cattle range relations on summer range in Utah. Transactions of the North American Wildlife and Natural Resources Conference 29:404–414.

Keller, S. A. C. 1986. Mount St. Helens: five years later. Eastern Washington University Press, Cheney.

Kie, J. G., V. C. Bleich, A. L. Medina, J. D. Yoakum, and J. W. Thomas. 1994. Managing rangelands for wildlife. Pages 663–688 in T. A. Bookhout, editor. Research and management techniques for wildlife and habitats. The Wildlife Society, Bethesda, Maryland.

Kimmins, J. P., and M. C. Feller. 1976. Effect of clear-cutting and broadcast slash-burning on nutrient budgets, streamwater chemistry and productivity in western Canada. Proceedings of the XVI IUFRO World Congress, Oslo, Div. 1:186–197.

Knopf, F. L. 1992. Faunal mixing, faunal integrity, and the biopolitical template for diversity conservation. Transactions of the North American Wildlife and Natural Resources Conference 57:330–342.

Knopf, F. L., and F. B. Samson. 1996. Scale perspectives on avian diversity in western riparian ecosystems. Conservation Biology (in press).

Knopf, F. L., and M. L. Scott. 1990. Altered flows and created landscapes in the Platte River headwaters, 1840–1990. Pages 49–70 in J. M. Sweeney, editor. Management of dynamic ecosystems. North Central Section, The Wildlife Society, West Lafayette, Indiana.

Knopf, F. L., J. A. Sedgwick, and R. W. Cannon. 1988. Guild structure of a riparian avifauna relative to seasonal cattle grazing. Journal of Wildlife Management 52:280–290.

Knutson, M. G. 1995. Birds of large floodplain forests: local and regional habitat associations of the Upper Mississippi River. Ph.D diss., Iowa State University, Ames.

Laws, R. M. 1968. Interactions between elephants and hippopotamus populations and their environments. East African Agriculture and Forestry Journal 33:140–147.

———. 1981. Experiences in the study of large mammals. Pages 19–45 in C. W. Fowler and T. D. Smith, editors. Dynamics of large mammal populations. John Wiley, New York, New York.

Leopold, A. 1933. Game management. Charles Scribner's Sons, New York, New York.

Malanson, G. P. 1993. Riparian landscapes. Cambridge University Press, Cambridge, England.

Megahan, W. F., and P. N. King. 1985. Identification of critical areas on forest lands for control of nonpoint sources of pollution. Environmental Management 9:7–18.

Minshall, G. W., and C. T. Robinson. 1992. Effects of the 1988 wildfires on stream systems of Yellowstone National Park. University of Wyoming–National Park Service Research Center, Annual Report 16:191–198.

Nolet, B. A., A. Hoekstra, and M. M. Ottenheim. 1994. Selective foraging on woody species by the beaver, *Castor fiber*, and its impact on a riparian willow forest. Biological Conservation 70:117–126.

Noss, R. F. 1987. Corridors in real landscapes: a reply to Simberloff and Cox. Conservation Biology 1:159–164.

Oakley, A. L., J. A. Collins, L. B. Everson, D. A. Heller, J. C. Howerton, and R. E. Vincent. 1985. Riparian zones and freshwater wetlands. Pages 57–80 *in* E. R. Brown, editor. Management of wildlife and fish habitats in forests of western Oregon and Washington. Part 1, chapter narratives. U.S. Forest Service, Portland, Oregon.

Olmart, R. D., and B. W. Anderson. 1986. Riparian habitat. Pages 169–199 *in* A. Y. Cooperrider, R. J. Boyd, and H. R. Stuart, editors. Inventory and monitoring of wildlife habitat. U.S. Bureau of Land Management, Denver, Colorado.

Parker, M., F. J. Wood, B. H. Smith, and R. G. Elder. 1985. Erosional downcutting in lower order riparian ecosystems: have historical changes been caused by removal of beaver? Pages 35–38 *in* R. R. Johnson, C. D. Ziebell, D. R. Patton, P. F. Ffolliott, and R. H. Hamre, editors. Riparian ecosystems and their management: reconciling conflicting uses. U. S. Forest Service General Technical Report RM-120.

Payne, N. F. 1992. Techniques for wildlife habitat management of wetlands. McGraw-Hill, New York, New York.

Payne, N. F., and F. C. Bryant. 1994. Techniques for wildlife habitat management of uplands. McGraw-Hill, New York, New York.

Pimentel, D., C. Harvey, P. Resosudarmo, K. Sinclair, D. Kurz, M. McNair, S. Crist, L. Shpritz, L. Fitten, R. Saffouri, and B. Blair. 1995. Environmental and economic costs of soil erosion and conservation benefits. Science 267:1117–1123.

Reed, C. E. M., D. P. Murray, and D. J. Butler. 1993. Black stilt recovery plan *(Himantopus novaezealandiae)*. Department of Conservation, Wellington, New Zealand.

Robinson, W. L., and E. G. Bolen. 1989. Wildlife ecology and management. 2d edition. Macmillan, New York, New York.

Scott, J. M., F. Davis, B. Csuti, R. Noss, B. Butterfield, C. Groves, H. Anderson, S. Caicco, F. D'Erchia, T. S. Edwards, Jr., J. Ulliman, and R. G. Wright. 1993. Gap analysis: a geographic approach to protection of biological diversity. Wildlife Monographs 123:1–41.

Sedell, J. R., F. H. Everest, and D. R. Gibbons. 1989. Streamside vegetation management for aquatic habitat. Pages 115–125 *in* Proceedings of the national silvicultural workshop: silviculture for all resources. Timber Management, U.S. Forest Service, Washington, D.C.

Sidle, J. J., D. E. Carlson, E. M. Kirsch, and J. J. Dinan. 1992. Flooding mortality and habitat renewal for least terns and piping plovers. Colonial Waterbirds 15:132–136.

Simberloff, D., J. A. Farr, J. Cox, and D. W. Mehlman. 1992. Movement corridors: conservation bargains or poor investments? Conservation Biology 6:493–504.

Sinclair, A. R. E. 1983. Management of conservation areas as ecological baseline controls. Pages 13–22 *in* R. N. Owen-Smith, editor. Management of large mammals in African conservation areas. Haum, Pretoria, South Africa.

Smith, S. D., A. B. Weelington, J. L. Nachlinger, and C. A. Fox. 1991. Functional responses of riparian vegetation to streamflow diversion in the eastern Sierra Nevada. Ecological Applications 1:89–97.

Sousa, W. P. 1984. The role of disturbance in natural communities. Annual Review of Ecology and Systematics 15:353–391.

Sparks, R. E. 1995. Need for ecosystem management of large rivers and their floodplains. BioScience 45:168–182.

Swanson, F. J., and G. W. Lienkaemper. 1978. Physical consequences of large organic debris in Pacific Northwest streams. U.S. Forest Service General Technical Report PNW-69.

Swift, B. L. 1984. Status of riparian ecosystems in the United States. Water Resources Bulletin 20:223–228.

Walters, C. J. 1986. Adaptive management of renewable resources. John Wiley, New York, New York.

Wells, R. W. 1968. Fire at Peshtigo. Prentice Hall, Englewood Cliffs, New Jersey.

Wilcove, D. S. 1985. Nest predation in forest tracts and the decline of migratory songbirds. Ecology 66:1211–1214.

———. 1987. From fragmentation to extinction. Natural Areas Journal 7:23–29.

Chapter 8 **Managing Forested Wetlands**
Leigh H. Fredrickson

Worldwide forested wetlands are represented by many diverse types, including bogs, swamps, and floodplain forests (Lugo et al. 1990). The structure and function of these wetlands are driven by hydrologic and climatic characteristics. Bogs are peat-rich ecosystems at high latitudes with water-logged soils; swamps tend to be seasonally flooded, peat-poor systems in tropical and subtropical regions; and floodplain forests are associated with alluvial areas that are intermittently flooded during periods of high flow. Forested wetlands contribute about 3.4×10^{12} m² to the estimated global wetland area of 5.3×10^{12} m² (Lugo et al. 1990). Thus on a global scale, forested wetlands are of great importance, providing many important benefits to society and contributing to global biodiversity.

Many different types of forested wetlands have been described by the dominant vegetation in North America (table 8.1). Bog-type wetlands are widely distributed between 45° and 75° north latitude (Matthews 1990; fig. 8.1), whereas southern forested wetlands are distributed mainly below 45° N (fig. 8.1). Different descriptors, reflecting colloquial as well as scientific perspectives, have been used for somewhat similar wetlands (table 8.1). In North America, conversion of temperate and subtropical wetland forests to other uses in combination with perturbations that influence functions and values of these habitats have been ongoing since the arrival of Europeans (Korte and Fredrickson 1977, MacDonald et al. 1979, Tiner 1984). In contrast, northern forested wetlands have fewer modifications and perturbations than those in more temperate zones. Nevertheless these northern wetlands also have been impacted by water projects, tree harvest, and such subtle factors as air pollution (Environmental Defense Fund 1993).

Changes in forested wetland area and disruptions in the distribution and functions of forested wetlands require strategies at many different scales to maintain their productivity. Protecting and managing forested wetlands is a problem because certain perturbations might change the structure and composition of the forest within a year. Once the structure is altered, restoration of forested wetland structure and function might take decades or even centuries. Thus management decisions in forested wetland ecosystems require careful planning to prevent disruptions of current values and to assure the restoration of functions and values for future generations.

A full discussion of these problems and opportunities relative to the many different forested wetlands on this continent is beyond the scope of this chapter. My discussion focuses on southern forested wetlands of the Mississippi alluvial valley. Conceptually, the challenges facing managers of these southern wetlands are not unlike those for other forested wetlands in North America and elsewhere. Thus the framework for making decisions and the strategies required to protect and maintain functions and biodiversity in southern forested wetlands can be used as a model for efforts to maintain viable forested wetland ecosystems anywhere. This chapter identifies key factors associated with wetlands and describes characteristics of southern forested wetlands and their current status. Current management, strategies to restore ecosystem functions and values, and the associated biodiversity are discussed.

WETLAND ECOSYSTEMS

Wetlands are transitional habitats with indistinct boundaries between truly aquatic and terrestrial systems. They are characterized by the presence of plants that are predominantly hydrophytes and have soils that are saturated or covered with shallow water (Cowardin et al. 1979). Daily, seasonal, or long-term dynamics of wetlands further complicate the study and management of wetland habitats. Dynamics of wetlands and their continuing creation and destruction within riverine systems have led to confusion and debate over wetland delineation.

The structure and function of wetlands are related to two groups of factors commonly described as biotic and abiotic. The abiotic factors include soils,

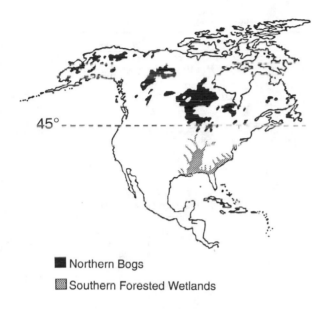

■ Northern Bogs
▨ Southern Forested Wetlands

Fig. 8.1. Distribution of forested wetlands in North America.
Modified from Lugo et al. 1990.

fire, climate, groundwater table, water quality, and hydrology (fig. 8.2). Fore-most among these factors is hydrology (Mitsch and Gosselink 1993). Sub-surface substrates in wetlands also can be diverse and patchy. Subsurface layers that restrict water movement vary in depth; different sediments influence soil moisture conditions. These subsurface and sediment conditions have a profound influence on plant species composition, plant growth rates, and forest productivity (Richardson et al. 1992). Some wetland managers have an excellent understanding of hydrology and how water can be manipulated to restore hydrologic functions and the associated wetland benefits. Although most managers have an appreciation for these hydrologic constraints, they often lack a comprehensive understanding of the interactions among abiotic factors and the implications that these factors have on biological processes.

Some important biotic components are external to the wetland basin be-cause some species that use wetlands are not truly adapted to wetland life or they can have an inconsistent role or effect on wetland organisms. Predators (for example, raccoons *[Procyon lotor]*) often exploit wetland wildlife as prey. Pathogens (for example, fowl cholera *[Pasteurella multocida]* and bot-ulism *[Clostridium botulinum]*) can have a profound influence on the mor-tality of wetland organisms. Some upland species (for example, white-tailed deer *[Odocoileus virginianus]* and wild turkey *[Meleagris gallopavo]*) use

Table 8.1 Selected types of forested wetlands in North America

Wetland type	Location
Riverside swampland	Hay River, Northwest Territories
Fan palm oases	San Andreas fault, California
Phreatophyte systems	Rio Grand, New Mexico
Bog forests, moors, and bird bogs	Alberta
Tamarack-dominated swamps, slowly drained swamps	Northern Lake Michigan to Chicago, Illinois
White cedar swamp	Central Florida, Maine, Wisconsin
Cypress heads or domes	Central Florida
Deep water swamps, shallow water swamps, peaty-freshwater swamps	Southern United States
Tupelo gum swamp	Alabama
Bottomland forests	North Central Oklahoma
Cypress-gum, cedar, maple-gum, and mixed hardwood swamp	Great Dismal Swamp, Virginia
Cypress swamp	Southern United States
Alder swamp	Finger Lakes, New York
Low floodplain	Central Oklahoma
Bog forest	Wisconsin–Michigan
Mangrove	Florida
Pocosin	Carolinas

Sources: see Lugo et al. 1990, Richardson 1981, and Chapman 1977.

wetland resources as food and for protective cover, but they are not wetland-dependent species.

The biotic components within wetlands have complex interactions and include microbes (for example, algae, fungi, and bacteria) that colonize detritus and play an important role in the decomposition process. Conditioning of litter by the microbial component is a key factor in nutrient dynamics and in stimulating invertebrate responses. Algal biomass can be considerable over the course of the annual cycle and might be more important to the wetland

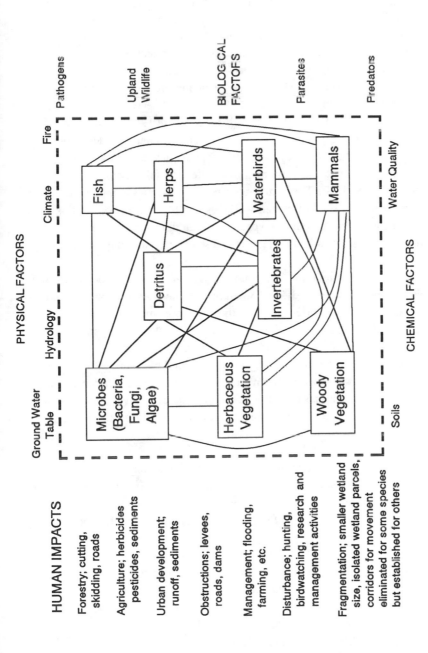

Fig. 8.2. Conceptual wetland model. The dashed line suggests the transitional nature of wetlands between aquatic and terrestrial systems and the flow of energy into and from wetlands.

food web than productivity indicates (Zedler 1980, Wiegert and Freeman 1990). Filamentous algae tie up and retain nutrients in wetlands because these plants remain attached to vascular vegetation and other wetland substrates as water recedes. Vascular plants provide structure, protective cover, nest sites, and foods (Weller 1994). Woody species dominate the structure in forested wetlands and are represented by many plant families. Invertebrates are important in the decomposition process (fig. 8.3) and serve as important food for vertebrates (Reid 1985, Batema et al. 1985, White 1985, Heitmeyer 1985). Forested systems are particularly dependent on shredders for decomposition (fig. 8.3). Many shredders have life-history strategies that allow them to survive summer drought and respond immediately to fall flooding (Batema et al. 1985, White 1985). These invertebrates are essential components in the food web associated with forested wetlands and form an important link with waterbirds. Of more than eighty species of waterbirds that use southern forested wetlands, nearly all either require or use invertebrates during some stage of the annual cycle.

Vertebrates are important consumers in wetlands and include fishes, amphibians, reptiles, waterbirds, and mammals (table 8.2). Although fish, amphibians, and reptiles are consumers, they also are important prey for birds and mammals. Birds are the most obvious vertebrates in wetlands because many have a large body size, and most have diurnal habits as well as interesting behavior. Furthermore, birds are represented by many different forms and often are present in large numbers. Few mammals have adapted to life in wetlands, although many exploit the more ephemeral wetland zones (table 8.2). In contrast to birds, most mammals are inconspicuous and nocturnal.

CHARACTERISTICS OF SOUTHERN FORESTED WETLANDS

Southern forested wetlands typically have a predominance of seasonal flooding during the dormant season. The dynamic flooding regimes are driven by local precipitation and overflows from streams (Heitmeyer et al. 1989). Flooding is dynamic within and among years; the timing, depth, and duration of seasonal flooding varies annually and over the course of long-term flooding cycles. The flooding gradient determines the distribution of woody and herbaceous plant communities as well as animal communities. Tree composition is driven mainly by tolerance to flooding within and among years, in combination with soils and shade (Bedinger 1979, Fredrickson 1979, Conner et al. 1981, Larson et al. 1981, McKnight et al. 1981, Heitmeyer et al. 1989; tables 8.3, 8.4). The most water-tolerant trees, such as bald cypress (*Taxodium distichum*) and water tupelo (*Nyssa aquatica*), occur where deep flooding is extensive for much of the year. Overcup oak (*Quercus lyrata*), nuttall

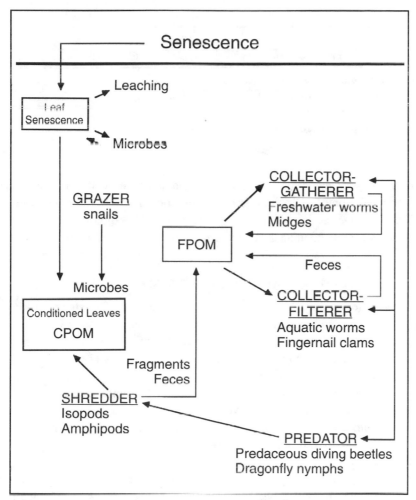

Fig. 8.3. Decomposition process in forested wetlands. CPOM = coarse particulate organic matter; FPOM = fine particulate organic matter.

oak *(Q. nuttalli)*, pin oak *(Q. palustris)*, willow oak *(Q. phellos)*, and cherrybark oak *(Q. falcata* var. *pagodaefolia)* occur along a continuum as depth and duration of flooding decrease. Many plant families have members adapted to flooding, which allows survival somewhere along the flooding gradient (table 8.3). Variability in surface topography can result in patchy flooding patterns, yielding a patchy distribution of forest vegetation.

The large size and great height of many trees, combined with a rich association of midcanopy trees and understory vegetation, forms a highly lay-

Table 8.2 Number of species regularly using lowland hardwood wetlands during fall–spring

Flooding regime and plant community	Amphibians/ reptiles	Fish	Mammals	Birds
Permanently flooded Cypress/Tupelo	10	>50	8	20
Semipermanently flooded Overcup/Red maple	6	35	14	15
Seasonally flooded Red oaks	12	25	22	>60
Temporarily flooded Red oak/Hickory	15	5	25	>50

Sources: Wharton et al. 1982, Fredrickson and Batema 1992.

ered system with many niches. The rich fauna associated with these niches is distributed along the flooding gradient (Fredrickson 1979, Wharton et al. 1982). Some species are adapted to a wide range of flooding depths; others are less tolerant of flooding and have a more restricted distribution.

CURRENT STATUS OF SOUTHERN FORESTED WETLANDS

The pristine forested wetland ecosystem has been severely disrupted by many human activities, with resulting extinctions, extirpations, and decreasing bio-diversity (table 8.5). Originally, forested wetlands of several different types were distributed over 10 million ha in the Mississippi alluvial valley. In the late 1800s, only a small amount of this area was converted to other uses, largely because the capital and technology to drain the southern swamps were not available or had limited effectiveness. All of this changed at the end of the nineteenth century with the development of the dipper dredge and with the passage of legislation that transferred swamplands to the states and then to the counties and that allowed the sale of bonds to finance drainage projects (Korte and Fredrickson 1977). The initial conversion from forested wetlands to other uses was promoted because of economic opportunities coupled with an attitude to conquer the wilderness. Early settlers often described southern forested wetlands as snake- and mosquito-infested swamps that were imped-iments to progress (Nolen 1913). This attitude, coupled with economic op-portunities and incentives, culminated with massive land clearing that reached a peak following World War II. Crops such as soybeans, which are adapted to the soils and climate in this region, were largely responsible for forest conversion during this period. Near the end of the 1900s, these activities reduced the forested wetland area in the Mississippi alluvial valley to about 20% of the original area or about 2 million ha (Tiner 1984). This loss resulted from a combination of constantly changing conditions over the past 150 years rather than from a single factor (table 8.5). The insidious process has been

Table 8.3 Flood tolerance scale for major tree species occurring in bottomland hardwood wetlands

Intolerant	Weakly tolerant	Moderately tolerant	Tolerant
Red elm *Ulmas rubra*	Cherrybark oak *Quercus falcata*	Green ash *Fraxinus pennsylvanica*	Pumpkin ash *Fraxinus profunda*
American beech *Fagas grandiflora*	Water oak *Quercus nigra*	Swamp Cottonwood *Populus heterophylla*	Buttonbush *Cephalanthus occidentalis*
Black cherry *Prunus serotina*	Willow oak *Quercus phellos*	Sweetgum *Liquidambar styraciflua*	Bald cypress *Taxodium distichum*
Flowering dogwood *Cornus florida*	White ash *Fraxinus americana*	Sugarberry *Celtis laevigata*	Pond cypress *Taxodium ascendens*
Eastern hophornbean *Ostrya virginiana*	Eastern cottonwood *Populas deltoides*	Red maple *Acer rubrum*	Water elm *Planera aquatica*
Paw paw *Asimina triloba*	Winged elm *Ulmas alata*	Nuttall oak *Quercus nuttallii*	Water tupelo *Nyssa aquatica*
Sassafras *Sassafras albidum*	Black tupelo *Nyssa sylvatica*	Overcup oak *Quercus lyrata*	Swamp privet *Forestiera acuminata*
	Shellbark hickory *Carya lacinosa*	Pin oak *Quercus palustris*	Black willow *Salix nigra*
	Pecan *Carya illinoensis*	River birch *Betula nigra*	
	Black walnut *Juglans nigra*	American elm *Ulnus americana*	
	Live oak *Quercus virginiana*	Hawthorne *Crateagus spp.*	
	American holly *Ilex opaca*	Water hickory *Carya aquatica*	
	American hornbeam *Carpinus caroliniana*	Laurel oak *Quercus laurifolia*	
	Southern magnolia *Magnolia grandifolia*	Water locust *Gleditsia aquatica*	
	Red mulberry *Morus rubra*	Loblolly pine *Pinus taeda*	
		Persimmon *Diospyros virginiana*	

Table 8.3 (cont.)

Intolerant	Weakly tolerant	Moderately tolerant	Tolerant
		Honey locust	
		Gledisia triacanthos	
		Sycamore	
		Platanus occidentalis	
		Pond pine	
		Pinus serotina	
		Sweetbay	
		Magnolia virginiana	
		Possumhaw	
		Ilex decidua	

Source: McKnight et al. 1980.

Note: Flood tolerance scale defined as follows:

Intolerant. Cannot survive even short periods of flooding into the growing season and has no adaptations for low oxygen levels.

Weakly tolerant. Survives flooding into growing season for a few days or weeks, but trees do not develop adaptations for low oxygen levels. Not tolerant of any flooding above the root crowns during the growing season, but flooding for one month annually may occur regularly during the dormant season.

Moderately tolerant. Survives and grows in saturated conditions for several months of the growing season, but mortality ensues if flooding persists. Some adaptations for living in low oxygen. Many tolerate flooding for one to three months, part of which occurs during the growing season.

Tolerant. Survives and grows in saturated soil conditions for long periods of time and has adaptations for living in low-oxygen environments. Some species can survive continuous flooding, but they regularly tolerate six months of inundation annually, including parts of the growing season.

the demise of the once vast Mississippi alluvial valley floodplain forest.

The area of forests lost is reasonably well documented, but effects related to biodiversity and ecosystem functions are often overlooked. Foremost among these changes is the total disruption of the historic hydrology. In addition to the mainstem dams in the upper reaches of the drainage system that modify river flows, levees along the Mississippi and other major tributaries and an extensive drainage system within each county have changed the timing, duration, and pattern of flooding. For example, in Mississippi County, Arkansas, the length of the drainage system extends for about 1,600 km (C.

Klimas, U.S. Army Corps of Engineers, pers. comm. 1987) and is similar to drainage developments required for successful agriculture throughout the region. The effectiveness of the drainage system is apparent in forests close to ditches (fig. 8.4). Surface water does not accumulate in forests immediately next to ditches; flood duration and depth is very different from flooding on sites less influenced by drainage.

The vast old-growth forest of giant trees has been replaced by a young second-growth forest. Another compromising factor affecting this ecosystem is the current area of different wetland types, compared with their historic distribution. Seasonally flooded sites with shallowly flooded oak communities are less abundant today than in the pristine wetland forest. Remnant habitats often are those at low elevations that usually are difficult to drain. These wetter sites have plant communities representing more water-tolerant vegetation (table 8.3). Plant communities associated with short-duration seasonal flooding were the first to be affected by drainage and land clearing because of ease in drainage.

Remnant forests are about equally distributed between lands within the batture (area between mainstem levees) of the Mississippi River and its tributaries and lands outside the batture (fig. 8.5). The largest contiguous blocks of forest occur within the batture, but these sites are subjected to high flow velocities, higher and immediate flood peaks, and deeper flooding for more extended periods. Remnant forests outside the batture are often isolated and may be small parcels of less than 50 ha. Hydrology has been severely modified in many of these small forested blocks because flood frequency is less, flood depths are shallower, and flood duration is shorter.

Other developments in this ecosystem have compromised historic functions and values (table 8.5). Overland dispersal by terrestrial animals has been compromised because most forest remnants are discontinuous. Barriers to movements, such as deep channels, highways, railroads, powerline right-of-ways, and intensive farming, are plentiful. Highways that are barriers for some animals become routes of dispersal for exotic vegetation. Changes in drainage flow patterns also may affect mobility of some aquatic species and preclude their dispersal among remnant forest parcels. In contrast, the extensive drainage system provides conduits for movement of some exotic aquatic plant and animal species. Exotics, such as carp *(Cyprinus carpio),* have easy access to most water bodies through the extensive drainage infrastructure.

The combination of impacts, including change in size of forested area, forest fragmentation, loss of old-growth forest, and modified hydrology, among other perturbations, not only has an impact on the distribution and abundance of wetland wildlife but has caused extinctions and extirpations of

Table 8.4 Shade tolerance scale for major tree species occurring in bottomland hardwood wetlands

Very intolerant	Intolerant	Moderately tolerant	Tolerant	Very tolerant
E. cottonwood *Populus deltoides*	Sweetgum *Liquidambar styraciflua*	Green ash *Fraxinus pennsylvanica*	Red elm *Ulmus rubra*	Sugarberry *Celtis laevigata*
Black willow *Salix nigra*	Nuttall oak *Quercus nuttallii*	S. cottonwood *Populus heterophylla*	Red maple *Acer rubrum*	American beech *Fagus grandiflora*
Sandbar willow *Salix interior*	Pin oak *Quercus palustris*	Overcup oak *Quercus lyrata*	Buttonbush *Cephalanthus occidentalis*	American holly *Ilex opaea*
	Water oak *Quercus nigra*	Pumpkin ash *Fraxinus profunda*	American elm *Ulmas americana*	Possumhaw *Ilex decidua*
	Willow oak *Quercus phellos*	White ash *Fraxinus americana*	Winged elm *Ulmas alata*	Hornbeam *Carpinus carolinius*
	Cherrybark oak *Quercus falcata*	Bald cypress *Taxodium distichium*	Water elm *Planera aquatica*	Mulberry *Morus spp.*
	River birch *Betula nigra*	Pond cypress *Taxodium ascendens*	Persimmon *Diospyros virginiana*	Paw paw *Asimina triloba*
	Honey locust *Gleditsia triaconthos*	Black tupelo *Nyssa sylvatica*	Swamp privet *Forestiera acuminata*	
	Water locust *Gleditsia aquatica*	Silver maple *Acer saccharinum*		

Live oak	Water hickory
Quercus virginiana	*Carya aquatica*
Pond pine	Sweet bay
Pinus serotina	*Magnolia virginiana*
Sassafras	Sweet bay
Sassafras albidums	*Magnolia virginiana*
Black walnut	Swamp chestnut oak
Juglans nigra	*Quercus bicolor*
	Laurel oak
	Quercus laurifolia
	Loblolly pine
	Pinus taeda
	Sycamore
	Platanus occidentalis

Source: McKnight, et al. 1980.

Note: Shade tolerance scale is defined as follows:

Very intolerant. Cannot survive under full shade. No adaptations for low-light conditions.

Intolerant. Decreased growth under full shade; will not survive if conditions persist. Few adaptations for low-light conditions.

Moderately intolerant. Survives under full shade, but does best with partial shade. Some adaptations to low-light conditions.

Tolerant. Survives and grows under full shade. Adapted to low-light conditions.

Very tolerant. Optimal growth under full shade. Fully adapted to low-light conditions.

Table 8.5 Effects of human activities on biodiversity of southern forested wetlands

			Potential response			
Effect	Plants	Invertebrates	Fish	Herps	Birds	Mammals
Timber harvest	Change in structure and composition	Change in composition	—	Population decline	Extirpation or extinction of some species, population decline	Extirpation of some species, population decline
Reduced size, fragmentation	Less habitat	Reduced species richness and population size	Reduced species richness and population size	Reduced species richness and population size	Loss of species with large home range; reduced species richness and population size	Loss of largest species, reduced species richness and population size
Disruption of wetland complexes	Reduced species richness	Extirpation of some species, reduced population size	Extirpation of some species, life history events compromised	Extirpation of some species, population decline	Extirpation of some species; compromise life history events; population decline	Extirpation of some species, population decline
Habitats no longer present in historic proportions[a]	Reduced species richness	Species richness and population decline	Species richness and population decline	Species richness and population decline	Species richness and population decline	Extirpation of some species, population decline

Modified hydrology	Reduced species richness, change in structure and distribution	Reduced species richness and biomass	Reduced species richness and population size	Reduced species richness and population size	Reduced species richness and population size	Reduced species richness and population size
Herbicides	Reduced diversity and biomass	Reduced species richness, population size, and biomass	Reduced species richness and population size	Reduced species richness and population size	Reduced population size	—
Pesticides	—	Mortality, reduced species richness	Mortality, reduced species richness	Mortality	Mortality	Mortality
Fertilizers	Increased production; shift to monotypic composition; change in structure	Reduced species richness; increase in some species and decrease in others	Reduced species richness	Reduced species richness	Reduced species richness	—
Sedimentation	Reduced species; shift to monotypic composition; change in structure	Reduced species richness and population size	Reduced species richness and population size	Reduced species richness and population size	Reduced species richness and population size	Reduced species richness and population size

aE.g., originally oakflats accounted for 50% of total wetland area, but today they only account for 10% of remaining wetland area.

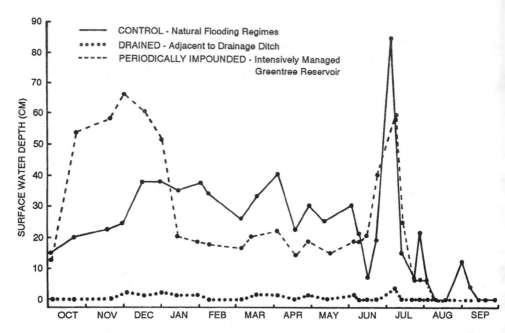

Fig. 8.4. Comparison of surface flooding in southern forested wetlands that have not been drained, those subjected to periodic impoundment for waterfowl management, and those with extensive drainage systems.

nonwetland wildlife as well. The passenger pigeon *(Ectopistes migratorius)*, Carolina parakeet *(Conuropsis carolinensis)*, and ivory-billed woodpecker *(Campephilus principalis)* are extinct; the black bear *(Ursus americanus)*, mountain lion *(Felis concolor)*, and red wolf *(Canis rufus)* are extirpated from most of this region. Other forest-dependent species, as well as neotropical migrants, have been adversely affected by the loss and perturbations of forested wetlands, adding to the concern over biodiversity.

CURRENT MANAGEMENT

Current wetland management follows a pattern established during the 1930s and 1940s. The main focus lies with federal and state agencies on site-specific projects. Properties are held in fee title or under lease agreements; most holdings are widely scattered, and few are larger than 20,000 ha. The most notable exceptions are White River National Wildlife Refuge in Arkansas and the Atchafalaya Basin in Louisiana, which are over 50,000 ha each. As more and

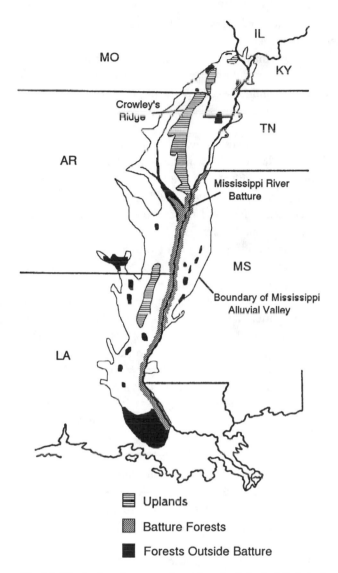

Fig. 8.5. Distribution of remnant forested wetlands in the Mississippi
alluvial valley.

more private land has been converted to agriculture and other uses, public
parcels have become small, isolated islands of habitat in a sea of agriculture.

Cooperative efforts are lacking among and within agencies with a focus
on regional or continental populations and on maintaining functions and values of the larger wetland forested ecosystem. Until recently, the importance

of private lands in providing important resources has been poorly documented or understood. Effective management of the entire forested wetland system is compromised because few management activities are coordinated either among different agencies or within the same agency. Furthermore, management goals often are strongly influenced by political pressure, local interest groups, navigation, or a focus on a few species or species groups of special interest.

Currently, the most commonly practiced wetland management in the Mississippi alluvial valley is site specific. Most managers assigned to a project, whether state, federal, or private, typically have a rather specific management goal related to forestry, a species, or a group of species. The most commonly identified groups are waterfowl, endangered and/or threatened species, or species with economic value (for example, crayfish). In some cases, management goals are broader and might include protection of a habitat or an effort to emulate natural hydrology. Even though these goals are broader than a single-species approach, little coordination or cooperation exists among or within agencies.

One management strategy often used in the Mississippi alluvial valley is to produce foods for some wetland-dependent target species in habitats other than forests. Foremost among groups that are adapted to exploit habitats other than flooded forests are waterfowl, especially mallards *(Anas platyrhynchos)* and Canada geese *(Branta canadensis)*. A full understanding of whether such waterfowl as mallards are as successful in nonforested habitats as in forested wetlands is lacking. Nevertheless, dabbling ducks consistently use many different wetland conditions and readily exploit agricultural fields as well as seasonally flooded wetlands (moist-soil impoundments), where native herbaceous vegetation is produced on large management units (Fredrickson and Heitmeyer 1988, Laubhan and Fredrickson 1993). Sites with native vegetation, diverse invertebrate populations, and some fish provide benefits for a much more diverse fauna than agricultural crops (Fredrickson and Reid 1986). Shorebirds, herons, rails, and grebes often use impoundments with native vegetation more than grain fields (Fredrickson and Taylor 1982). Nevertheless, many fishes, amphibians, and reptiles, as well as some birds, are not accommodated on these intensively managed habitats because they need forest cover or foods. Thus, although this approach is effective for some groups and/or species, extensive use of the technique can compromise life history success of other endemic forms (Laubhan and Fredrickson 1993).

Much progress has been made in understanding southern forested wetlands in the past two decades. Foremost among these findings is the dynamic nature of flooding regimes (Heitmeyer et al. 1989) and the effects this variation in seasonal and long-term flooding have on the biota (Gosselink and Turner 1978, Laubhan and Fredrickson 1993). Implementation of this hydrologic information into management strategies has been instrumental in protecting wetland forests from damage and mortality and in promoting regeneration (Fredrickson and Batema 1992). Development of the management infrastructure has included such features as multiple units no larger than 100 ha, contour levees, shallow flooding, stoplog water-control structures, independent supply and discharge of water, and constantly changing water manipulations within and between years (Fredrickson and Batema 1992, Fredrickson and Laubhan 1994a).

THE CHALLENGE OF ECOSYSTEM MANAGEMENT

The size and quality of remnant areas, distribution and composition of habitats, and modified hydrology have created a current condition where some intensive management is essential. When drainage is effective, dormant-season flooding may not occur or the frequency of flooding may be less than during historic times. Likewise, poorly designed or managed sites have a timing, speed, depth, duration, and extent of flooding different from historic regimes. Recently, the use of agricultural lands by wetland wildlife within the Mississippi alluvial valley has been compromised further because extensive areas have been laser leveled (a specified slope is established across the entire field) for flood irrigation. These leveled fields lack small depressions where water collects and where native wetland vegetation develops during wet seasons. Wetland wildlife have limited benefits from leveled fields unless they are diked and flooded, as in rice culture. Such changes, combined with changes in flooding regimes related to water projects and agricultural drainage, require the development of a management infrastructure to emulate historic hydrologic conditions. Thus, part of the approach to ecosystem management requires a thorough understanding of required habitat conditions at a specific time and location for a large number of species. The infrastructure for intensive wetland management is costly and requires extensive expertise in engineering and water manipulations to meet the biological needs of the diverse native biota (Laubhan and Fredrickson 1993, Fredrickson and Laubhan 1994b).

One great management challenge relates to a limited habitat base combined with conflicting habitat needs among different species. Because a single wetland type rarely provides all of the resources required for a single species or for a group of species for all life history events, different wetland types of adequate size and in close proximity are an essential ingredient for survival and reproduction of many forested wetland inhabitants (table 8.6). Of the wetland types commonly found in southern forests, each wetland has specific characteristics that provide resources for some species but not others. Oak flats flooded during the dormant season provide breeding habitat for salamanders and swamp fishes, as well as foraging habitat for waterfowl, white-tailed deer, turkey, and raccoons (table 8.6). These same habitats have an abundance of Neotropical migrants during spring and fall. During the summer, when surface flooding normally is absent, nonwetland wildlife use the area extensively while wetland-dependent species move to other sites with surface water.

The magnitude of this challenge is apparent from the number of species that require southern forested wetlands for life-history success (Wharton et al. 1982, Fredrickson and Batema 1992; table 8.2) and the great array of conditions that are required for a single species during the course of one annual cycle (Laubhan and Fredrickson 1993). Such factors as mobility and interconnected habitats become critical for successful completion of some life histories. Six common species that exploit southern forested wetlands provide insights into the complex requirements for life-history success and the associated challenge of implementing effective intensive management in a disrupted landscape (table 8.6). Many direct conflicts exist among species needs. Bantam sunfish *(Lepomis symmetricus)* and alligator snapping turtles *(Macrochelys temmincki)* need permanent water. Sites with permanent deep water do not provide the necessary protective cover and foods used by mallards and king rails *(Rallus elegans).* So little information is available for so many species that we cannot anticipate when manipulations to provide essential requirements for one species might compromise the needs of other species.

Recent findings identified interconnections among wetlands across wide geographic areas. These interconnections are most obvious for migratory birds that move great distances between wintering and breeding areas. These cross-seasonal effects have been documented for purple herons *(Ardea purpensis)* in Europe that winter in the Senegal River in Africa (den Held 1981) and for mallards that winter in southern forested wetlands and breed at more northern latitudes (Heitmeyer and Fredrickson 1981, Kaminski and Gluesing 1987). Ideal conditions during the winter in southern wetland forests consistently precede years when more mallard young are produced on northern habitats. Thus, the effective exploitation of one wetland type requires desirable wetland conditions at some distant wetland in order to assure the ideal physiological condition for breeding. Few species have been studied in enough detail to identify how extensive and to what degree cross-seasonal effects influence animal populations. Cross-seasonal needs are an aspect of management with important implications for an unknown number of wetland species that migrate across wide geographic areas.

The tremendous loss of forested wetlands has prompted agencies to develop methods for reforestation. Use of this approach is critical to ecosystem management of southern forested wetlands because remnant parcels often are not large enough to meet the needs of larger species, or interconnecting corridors must be developed between remnants to assure interchange among populations that maintain genetic diversity. Direct seeding and planting seedlings and saplings of dominant trees have been effective, but the technology needed to reestablish all components, including shrubs and herbaceous plants, has not been developed. Reasonable potential exists to restore the canopy structure, but duplicating the diverse plant communities of historic forested wetland will be difficult.

STRATEGIES FOR ECOSYSTEM MANAGEMENT

Currently, there is much emphasis on the concept of ecosystem management. Nonetheless, implementation of an ecosystem approach is untested, opinions vary widely, and the necessary cooperation for implementation must be developed. Although the concept is a worthy and necessary goal, assessment of the effectiveness of this landscape scale-approach remains in its infancy. Clearly, there must be much improvement in the use of biological information at the administrative as well as at the land management level to meet these important ecosystem goals.

Administrative Decisions

Historically, the purchase or failure to purchase property by state and federal agencies has been driven by the availability of property, funds for purchase, political demands by powerful congressional representatives or the presidential administration, or the politics of maintaining public support for an agency. Biotic and abiotic information relating to hydrology, biodiversity, island biogeography, and landscape ecology has not been used consistently. In addition, some competition occurs among agencies for purchase and management of remnant parcels. Thus, the first step in effective ecosystem management must be the implementation of administrative decisions that incorporate information relating to biodiversity and the key abiotic factors in ecosystem functions (fig. 8.6a, b). This is an ominous task because the decision-making process must cross the boundaries of agency responsibility and must overlook the traditions often deeply ingrained within agencies (Knopf 1992). Furthermore, success can be achieved only if the ecosystem goal is

Table 8.6 Habitat conditions required to complete annual cycle events that occur in southern forested wetlands

Annual cycle event	Wetland type[a]	Hydrology		Vegetation		Foods
		Type[b]	Depth	Type	Density	
Bantam sunfish						
resident	ss, f, ow, dt	sf, sp, p	>15 cm	perennial	medium	invertebrates, amphibians, fishes
Bullfrog						
resident	ss, mh, f, ow	p	>30 cm	perennial	medium	invertebrates, amphibians, fishes
Alligator snapping turtle						
resident	ss, ow	p	>70 cm	perennial	low	fishes
King rail						
spring migration	mh, ms	sf, sp, p	10 cm	perennial	high	crayfish, aquatic insects
nest/incubation	mh, ms	sf, sp, p	7.5 cm	perennial	high	crayfish
brood rearing						
young (1–3 weeks)	mh, ms	sf, sp, p	2.5 cm	perennial	low	
old (4–8 weeks)	mh, ms	sf, sp, p	<2.5 cm	perennial	low	leeches, beetles, oligochaetes
fall migration	mh, ms	sf, sp, p	7.5 cm	perennial	high	crayfish, aquatic insects

Mallard						
autumn migration/ prealternate molt	mh, ms, ss	sf, sp, p	<25.4 cm	annual	high	invertebrates, seeds, tubers
pair/courtship	ms, f, ss	sf, sp, p	<60 cm	perennial	low	seeds, acorns
prebasic molt	mh, ow, gtr, f, ss, ag	sf, sp, p	<25.4 cm	perennial	low	seeds, invertebrates, rowcrops
spring migration	dt, f, mh, ms	sf, sp, p	<25.4 cm	perennial/annual	moderate	invertebrates (crustaceans), acorns
Beaver						
resident	ss, f, dt, ow	sp, p	>60 cm	perennial	high	aquatic and woody plants

[a] mh = marsh; ms = moist-soil; ss = scrub/shrub; f = naturally flooded forest; ow = open water; gtr = greentree reservoir; dt = dead timber; ag = agricultural fields.

[b] sf = seasonally flooded; sp = semipermanent; p = permanent.

truly to provide natural resource benefits and not personal gain or satisfaction of agency agendas unrelated to biological issues.

Management Planning

Once lands are in public ownership, an ecosystem approach requires an inventory of properties, including the size, wetland types, juxtaposition with other wetlands, hydrology, soils, and species of concern, among others (fig. 8.6b). Similar information on private holdings is just as important because private property encompasses a much larger area than land held in fee-title or easements by public agencies. In many cases agencies hold property that has important benefits for biota associated with southern forested wetlands, but these benefits are compromised because of size, location, or conflicting interests. A cooperative effort among agencies and the private sector is needed to identify essential habitats that assure biodiversity. This is a major step toward ecosystem management.

Foremost in the planning process is to determine whether intensive management is reasonable and needed. This decision hinges mainly on whether the hydrology is somewhat natural or has been modified (fig. 8.6b). Decisions relating to protection or management can be very complicated. Some projects in public ownership need intensive development and substantial infrastructure to restore wetland functions and values. Other sites within the project area might need protection as a best-management strategy.

Today no single source identifies the area of wetlands in public ownership, the proportion of this ownership in different wetland types, or the location of different wetland types. This information is essential in the planning process for landscape-scale projects. Estimating the area and types of wetlands in private holdings is just as important. Currently, efforts are under way to identify all private holdings in the Mississippi alluvial valley that provide wetland benefits (K. Reinecke, National Biological Survey, and C. Baxter, U.S. Fish and Wildlife Service, pers. comm. 1994). Estimating total area, area in different types of habitats, and the duration and timing of flooding are included in these efforts.

Management Implementation

The best planning is often compromised during implementation. Such factors as unqualified personnel, inadequate monitoring or quality control, and lack of evaluation following the implementation are common problems. Often monies are available for development, but the personnel required for successful management are not provided.

Foremost among requirements for successful wetland manipulations is in-

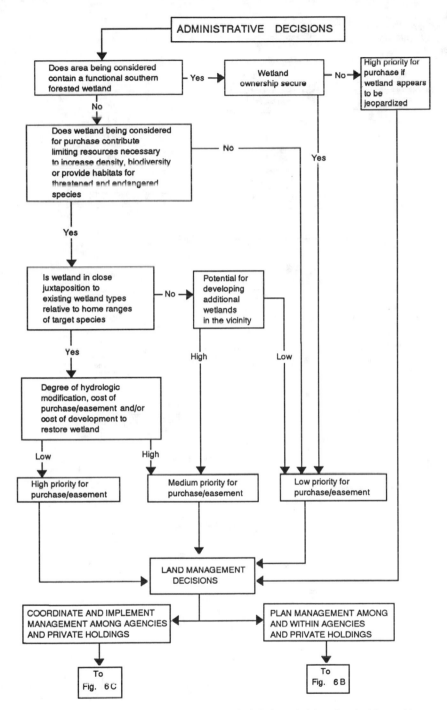

Fig. 8.6a. Flow chart showing key information required during administrative decision making for ecosystem management.

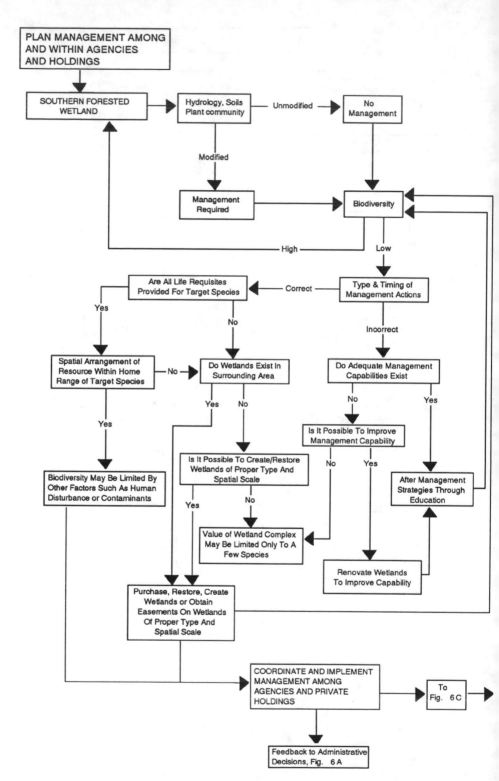

Fig. 8.6b. Flow chart showing key information required during the planning process within and among agencies to meet ecosystem management goals.

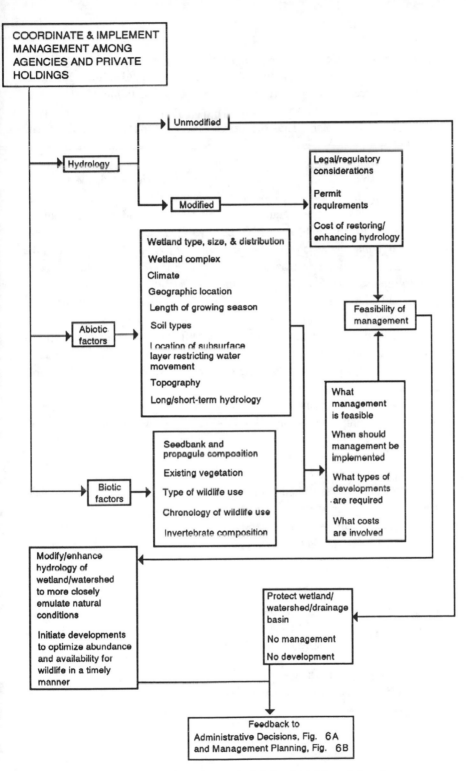

Fig. 8.6c. Flow chart showing key information required during the implementation of management to meet ecosystem management goals.

formation on abiotic factors, especially hydrology and soils (fig. 8.6c). Integration of information for abiotic and biotic components identifies approaches linked to success and cost effectiveness. Properly functioning forested wetland habitats are easily degraded and fragmented even when management intentions are exemplary. Thus successful ecosystem management of remnant forested wetlands requires careful consideration of current conditions, including an assessment of the forested wetland condition and identification of controlling abiotic factors. Among the most important aspects is identification of how hydrology has been modified. Foremost among management challenges is the lengthy time required for a forested wetland to reach maturity. Because this time period for many forests crosses several human generations, current-day management planning strategies are too ephemeral to match the long-term challenge. For example, the pictoral record from Custer's expedition to the Black Hills of South Dakota provides a baseline showing a more open and less extensive forest (Progulske 1974). Subtle changes occurring in this western forest during the past century (Progulske 1974) suggest the importance of developing monitoring mileposts for resource agencies. Unfortunately, the impatience of land managers, administrators, and their clients, coupled with our poor understanding of forested wetlands and lack of viable restoration strategies, provides a sobering challenge to current and future researchers and managers. The design of long-term studies that cross generations and the implementation of effective management strategies are in their infancy and will provide thought-provoking challenges far into the future.

Many individuals have shaped my thinking about wetland management and the challenges associated with meeting ecosystem goals in southern wetlands. Foremost among these are D. Batema, J. Boyles, P. Covington, M. Heitmeyer, C. Klimas, F. Reid, and M. Weller. I kindly thank S. McLeod and G. Pogue for thoughtful reviews.

LITERATURE CITED

Batema, D. L., G. S. Henderson, and L. H. Fredrickson. 1985. Wetland invertebrate distribution in bottomland hardwoods as influenced by forest type and flooding regime. Proceedings of the Central Hardwoods Forest Conference 5:196–202.

Bedinger, M. S. 1979. Relation between forest species and flooding. Pages 427–435 in P. C. Greeson, J. R. Clark, and J. E. Clark, editors. Wetland functions and values: the state of our understanding. American Water Resources Association Technical Publication 79-2.

Chapman, V. J. (editor). 1977. Ecosystems of the world, 1. West coastal ecosystems. Elsevier, New York, New York.

Conner, W. H., J. G. Gosselink, and R. T. Parrondo. 1981. Comparison of the vegetation of three Louisiana swamp sites with different flooding regimes. American Journal Botany 68:320–331.

Cowardin, L. M., V. Carter, F. C. Golet, and E. T. LaRoe. 1979. Classification of wetlands and deepwater habitats of the United States. U.S. Fish and Wildlife Service Publication FWS/OBS-79/31.

den Held, J. J. 1981. Population changes in the purple heron in relation to drought in the wintering area. Ardea 69:185–191.

Environmental Defense Fund. 1993. Large-scale pollution revealed throughout the Arctic. Report to Members of the Environmental Defense Fund. 24(5):1, 5.

Fredrickson, L. H. 1979. Lowland hardwood wetlands: current status and habitat values for wildlife. Pages 296–306 in P. E. Greeson, J. R. Clark, and J. E. Clark, editors. Wetland functions and values, the state of our understanding. American Water Resources Technical Publication 79-2.

Fredrickson, L. H., and D. L. Batema. 1992. Greentree reservoir management handbook. Gaylord Memorial Laboratory, Wetland Management Series 1. Gaylord Laboratory, Puxico, Missouri.

Fredrickson, L. H., and M. E. Heitmeyer. 1988. Wetland use of southern forested wetlands by waterfowl. Pages 307–343 in M. W. Weller, editor. Waterfowl in winter. University of Minnesota Press, Minneapolis.

Fredrickson, L. H., and M. K. Laubhan. 1994a. Managing wetlands for wildlife. Pages 623–647 in T. A. Bookhout, editor. Research and management techniques for wildlife and habitats. 5th edition. The Wildlife Society, Bethesda, Maryland.

———. 1994b. Intensive wetland management: a key to biodiversity. Transactions of the North American Wildlife and Natural Resources Conference 59:555–565.

Fredrickson, L. H., and F. A. Reid. 1986. Wetland and riparian habitats: a nongame management overview. Pages 59–96 in J. B. Hale, L. B. Best, and R. L. Clawson, editors. Management of nongame wildlife in the Midwest: a developing art. North Central Section of the Wildlife Society, West Lafayette, Indiana.

Fredrickson, L. H., and T. S. Taylor. 1982. Management of seasonally flooded impoundments for wildlife. Resource Publication 148, U.S. Fish and Wildlife Service, Washington, D.C.

Gosselink, J. G., and R. E. Turner. 1978. The role of hydrology in freshwater wetland ecosystems. Pages 63–78 in R. E. Good, D. F. Whigham, and R. L. Simpson, editors. Freshwater wetlands: ecological processes and management potential. Academic Press, New York, New York.

Heitmeyer, M. E. 1985. Wintering strategies of female mallards related to dynamics of lowland hardwood wetlands in the Upper Mississippi Delta. Ph.D. dissertation. University of Missouri, Columbia.

Heitmeyer, M. E., and L. H. Fredrickson. 1981. Do wetland conditions in the Mississippi Delta hardwoods influence mallard recruitment? Transactions of the North American Wildlife Natural Resources Conference 46:44–57.

Heitmeyer, M. E., L. H. Fredrickson, and G. F. Krause. 1989. Water and habitat dynamics of the Mingo Swamp in southeastern Missouri. U.S. Fish and Wildlife Resource Publication 180.

Kaminski, R. M., and E. A. Gluesing. 1987. Density and habitat related recruitment in mallards. Journal Wildlife Management 51:141–148.

Knopf, F. L. 1992. Faunal mixing, faunal integrity, and the biopolitical template for diversity conservation. Transactions of the North American Wildlife and Natural Resources Conference 57:330–342.

Korte, P. A., and L. H. Fredrickson. 1977. Loss of Missouri's lowland hardwood ecosystem. Transactions of the North American Wildlife Natural Resources Conference 42:31–41.

Larson, J. S., M. S. Bedinger, C. F. Bryan, S. Brown, R. T. Huffman, E. C. Miller, D. G. Rhodes, and B. A. Touchet. 1981. Transition from wetlands to uplands in southeastern bottomland hardwood forests. Proceedings of a workshop on bottomland hardwood forest wetlands of southeastern United States, Lake Lanier, Georgia. June 1–5, 1980. Elsevier, New York, New York.

Laubhan, M. K., and L. H. Fredrickson. 1993. Integrated wetland management: concepts and opportunities. Transactions of the North American Wildlife and Natural Resources Conference. 58:323–334.

Lugo, A. E., M. Brinson, and S. Brown. 1990. Ecosystems of the world 15, forested wetlands. Elsevier, New York, New York.

MacDonald, P. O., W. E. Frayer, and J. K. Clauser. 1979. Documentation, chronology, and future projections of bottomland hardwood losses in the lower Mississippi Alluvial Plain. Volume 1. U. S. Fish and Wildlife Service, Washington, D.C.

Matthews, E. 1990. Global distribution of forested wetlands. Addendum in A. E. Lugo, M. Brinson, and S. Brown. Ecosystems of the world 15, forested wetlands. Elsevier, New York, New York.

McKnight, J. S., D. D. Hook, O. G. Langdon, and R. L. Johnson. 1981. Flood tolerance and related characteristics of trees of bottomland forests of the southern United States. Pages 29–65 in J. R. Clark and J. Benforado, editors. Wetlands of bottomland forests: proceedings of a workshop on bottomland hardwood forest wetlands of the southeastern United States. Elsevier, New York, New York.

Mitsch, W. J., and J. G. Gosselink. 1993. Wetlands. 2d edition. Van Nostrand Reinhold, New York, New York.

Nolen, J. H. 1913. Missouri's swamp and overflowed lands and their reclamation. Report to 47th Missouri General Assembly. Hugh Stephens Printing Co., Jefferson, Missouri.

Progulske, D. R. 1974. Yellow ore, yellow hair, yellow pine. South Dakota State University, Agricultural Experiment Station, Bulletin 616.

Reid, F. A. 1985. Wetland invertebrates in relation to hydrology and water chemistry.

Pages 72–79 *in* M. D. Knighton, editor. Water impoundments for wildlife: a habitat management workshop. USDA Forest Service, St. Paul, Minnesota.

Reinecke, K. J., R. C. Barkley, and C. K. Baxter. 1986. Potential effects of changing water conditions on mallard wintering in the Mississippi alluvial valley. Pages 325–337 *in* M. W. Weller, editor. Waterfowl in winter. University of Minnesota Press, Minneapolis.

Richardson, C. J. (editor). 1981. Pocosin wetlands. Hutchinson Ross Publishing, Stroudsburg, Pennsylvania.

Richardson, J. L., L. P. Wilding, and R. B. Daniels. 1992. Recharge and discharge of groundwater in aquic conditions illustrated with flownet analysis. Geogerma 53: 65–78.

Tiner, R. W., Jr. 1984. Wetlands of the United States: current status and recent trends. National Wetlands Inventory, Washington, D.C.

Weller, M. W. 1994. Freshwater marshes: ecology and wildlife management. 3d edition. University of Minnesota Press, Minneapolis.

Wharton, C. H., W. M. Kitchens, and T. W. Sipe. 1982. The ecology of bottomland hardwood swamps of the southeast: a community profile. U.S. Fish and Wildlife Service, FWS/OBS-81/37

Wharton, C. H., V. W. Lambour, J. Newsom, P. V. Winger, L. L. Gaddy, and P. Mancke. 1981. The fauna of bottomland hardwoods in southeastern U.S. Pages 87–160 *in* J. R. Clark and J. Benforado, editors. Wetlands of bottomland hardwood forests. Proceedings of a workshop on bottomland hardwood forest wetlands of southeastern United States. Elsevier, New York, New York.

White, D. C. 1985. Lowland hardwood wetland invertebrate community and production in Missouri. Archives Hydrobiologia 103:509–533.

Wiegert, R. G., and B. J. Freeman. 1990. Tidal salt marshes of the Southeastern Atlantic coast: a community profile. U.S. Department of Interior, Fish and Wildlife Service, Washington, D.C. Biological Report 85(7.29).

Zedler, J. B. 1980. Algae mat productivity: comparisons in a salt marsh. Estuaries 3: 122–131.

TECHNIQUES AND CLASSIFICATION

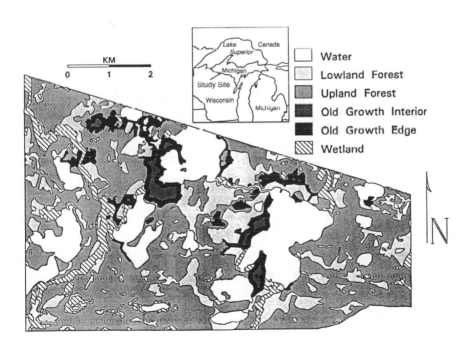

KM

0   1   2

Lake Superior   Canada
Michigan
Study Site
Wisconsin   Michigan

☐ Water

▨ Lowland Forest

▨ Upland Forest

▨ Old Growth Interior

■ Old Growth Edge

▨ Wetland

N

Across adequately large areas, ecosystem processes (such as disturbance, succession, evolution, natural extinction, recolonization, fluxes of materials, and other stochastic, deterministic, and chaotic events) that characterize the variability found in natural ecosystems should be present and functioning.
—Merrill Kaufmann and colleagues

Overleaf: Geographical information systems (GIS) have greatly facilitated our ability to manage natural resources on a large spatial scale, used here to examine the edge/interior patterns of old-growth forests in a 4,500 ha tract in northern Wisconsin. Shading delineates the penetration of a 100 m edge effect into the old-growth polygons and remaining old-growth interior (Mladenoff et al. 1994). Reprinted by permission of Blackwell Science, Inc.

Chapter 9 National Hierarchical Framework of Ecological Units

David T. Cleland, Peter E. Avers,
W. Henry McNab, Mark E. Jensen,
Robert G. Bailey, Thomas King,
and Walter E. Russell

To implement ecosystem management, we need basic information about the nature and distribution of ecosystems. To develop this information, we need working definitions of ecosystems and supporting inventories of the components that make up ecosystems. We also need to understand ecological patterns and processes and the interrelations of social, physical, and biological systems. To meet these needs, we must obtain better information about the distribution and interaction of organisms and the environments in which they occur, including the demographics of species, the development and succession of communities, and the effects of human activities and land use on species and ecosystems (Urban et al. 1987). Research has a critical role in obtaining this information.

This chapter presents a brief background of regional land classifications, describes the hierarchical framework for ecological unit design, examines underlying principles, and shows how the framework can be used in resource planning and management. The basic objective of the hierarchical framework is to provide a systematic method for classifying and mapping areas of the earth based on associations of ecological factors at different geographic scales. The framework is needed to improve our efforts in national, regional, and forest-level planning; to achieve consistency in ecosystem management across national forests and regions; to advance our understanding of the nature and distribution of ecosystems; and to facilitate interagency data planning. Furthermore, the framework will help us evaluate the inherent capabilities of land and water resources and the effects of management on them.

Ecological units delimit areas of different biological and physical potentials. Ecological unit maps can be coupled with inventories of existing vegetation, air quality, aquatic systems, wildlife, and human elements to characterize complexes of life and environment, or ecosystems. This information on ecosystems can be combined with our knowledge of various processes to facilitate a more ecologically sound approach to resource planning, management, and research.

Note that ecological classification and mapping systems are devised by humans to meet human needs and values. Ecosystems and their various components often change gradually, forming continua on the earth's surface which cross administrative and political boundaries. Based on their understanding of ecological systems, humans decide on ecosystem boundaries by using physical, biological, and social considerations.

We recognize that the exact boundaries for each level envisioned in this process and developed in map format may not fit every analysis and management need. Developing boundaries of areas for analysis, however, will not change the boundaries of ecological units. In some cases an ecological unit may be the analysis area. In other cases, watersheds, existing conditions, management emphasis, proximity to special features (for example, natural, wilderness, or urban areas), or other conditions may define an analysis area. In these cases, ecological units can be aggregated or divided if necessary to focus on relevant issues and concerns.

BACKGROUND: REGIONAL LAND CLASSIFICATIONS

Hierarchical systems using ecological principles for classifying land have been developed for geographical scales ranging from global to local. Using a bioclimatic approach at a global scale, several researchers have developed ecological land classifications: Holdridge (1967), Walter and Box (1976), Udvardy (1975), and Bailey (1989a, b). Wertz and Arnold (1972) developed land stratification concepts for regional and land unit scales. Other ecologically based classifications proposed at regional scales include those of Driscoll et al. (1984), Gallant et al. (1989), and Omernik (1987) in the United States and those of Wiken (1986) and the Ecoregions Working Group (1989) in Canada. Concepts have also been presented for ecological classification at subregional to local scales in the United States (Barnes et al. 1982), Canada (Jones et al. 1983, Hills 1952), and Germany (Barnes 1984).

But no single system has the structure and flexibility necessary for developing ecological units at continental to local scales. Each of these systems has strong points that contribute to the strength of the national hierarchy. The concepts and terminology of the national system draw upon this work to devise a consistent framework for application throughout the United States.

ECOLOGICAL UNIT DESIGN

The primary purpose for delineating ecological units is to identify land and water areas at different levels of resolution that have similar capabilities and potentials for management. Ecological units are designed to exhibit similar patterns in: (1) potential natural communities, (2) soils, (3) hydrologic function, (4) landform and topography, (5) lithology, (6) climate, (7) air quality, and (8) natural processes for cycling plant biomass and nutrients (for example, succession, productivity, and fire regimes).

It should be noted that climatic regime is an important boundary criterion for ecological units, particularly at broad scales. In fact, climate, as modified by topography, is the dominant criterion at upper levels. Other factors, such as geomorphic process, soils, and potential natural communities, take on equal or greater importance than climate at lower levels.

It follows, then, that ecological map units are differentiated and designed by multiple components, including climate, physiography, geology, soils, water, and potential natural communities. These components may be analyzed individually and then combined, or multiple factors may be simultaneously evaluated to classify ecological types, which are then used in ecological unit design. The first option may be increasingly used as geographic information systems (GIS) become more available. The interrelations among independently defined components, however, will need to be carefully evaluated, and the results of layering component maps may need to be adjusted to identify units that are both ecologically significant and meaningful to management. When various disciplines cooperate in devising integrated ecological units, products from existing resource component maps can be modified, and integrated interpretations can be developed (Avers and Schlatterer 1991).

Ecological unit inventories are generally designed and conducted in cooperation with the Soil Conservation Service, Agricultural Experiment Stations of Land Grant Universities, Bureau of Land Management, and other appropriate state and federal agencies. Mapping conventions and soil classification meet standards of the National Cooperative Soil Survey.

Table 9.1 National hierarchy of ecological units

Planning and analysis scale	Ecological units	Purpose, objectives, and general use
Ecoregion		
Global	Domain	Broad applicability for modeling and sampling.
Continental	Division	Strategic planning and assessment. International planning.
Regional	Province	
Subregion	Section	Strategic, multiforest, statewide, and multiagency
	Subsection	analysis and assessment.
Landscape	Landtype association	Forest or areawide planning, and watershed analysis.
Land unit	Landtype	Project and management area planning and analysis.
	Landtype phase	
Hierarchy can be expanded by user to smaller geographical areas and more detailed ecological units if needed.		Very detailed project planning.

CLASSIFICATION FRAMEWORK

The National Ecological Unit Hierarchy is presented in tables 9.1, 9.2, and 9.3. The hierarchy is based on concepts and terminology developed by numerous scientists and resource managers (Hills 1952, Crowley 1967, Wertz and Arnold 1972, Rowe 1980, Allen and Starr 1982, Barnes et al. 1982, Forman and Godron 1986, Bailey 1987, Meentemeyer and Box 1987, Gallant et al. 1989, Cleland et al. 1992). The following is an overview of the differentiating criteria used in the development of the ecological units. Table 9.2 summarizes the principal criteria used at each level in the hierarchy.

Ecoregion Scale

At the ecoregion scale, ecological units are recognized by differences in global, continental, and regional climatic regimes and gross physiography. The basic assumption is that climate governs energy and moisture gradients, thereby acting as the primary control over more localized ecosystems. Three levels of ecoregions, adapted from Bailey (1980), are identified in the hierarchy:

Table 9.2 Principal map unit design criteria of ecological units

Ecological unit	Principal map unit design criteria
Domain	Broad climatic zones or groups (e.g., dry, humid, tropical)
Division	Regional climatic types (Koppen 1931, Trewartha 1968) Vegetational affinities (e.g., prairie or forest) Soil order
Province	Dominant potential natural vegetation (Kuchler 1964) Highlands or mountains with complex vertical climate-vegetation-soil zonation
Section	Geomorphic province, geologic age, stratigaphy, lithology Regional climatic data Phases of soil orders, suborders, or great groups Potential natural vegetation Potential natural communities (PNC) (FSH 2090)
Subsection	Geomorphic process, surficial geology, lithology Phases of soil orders, suborders, or great groups Subregional climatic data PNC—formation or series
Landtype association	Geomorphic process, geologic formation, surficial geology, and elevation Phases of soil subgroups, families, or series Local climate PNC—series, subseries, plant associations
Landtype	Landform and topography (elevation, aspect, slope gradient, and position) Phases of soil subgroups, families, or series Rock type, geomorphic process PNC—plant associations
Landtype phase	Phases of soil families or series Landform and slope position PNC—plant associations or phases

Note: The criteria listed are broad categories of environmental and landscape components. The actual classes of components chosen for designing map units depend on the objectives for the map.

Table 9.3 Map scale and polygon size of ecological units

Ecological unit	Map scale range	General polygon size
Domain	1:30,000,000 or smaller	1,000,000s of square miles
Division	1:30,000,000 to 1:7,500,000	100,000s of square miles
Province	1:15,000,000 to 1:5,000,000	10,000s of square miles
Section	1:7,500,000 to 1:3,500,000	1,000s of square miles
Subsection	1:3,500,000 to 1:250,000	10s to low 1,000s of square miles
Landtype association	1:250,000 to 1:60,000	1,000s to 10,000s of acres
Landtype	1:60,000 to 1:24,000	100s to 1,000s of acres
Landtype phase	1:24,000 or larger	<100 acres

1. *Domains,* subcontinental divisions of broad climatic similarity, such as lands that have the dry climates defined by Koppen (1931), which are affected by latitude and global atmospheric conditions. For example, the climate of the polar domain is controlled by arctic air masses, which create cold, dry environments where summers are short. In contrast, the climate of the humid tropical domain is influenced by equatorial air masses, and there is no winter season. Domains are also characterized by broad differences in annual precipitation, evapotranspiration, potential natural communities, and biologically significant drainage systems. The four domains are named according to the principal climatic descriptive features: polar, dry, humid temperate, and humid tropical.

2. *Divisions,* subdivisions of domains determined by isolating areas of definite vegetational affinities (prairie or forest) that fall within the same regional climate, generally at the level of the basic types of Koppen (1931) as modified by Trewartha (1968). Divisions are delineated according to: (a) the amount of water deficit, which subdivides the dry domain into semi-arid, steppe, or arid desert, and (b) the winter temperatures, which have an important influence on biological and physical processes and the duration of any snow cover. This temperature factor is the basis of distinction between temperate and tropical/subtropical dry regions. Divisions are named for the main climatic regions they delineate, such as steppe, savannah, desert, Mediterranean, marine, and tundra.

3. *Provinces,* climatic subzones controlled primarily by such continental weather patterns as length of dry season and duration of cold temperatures. Provinces are also characterized by similar soil orders. The climatic subzones are evident as extensive areas of similar potential natural vegetation, as mapped by Kuchler (1964). Provinces are named typically using a

binomial system consisting of a geographic location and vegetative type: Bering tundra, California dry steppe, and eastern broadleaf forests.

Highland areas that exhibit altitudinal vegetational zonation and that have the climatic regime (seasonality of energy and moisture) of adjacent lowlands are classified as provinces (Bailey et al. 1985). The climatic regime of the surrounding lowlands can be used to infer the climate of the highlands. For example, in the Mediterranean division along the Pacific Coast, the seasonal pattern of precipitation is the same for the lowlands and highlands except that the mountains receive about twice the quantity. The provinces are named for the lower-elevation and upper-elevation (subnival) belts, for example, Rocky Mountain forest-alpine meadows.

Subregional Scale

Subregions are characterized by combinations of climate, geomorphic process, topography, and stratigraphy that influence moisture availability and exposure to radiant solar energy, which in turn directly control hydrologic function, soil-forming processes, and potential plant community distributions. Sections and subsections are the two ecological units mapped at this scale.

1. *Sections*, broad areas of similar geomorphic process, stratigraphy, geologic origin, drainage networks, topography, and regional climate. Such areas are often inferred by relating geologic maps to potential natural vegetation "series" grouping, as mapped by Kuchler (1964). Boundaries of some sections approximate geomorphic provinces (for example, Blue Ridge) as recognized by geologists. Section names generally describe the predominant physiographic feature upon which the ecological unit delineation is based, such as Flint hills, Great Lakes morainal, bluegrass hills, Appalachian piedmont.

2. *Subsections*, smaller areas of sections with similar surficial geology, lithology, geomorphic process, soil groups, subregional climate, and potential natural communities. Names of subsections are usually derived from geologic features, such as Plainfield sand dune, Tipton till plain, and granite hills.

Landscape Scale

At the landscape scale, ecological units are defined by general topography, geomorphic process, surficial geology, soil, and potential natural community patterns and local climate (Forman and Godron 1986). These factors affect biotic distributions, hydrologic function, natural disturbance regimes, and general land use. Local landform patterns become apparent at this level in the hierarchy, and differences among units are usually obvious to on-the-ground observers. At this level, terrestrial features and processes may also have a strong influence on ecological characteristics of aquatic habitats (Platts 1979, Ebert et al. 1991).

Landtype association ecological units represent this scale in the hierarchy. These are groupings of landtypes or subdivisions of subsections based upon similarities in geomorphic process, geologic rock types, soil complexes, stream types, lakes, wetlands, and series, subseries, or plant association vegetation communities. Repeatable patterns of soil complexes and plant communities are useful in delineating map units at this level. Names of landtype associations are often derived from geomorphic history and vegetation community.

Land Unit Scale

At the basic land unit scale, ecological units are land mapped in the field based on properties of local topography, rock types, soils, and vegetation. These factors influence the structure and composition of plant communities, hydrologic function, and basic land capability. Landtypes and landtype phases are the ecological units mapped at this scale.

1. *Landtypes,* subdivisions of landtype associations or groupings of landtype phases based on similarities in soils, landform, rock type, geomorphic process, and plant associations. Land surface form that influences hydrologic function (for example, drainage density, dissection relief) is often used to delineate different landtypes in mountainous terrain. Valley bottom characteristics (for example, confinement) are commonly used in establishing riparian landtype map units. Names of landtypes include an abiotic and biotic component (USDA Forest Service Handbook 2090.11).

2. *Landtype phases,* more narrowly defined landtypes based on topographic criteria (for example, slope shape, steepness, aspect, position, hydrologic characteristics, associations and consociations of soil taxa, and plant associations and phases). These factors influence or reflect the microclimate and productivity of a site. Landtype phases are often established based on interrelations between soil characteristics and potential natural communities. In riparian mapping, landtype phases may be established to delineate different stream-type environments (Herrington and Dunham 1967). Naming is similar to landtypes (USDA Forest Service Handbook 2090.11).

The landtype phase is the smallest ecological unit recognized in the hierarchy. However, even smaller units may need to be delineated for very detailed project planning at large scales (table 9.1). Map design criteria depend on project objectives.

Plot Data

Point or plot sampling units are used to gather ecological data for inventory, monitoring, and quality control and for developing classifications of vegetation, soils, or ecological types. These plot data feed into databases for analysis, description, and interpretation of ecological units (Keane et al. 1990). Specific standards to be followed are in the Standards for Data and Data Structures Handbook. Other directives may also apply, such as those in the Permanent Plot Handbook. The plots can serve as reference sites for ecological types. Plots, while not mappable, can be shown on maps as point data.

In summary, the national framework has an extensive scientific basis and provides a hierarchical system for mapping ecological units ranging in size from global to local. At each level, abiotic and biotic components are integrated to classify and delineate geographical areas with similar ecological potentials. These ecological units, combined with information on existing conditions and ecological processes, provide a basis for managing ecosystems.

UNDERLYING PRINCIPLES

Ecosystem Concept

Ecosystems are places where life-forms and environment interact; they are three-dimensional segments of the earth (Rowe 1980). Tansley introduced the term *ecosystem* in 1935, first articulating the idea of ecological systems composed of multiple abiotic and biotic factors (Major 1969). The ecosystem concept brings the biological and physical worlds together into a holistic framework within which ecological systems can be described, evaluated, and managed (Rowe 1992). The structure and function of ecosystems are largely regulated along energy, moisture, nutrient, and disturbance gradients. These gradients are affected by climate, physiography, soils, hydrology, flora, and fauna (Barnes et al. 1982, Jordan 1982, Spies and Barnes 1985), and these factors change at different spatial and temporal scales. Ecological systems therefore exist at many spatial scales, from the global ecosphere down to regions of microbial activity.

At global, continental, and regional scales, ecosystem patterns correspond with climatic regions, which change mainly due to latitudinal, orographic, and maritime influences (Bailey 1987, Denton and Barnes 1988). Within climatic regions, physiography or landforms modify macroclimate (Rose 1984, Smalley 1986, Bailey 1987), and affect the movement of organisms, the flow and orientation of watersheds, and the frequency and spatial pattern of disturbance by fire and wind (Swanson et al. 1988). Within climatic-physiographic regions, water, plants, animals, soils, and topography interact to form ecosystems at land unit scales (Pregitzer and Barnes 1984). The challenge of ecosystem classification and mapping is to distinguish natural associations of ecological factors at different spatial scales, and to define ecological types and map ecological units that reflect these different levels of organization.

While the association of multiple biotic and abiotic factors is all important in defining ecosystems, all factors are not equally important at all spatial scales. At coarse scales, the important factors are largely abiotic, while at finer scales both biotic and abiotic factors are important. Furthermore, the level of discernible detail, the number of factors contributing to ecosystems, and the number of variables used to characterize these factors progressively increase at finer scales. Hence the data and analysis requirements, as well as the investments for ecosystem classification and mapping, also increase for finer-scaled activities.

The conditions and processes occurring across larger ecosystems affect and often override those of smaller ecosystems, and the properties of smaller ecosystems emerge in the context of larger systems (Rowe 1984). Moreover, environmental gradients change due to climatic, physiographic, and edaphic variations that affect ecological patterns and processes at different spatial scales. Thus it is useful to conceive of ecosystems as occurring in a nested geographic arrangement, with smaller ecosystems embedded in larger ones (Allen and Starr 1982, O'Neill et al. 1986, Albert et al. 1986). This spatial hierarchy is organized in decreasing orders of scale by the dominant factors affecting ecological systems. Ecosystems become networked, however, when nonadjacent systems exhibit similar structure and function with respect to specific biota (for example, sedentary plants as opposed to wide-ranging animals) and various processes; hence the networking of ecological systems is scale dependent (Allen and Hoekstra 1992). Networking of ecosystems occurs most often at lower levels of the hierarchy and depends upon requirements, environmental tolerances, and dispersion mechanisms of biota, as well as other factors that affect biotic-abiotic interactions occurring within and across local, landscape, and regional ecosystems.

Life and Environmental Interactions

Life-forms and environment have interacted and codeveloped at all spatial and temporal scales, one modifying the other through feedback. Appreciating these interactions is integral to understanding ecosystems. At a global scale, scientists have theorized that the evolution of cyanobacteria, followed by terrestrial plants capable of photosynthesis, carbon fixation, and oxygen production, converted the earth's atmosphere from hydrogen based to oxygen based and still sustain it today. At a continental scale, the migration of species in response to climate change and the interaction of their environmental tolerances and dispersal mechanisms with landform-controlled migration routes formed today's patterns in the distributions of species. At a landscape scale, life-forms, environment, and disturbance regimes have interacted to form patterns and processes. For example, pyrophilic communities tend to occupy droughty soils in fire-prone landscape positions, produce volatile substances, and accumulate litter, thereby increasing their susceptibility to burning. At yet finer scales, vegetation has induced soil development over time through carbon and nutrient cycling, enabling succession to proceed to communities with higher fertility requirements.

In each of these examples, life-forms and environment have modified one another through feedback to form ecological patterns and processes. These types of relations underscore the need to consider both biotic and environmental factors while classifying, mapping, and managing ecological systems.

Spatial and Temporal Variability

The structure and function of ecosystems change through space and time. Consequently, we need to address both spatial and temporal sources of variability while evaluating, classifying, mapping, or managing ecosystems (Delcourt et al. 1983, Forman and Godron 1986). At a land unit scale, for example, the fertility of particular locations changes through space because of differences in soil properties or hydrology, and at ecoregion scales, conditions vary from colder to warmer because of changes in macroclimate. These relatively stable conditions favor certain assemblages of plants and animals while excluding others because of biotic tolerances and such processes as competition. These environmental conditions are classified as ecological types and mapped as ecological units.

Within ecological units, ecosystems may support vegetation that is young, mature, or old, and they may be composed of communities that are early, mid-, or late successional. These relatively dynamic conditions also benefit certain plant and animal species and assemblages. Conditions that vary temporally are classified and mapped as existing vegetation, wildlife, water quality, and so forth.

These examples illustrate that ecological units do not contain all the information needed to classify, map, and manage ecosystems. Ecological units address the spatial distributions of relatively stable associations of ecological factors that affect ecosystems. When combined with information on existing conditions, the National Hierarchy of Ecological Units provides a means of addressing spatial and temporal variations that affect the structure and function of ecosystems. Adding our knowledge of processes to this information will enable us to evolve better ecosystem management.

USE OF ECOLOGICAL UNITS

Ecological units provide basic information for natural resource planning and management. Ecological unit maps may be used for such activities as delineating ecosystems, assessing resources, conducting environmental analyses, establishing desired future conditions, and managing and monitoring natural resources.

Ecosystem Mapping

To map ecosystems, or places where life and environment interact, we need to combine two types of maps: maps of existing conditions that change readily through time, and maps of potential conditions that are relatively stable. Existing conditions change due to particular processes that operate within the bounds of biotic and environmental, or ecological, potentials. Existing conditions are inventoried as current vegetation, wildlife, water quality, and so forth. Potential conditions are inventoried as ecological units. When these maps are combined, biotic distributions and ecological processes can be evaluated, and results can be extrapolated to similar ecosystems. The integration of multiple biotic and abiotic factors, then, provides the basis for defining and mapping ecosystems.

Fundamental base maps are key to mapping ecosystems and integrating resource inventories. These maps include the primary base map series, showing topography, streams, lakes, ownership, political boundaries, cultural features, and other layers in the cartographic features file. On this base, the next set of layers could include ecological units, watersheds, and inventories of aquatic systems at appropriate spatial scales. Next would be layers of information on existing vegetation, wildlife populations, fish distribution, demographics, cultural resources, economic data, and other information needed to delineate ecosystems to meet planning and analysis needs.

GIS will provide a tool for combining these separate themes of information and representing the physical, biological, and social dimensions to define and map ecosystems. But scientists and managers using this technology must actually integrate information themes, comprehend processes, and formulate management strategies. These tasks will not be accomplished mechanically.

Resource Assessments

The hierarchical framework of ecological units can provide a basis for assessing resource conditions at multiple scales. Broadly defined ecological units (for example, ecoregions) can be used for general planning assessments of resource capability. Intermediate scale units (for example, landtype associations) can be used to identify areas with similar natural disturbance regimes (for example, mass wasting, flooding, fire potential). Narrowly defined land units can be used to assess site-specific conditions, including distributions of terrestrial and aquatic biota; forest growth, succession, and health; and various physical conditions (for example, soil compaction and erosion potential, water quality).

High-resolution information obtained for fine-scale ecological units can be aggregated for some types of broader-scale resource assessments. Resource production capability, for example, can be estimated based on potentials measured for landtype phases, and estimates can be aggregated to assess ranger district, national forest, regional, and national capabilities.

Environmental Analyses

Ecological units provide a means of analyzing the feasibility and effects of management alternatives. To discern the effects of management on ecosystems, we often need to examine conditions and processes occurring above and below the level under consideration (Rowe 1980). For example, the effects of timber harvesting are manifest not only at a land unit scale but also at micro-site and landscape scales. Although the direct effects of management are assessed at the land unit scale, indirect and cumulative effects take place at different points in space or time, often at higher spatial scales. Ecological units defined at different hierarchical levels will be useful in conducting multiscaled analyses for managing ecosystems and documenting environmental effects (Jensen et al. 1991).

Watershed Analysis

The national hierarchy provides a basis for evaluating the linkages between terrestrial and aquatic systems. Because of the interdependence of geographic components, aquatic systems are linked or integrated with surrounding terrestrial systems through the processes of runoff, sedimentation, and migration of biotic and chemical elements. Furthermore, the context of water bodies affects their ecological significance. A lake embedded within a landscape containing few lakes, for example, functions differently from one embedded within a landscape composed of many lakes for wildlife, recreation, and other ecosystem values. Aquatic systems delineated in this indirect way have many characteristics in common, including hydrology and biota (Frissell et al. 1986). Overlays of hierarchical watershed boundaries on ecological mapping units are useful for most watershed analysis efforts. In this case, the watershed becomes the analysis area, which is both superposed by and composed of a number of ecological units which affect such hydrologic processes as water runoff and percolation, water chemistry, and ecological function due to context.

Desired Future Conditions

Desired future conditions (DFCs) portray the land or resource conditions expected if goals and objectives are met. Ecological units will be useful in establishing goals and methods to meet DFCs. When combined with information on existing conditions, ecological units will help us project responses to various treatments.

Ecological units can be related to past, present, and future conditions. Past conditions serve as a model of functioning ecosystems and provide insight into natural processes. It is unreasonable, for example, to attempt to restore systems like oak savannas or old-growth forests in areas where they did not occur naturally. Moreover, natural processes like disturbance or hydrologic regimes are often beyond human control. Ecological units will be helpful in understanding these processes and in devising DFCs that can be attained and perpetuated.

Desired future conditions can be portrayed at several spatial scales. We can minimize conflicting resource uses (for example, remote recreational experiences versus developed motorized recreation, habitat management for area-sensitive species versus edge species) if we consider the effects of projects at several scales of analysis. Ecological units will be useful in delineating land units at relevant analysis scales for planning DFCs.

Resource Management

Information on ecological units will help establish management objectives and will support such management activities as the protection of habitats of sensitive, threatened, and endangered species or the improvement of forest and rangeland health to meet conservation, restoration, and human needs. Information on current productivity can be compared to potentials determined for landtype phases, and areas producing less than their potential can be identified (Host et al. 1988). Furthermore, long-term sustained yield capability can be estimated based on productivity potentials measured for fine-scale ecological units.

Monitoring

Monitoring the effects of management requires baseline information on the condition of ecosystems at different spatial scales. Through the ecological unit hierarchy, managers can obtain information about the geographic patterns in ecosystems. They are thus in a position to design stratified sampling networks for inventory and monitoring. Representative ecological units can be sampled and information can then be extended to analogous unsampled ecological units, thereby reducing cost and time in inventory and monitoring.

By establishing baselines for ecological units and monitoring changes, we can protect landscape-, community-, and species-level biological diversity and other resource values, such as forest productivity, and air and water quality. The results of effectiveness and validation monitoring can be extrapolated to estimate effects and set standards in similar ecological units.

Evaluation of air quality is an example of how the National Hierarchical Framework of Ecological Units can be used for baseline data collection and monitoring. The Forest Service is developing a National Visibility Monitoring Strategy that addresses protection of air quality standards as mandated by the Clean Air Act, along with other concerns (USDA Forest Service 1993). Key to this plan is stratification of the United States at the subregion level of the national hierarchy into areas that have similar climatic, physiographic, cultural, and vegetational characteristics. Other questions dealing with effects of specific airborne pollutants on forest health, such as correlation of ozone with decline of ponderosa pine and other trees in mixed conifer forest ecosystems in the San Bernardino Mountains of southern California, will require establishment of sampling networks in smaller ecological units at landscape or lower levels.

Contemporary and Emerging Issues

The National Hierarchical Framework of Ecological Units is based on natural associations of ecological factors. These associations will be useful in responding to contemporary and emerging issues, particularly those that cross administrative and jurisdictional boundaries. Concerns regarding biological diversity, for example, can be addressed using the ecological unit hierarchy (Probst and Crow 1991). Conservation strategies can be developed using landscape-level units as coarse filters, followed by detailed evaluations and monitoring conducted to verify or adjust landscape designs. We can rehabilitate ecosystems and dependent species that have been adversely affected through fire exclusion, fragmentation, or other results of human activities if we grow to understand the natural processes that species and ecosystems codeveloped with, and then mimic those processes through ecosystem management.

Species may become rare, threatened, or endangered because their habitat is being lost or degraded, because they are endemic to a particular area, or because they are at the edge of their natural range. In the first two instances, protection or recovery efforts are warranted. In the latter case, however, it may be futile to try to maintain biota in environments where they are predisposed to decline. At a minimum, populations at the edge of their range can be evaluated for genetic diversity, and recovery programs can be administered accordingly. Species and community distributions can often be related to ecological units, which can be useful in their inventory and protection.

The new emphasis on sustaining and restoring the integrity of ecosystems may aid in arresting the decline of biological diversity and preempt the need for many future protection and recovery efforts. Developing basic information on the nature and distribution of ecosystems and their elements will enable us to better respond to issues like global warming, forest health, and biological diversity.

The hierarchical framework of ecological units was developed to improve our ability to implement ecosystem management. This framework, in combination with other information sources, is playing an important role in national, regional, and forest planning efforts; in the sharing of information between forests, stations, and regions; and in interregional assessments of ecosystem conditions.

Regions and stations, with national guidance, are coordinating their design of ecological units at higher levels of the national hierarchy. Development of landscape and land unit maps is being coordinated by appropriate regional, station, forest, and ranger district–level staff. As appropriate, new technologies (for example, remote sensing, GIS, expert systems) should be used in the design, testing and refinement of ecological unit maps.

The classification of ecological types and mapping of ecological units pose a challenge to integrate not only information but also the concepts and tools traditionally used by various disciplines. The effort brings together the biological and physical sciences that have too often operated independently. Specialists like foresters, fishery and wildlife biologists, geologists, hydrologists, community ecologists, and soil scientists will need to work together to develop and implement this new classification and mapping system. The results of these concerted efforts will then need to be applied in collaboration with planners, social scientists, economists, archaeologists, and the many other specialists needed to achieve a truly ecological approach to the management of our nation's national forests and grasslands.

LITERATURE CITED

Albert, D. A., S. R. Denton, and B. V. Barnes. 1986. Regional landscape ecosystems of Michigan. School of Natural Resources, University of Michigan, Ann Arbor.

Allen, T. F. H., and T. W. Hoekstra. 1992. Toward a unified ecology. Columbia University Press, New York, New York.

Allen, T. F. H., and T. B. Starr. 1982. Hierarchy: perspectives for ecological complexity. University of Chicago Press, Chicago, Illinois.

Avers, P. E., and E. F. Schlatterer. 1991. Ecosystem classification and management on national forests. *In* Proceedings of the symposium on systems analysis in forest resources, March 3–6, Charleston, South Carolina.

Bailey, R. G. 1980. Descriptions of the ecoregions of the United States. USDA Forest Service. Miscellaneous Publication 1391. U.S. Government Printing Office, Washington, D.C.

———. 1987. Suggested hierarchy of criteria for multiscale ecosystem mapping. Landscape and Urban Planning 14:313–319.

———. 1989a. Ecoregions of the continents (map). Scale 1:30,000,000. USDA Forest Service, Washington, D.C.

———. 1989b. Explanatory supplement to the ecoregions map of the continents. Environmental Conservation 15:307–309.

Bailey, R. G., S. C. Zoltai, and E. B. Wiken. 1985. Ecological regionalization in Canada and the United States. Geoforum 16(3):265–275.

Baldwin, J. L. 1973. Climates of the United States. U.S. Department of Commerce. National Oceanic and Atmospheric Administration. U.S. Government Printing Office, Washington, D.C.

Barnes, B. V. 1984. Forest ecosystem classification and mapping in Baden-Wurttemberg, West Germany. Pages 49–65 *in* Forest land classification: experience, problems, perspectives. Proceedings of the symposium, March 18–20, Madison, Wisconsin.

Barnes, B. V., K. S. Pregitzer, T. A. Spies, and V. H. Spooner. 1982. Ecological forest site classification. Journal of Forestry 80:493–498.

Brenner, R. N., and J. K. Jordan. 1991. The role of an ecological classification system in forest plan development and implementation. *In* Proceedings of the symposium on systems analysis in forest resources, March 3–6, Charleston, South Carolina.

Cleland, D. T., T. R. Crow, P. E. Avers, and J. R. Probst. 1992. Principles of land stratification for delineating ecosystems. *In* Proceedings of taking an ecological approach to management national workshop, April 27–30, Salt Lake City, Utah.

Cowardin, L. M., V. Carter, F. C. Golet, and E. T. LaRoe. 1979. Classification of wetlands and deepwater habitats of the United States. U.S. Department of the Interior, U:S. Fish and Wildlife Service, Washington, D.C.

Crowley, J. M. 1967. Biogeography in Canada. Canadian Geographer 11:312–326.

Delcourt, H. R., T. A. Delcourt, and T. Webb. 1983. Dynamic plant ecology: the spectrum of vegetative change in space and time. Quantitative Science Review 1: 153–175.

Denton, D. R., and B. V. Barnes. 1988. An ecological climatic classification of Michigan: a quantitative approach. Forest Science 34:119–138.

Driscoll, R. S., et al. 1984. An ecological land classification framework for the United States. USDA Forest Service. Miscellaneous Publication 1439. Washington, D.C.

Ebert, D. J., T. A. Nelson, and J. L. Kershner. 1991. A soil-based assessment of stream fish habitats in coastal plains streams. Proceedings of warmwater fisheries symposium, June 4–8, Phoenix, Arizona.

Ecoregions Working Group. 1989. Ecoclimatic regions of Canada, first approximation. Ecological Land Classification Series No. 23. Environment Canada, Ottawa. Includes map at 1:7,500,000.

Fenneman, N. M. 1938. Physiography of eastern United States. McGraw-Hill, New York, New York.

Forman, R. T. T., and M. Godron. 1986. Landscape ecology. John Wiley and Sons, New York, New York.

Frissell, C. A., W. J. Liss, C. E. Warren, and M. C. Hurley. 1986. A hierarchical framework for stream habitat classification: viewing streams in a watershed context. Environmental Management 10:199–214.

Gallant, A. L., T. R. Whittier, D. P. Larsen, J. M. Omernik, and R. M. Hughes. 1989. Regionalization as a tool for managing environmental resources. EPA/600/3-89/ 060. U. S. Environmental Protection Agency, Corvallis, Oregon.

Herrington, R. B., and D. K. Dunham. 1967. A technique for sampling general fish habitat characteristics of streams. USDA Forest Service. Research Paper INT-41. Intermountain Forest and Range Experiment Station, Ogden, Utah.

Hills, G. A. 1952. The classification and evaluation of site for forestry. Ontario Department of Lands and Forests. Resource Division Report 24.

Hix, D. M. 1988. Multifactor classification of upland hardwood forest ecosystems of the Kickapoo River watershed, southwestern Wisconsin. Canadian Journal of Forest Research 18:1405–1415.

Holdridge, L. R. 1967. Life zone ecology. Tropical Science Center. San Jose, California.

Host, G. E., K. S. Pregitzer, C. W. Ramm, J. B. Hart, and D. T. Cleland. 1987. Landform mediated differences in successional pathways among upland forest ecosystems in northwestern lower Michigan. Forest Science 33:445–457.

Host, G. E., K. S. Pregitzer, C. W. Ramm, D. T. Lusch, and D. T. Cleland. 1988. Variations in overstory biomass among glacial landforms and ecological land units in northwestern lower Michigan. Canadian Journal of Forest Research 18:659–668.

Jensen, M. C., C. McMicoll, and M. Prather. 1991. Application of ecological classification to environmental effects analysis. Journal of Environmental Quality 20: 24–30.

Jones, R. K., et al. 1983. Field guide to forest ecosystem classification for the clay belt, site region 3e. Ministry of Natural Resources, Ottawa, Ontario.

Jordan, J. K. 1982. Application of an integrated land classification. Pages 65–82 *in* Proceedings, artificial regeneration of conifers in the Upper Lakes Region, October 26–28, Green Bay, Wisconsin.

Keane, R. E., M. E. Jensen, and W. J. Hann. 1990. Ecodata and Ecopac: analytical tools for integrated resource management. The Compiler. 8(3):24–37.

Koppen, J. M. 1931. Grundriss der Klimakunde. Walter de Grayter, Berlin.

Kuchler, A. W. 1964. Potential natural vegetation of the coterminous United States. American Geographic Society Special Publication 36.

Major, J. 1969. Historical development of the ecosystem concept. Pages 9–22 *in* G. M. Van Dyne, ed. The ecosystem concept in natural resource management. Academic Press, New York, New York.

McNab, W. H. 1987. Rationale for a multifactor forest site classification system for the southern Appalachians. Pages 283–294 *in* Proceedings of 6th Central Hardwood Forest Conference, February 24–26, Knoxville, Tennessee.

Meentemeyer, V., and E. O. Box. 1987. Scale effects in landscape studies. Pages 15–34 *in* M. G. Turner, ed. Landscape heterogeneity and disturbance. Springer-Verlag, New York, New York.

Miller, P. R., O. C. Taylor, and R. G. Wilhour. 1982. Oxidant air pollution effects on a western coniferous forest ecosystem. Environmental Research Brief. EPA-600-/D-82-276. U.S. Government Printing Office, Washington, D.C.

Omernik, J. M. 1987. Ecoregions of the coterminous United States. Annals of the Association of American Geographers 77:118–125.

O'Neill, R. V., D. L. DeAngelis, J. B. Waide, and T. F. H. Allen. 1986. A hierarchical concept of ecosystems. Princeton University Press, Princeton, New Jersey.

Platts, W. S. 1979. Including the fishery system in land planning. USDA Forest Service. Gen. Tech. Rep. INT-60. Intermountain Forest and Range Experiment Station, Ogden, Utah.

———. 1980. A plea for fishery habitat classification. Fisheries 5(1):2–6.

Pregitzer, K. S., and B. V. Barnes. 1984. Classification and comparison of upland hardwood and conifer ecosystems of the Cyrus H. McCormick Experimental Forest, upper Michigan. Canadian Journal of Forest Research 14:362–375.

Probst, J. R., and T. R. Crow. 1991. Integrating biological diversity and resource management. Journal of Forestry 89:12–17.

Rowe, J. S. 1980. The common denominator in land classification in Canada: an ecological approach to mapping. Forest Chronicle 56:19–20.

———. 1984. Forest Land Classification: limitations of the use of vegetation. Pages 132–147 *in* Proceedings of the symposium on forest land classification, March 18–20, Madison, Wisconsin.

————. 1992. The ecosystem approach to forest management. Forest Chronicle 68: 222–224.

Russell, W. E., and J. K. Jordan. 1991. Ecological classification system for classifying land capability in midwestern and northeastern U.S. national forests. *In* Proceedings of the symposium, ecological land classification: applications to identify the productive potential of southern forests, January 7–9, Charlotte, North Carolina. USDA Forest Service Gen. Tech. Report SE-68, Asheville, North Carolina.

Smalley, G. W. 1986. Site classification and evaluation for the interior uplands. USDA Forest Service. Tech. Pub. R8-TP9. Southern Region, Atlanta, Georgia.

Spies, T. A., and B. V. Barnes. 1985. A multifactor ecological classification of the northern hardwood and conifer ecosystems of Sylvania Recreation Area, Upper Peninsula, Michigan. Canadian Journal of Forest Research 15:949–960.

Swanson, F. J., T. K. Kratz, N. Caine, R. G. Woodmansee. 1988. Landform effects on ecosystem patterns and processes. BioScience 38:92–98.

Trewartha, G. T. 1968. An introduction to climate. 4th edition. McGraw-Hill, New York, New York.

Udvardy, M. D. F. 1975. A classification of the biogeographical provinces of the world. Occasional Paper 18. International Union for Conservation of Nature and Natural Resources, Morges, Switzerland.

USDA. 1981. Land resource regions and major land resource areas of the United States. Agricultural Handbook 296. U.S. Government Printing Office, Washington, D.C.

USDA Forest Service. 1976. Land systems inventory guide. Northern Region.

————. 1993. Draft national visibility monitoring strategy. Washington, D.C.

USDA Forest Service. Handbook 2090. U.S. Government Printing Office, Washington, D.C.

USDA Forest Service. Manual 2060. U.S. Government Printing Office, Washington, D.C.

Urban, D. L., D. O'Neil, and H. H. Shugart. 1987. Landscape ecology. BioScience 37:119–127.

Walter, H., and E. Box. 1976. Global classification of natural terrestrial ecosystems. Vegetatio 32:75–81.

Wertz, W. A., and J. A. Arnold. 1972. Land systems inventory. USDA Forest Service, Intermountain Region, Ogden, Utah.

Wiken, E. B. 1986. Terrestrial ecozones of Canada. Ecological Land Classification Series No. 19. Environment Canada, Hull, Hull, Quebec.

Chapter 10 Geographic Information Systems and Remote Sensing Applications for Ecosystem Management

Frank D'Erchia

As resource managers recognize the importance of adopting an ecosystem approach in managing natural resources, the use of geographic information systems (GIS) and remote sensing technologies is increasing. These technologies provide tools for resource managers to use in analyzing and understanding an ecosystem, allowing decision makers to better visualize, integrate, and quantify available resource data.

The role of GIS in ecosystem studies is expanding as researchers exploit the increasingly sophisticated capabilities of GIS technology. Recent advances include the ability to store and manage large data sets and to perform spatial and statistical analyses. GIS can also provide input to both static and dynamic ecosystem models. For example, a static model may be used to make erosion estimates based on soil type and terrain characteristics, whereas a dynamic GIS model could be used to represent a spatial landscape at different time periods (Stow 1993). I will present three case studies demonstrating applications of GIS and remote sensing technologies.

The first case study discusses GIS applications in the Long Term Resource Monitoring Program for the Upper Mississippi River System. The long-term goals of the monitoring program are to understand the ecosystem, determine resource trends and impacts, develop management alternatives, and provide information and technology transfer (USFWS 1992). The monitoring program is responsible for collection of data on parameters including water quality, fisheries, vegetation, and invertebrates. Land cover/use spatial databases are developed, stored, and managed in a GIS for analysis and dissemination to river resource managers.

The second case study is a pilot project that uses GIS to analyze migratory bird habitats within the Upper Mississippi River corridor. This pilot project evaluated GIS technology as a tool to assist national wildlife refuge staff and other resource managers in making management decisions. Detailed land cover/use spatial data were used to predict habitat ranges for migratory birds by associating land cover classes with life-cycle habitat preferences (Lowenberg 1996).

The final case study discusses the Gap Analysis Program, a nationally instituted GIS effort to identify "gaps" in biodiversity protection, in the context of a tri-state effort in the upper Midwest. Gap analysis uses GIS to combine the distribution of natural vegetation, mapped from satellite imagery and other data sources, with distributions of vertebrate and other taxa as indicators of biodiversity (Scott et al. 1993). Maps of species-rich areas, individual species of concern, and overall vegetation types are generated and compared with land ownership and protection status.

CASE STUDY 1: LONG TERM RESOURCE MONITORING PROGRAM

From 1930 to 1950, the U.S. Army Corps of Engineers constructed twenty-nine locks and dams on the Upper Mississippi River System (UMRS) to maintain a 2.7 m channel for commercial navigation during periods of low flow. The locks and dams created a series of impounded pools, resulting in an initial boom in biological productivity. However, because natural alluvial river ecosystems are characterized by a floodplain and an annual flood pulse, this increase in aquatic habitats consequently brought loss of riparian habitat and reduced biological diversity, resulting in a net reduction in resource productivity. Floodplain encroachment and development, wetlands drainage, and channelization of the river have accelerated water velocities. Wave action resuspends lake bottom sediment and erodes islands and shorelines. Loss of islands creates greater wind fetches, resulting in greater wave action. The UMRS now experiences higher and more erratic river stages, which disrupt the natural functions of the floodplain ecosystem (Sparks et al. 1990).

Congress recognized the importance and uniqueness of the ecology of the area but also acknowledged the national economic significance of the navigation system. Therefore, the Water Resources Development Act of 1986 (Public Law 99-662) authorized the Upper Mississippi River System Long Term Resource Monitoring Program to provide decision makers with the information needed to maintain the UMRS as a sustainable multiple-use large river ecosystem.

The monitoring program is being implemented by the Environmental Management Technical Center, a National Biological Service Science Center, in cooperation with field stations staffed by the five UMRS states (Illinois, Iowa, Minnesota, Missouri, and Wisconsin). Guidance is provided by the U.S. Army Corps of Engineers, which has overall responsibility for the monitoring program (USFWS 1992).

The Environmental Management Technical Center is responsible for inventory and monitoring, research, analysis, and application of GIS and remote sensing technologies. Spatial data (including land cover/use, bathymetry, transportation, wetlands, and other spatial features) are processed and managed in a GIS at the center.

Study Region and Methods

The study area for the monitoring program encompasses the commercially navigable portions of the Mississippi River north of Cairo, Illinois, plus the entire Illinois River and waterway and four other midwestern rivers (fig. 10.1). Current research efforts focus on the floodplain of the UMRS. Land cover/use databases are developed through photo interpretation of aerial photography (Owens and Hop 1995). These databases are stored in a GIS at the center. Color-infrared photographs (1:15,000-scale) are taken annually of specific pools or reaches of the UMRS (a pool is defined as the impounded section of the river between locks and dams; e.g., Pool 8 is the impounded reach above Lock and Dam 8 to Lock and Dam 7). Detailed land cover/use spatial data coverages have been created for several pools within the study area from 1989 color-infrared aerial photography (D'Erchia 1993). The aerial photography is interpreted, transferred to base maps, and digitized to create spatial databases in an ARC/INFO GIS format (White and Owens 1991). ARC/INFO is a widely used commercial GIS software package developed by the Environmental Systems Research Institute, Redlands, California (Moorehouse 1985).

An additional source of land cover/use data is satellite imagery. The center has produced a classified coverage for the entire floodplain of the UMRS from Landsat Thematic Mapper (TM) satellite imagery (TM imagery has a pixel resolution of 30 m). The floodplain outline was extracted from the seven full scenes required to cover the study area.

Fig. 10.1. Long Term Resource Monitoring Program study area: the Upper Mississippi River System floodplain.

Finally, several historical databases were available for land cover/use. One was developed from aerial photography taken in 1975 under the auspices of the Great River Environmental Action Team (UMRBA 1982), and a preimpoundment perspective was obtained from the mapping efforts of the Mississippi River Commission Survey undertaken in the 1890s. The survey maps were found to be spatially and thematically accurate at a scale of 1:20,000; therefore, land cover/use features were digitized directly from the source base maps (fig. 10.2). These databases are used to analyze changes in the flood-plain ecosystem.

The following review of an analytical procedure conducted on Pool 8 demonstrates the use and applications of GIS and remote sensing technologies at the Environmental Management Technical Center.

A comparison of 1891 and 1989 land cover/use for Pool 8 reveals the postimpoundment land and water changes (fig. 10.3). Table 10.1 provides a summary of the spatial feature classes used, grouped for comparative purposes. As can be seen, there were increases in open water and marsh after impoundment and, conversely, decreases in woody terrestrial and grass and forb habitats. The area of woody terrestrial coverage not only decreased considerably by 1989, but the frequency increased substantially, resulting in a highly fragmented woody terrestrial ecosystem with many small polygons in 1989, compared with large homogeneous woody terrestrial areas in 1891. Urban areas expanded, with reduced use of the floodplain for agriculture in this section of the UMRS. Using GIS, these coverages can be combined to display changes in open water in the floodplain of Pool 8 (fig. 10.4). This analysis process provides a methodology to both visualize and quantify general poolwide changes pre- and postimpoundment.

Through analyses of physical effects on the system, GIS can be used to study how biological resources respond to change. A time-series display of interpreted aerial photography of the lower portion of Pool 8 from 1939 to 1989 reveals a dynamic island geomorphology (fig. 10.5). Erosion from wind and wave action has contributed considerably to island degradation. In a recent Environmental Management Technical Center study of island loss in lower Pool 8, bathymetric data were compared with historical water depth data collected in the 1930s (fig. 10.6).

Backwaters provide migrating waterfowl with critical habitats because of their highly productive submersed vegetation, which is an important food source for a number of species. However, in recent years, backwater vegetation productivity has been on the decline (Rogers 1994). GIS techniques are used to analyze the relations between increased sedimentation and decreased vegetation production. Areas that have experienced island loss, resulting in higher flows and increased sedimentation transport, have also displayed a decrease in submersed vegetation over time. Wind fetch models have been developed to assist in locating and evaluating artificial island placement to reduce sediment flow.

The U.S. Army Corp of Engineers (through the Environmental Management Program) is responsible for development of habitat rehabilitation projects to improve and prolong the longevity of the study area as a multiple-use

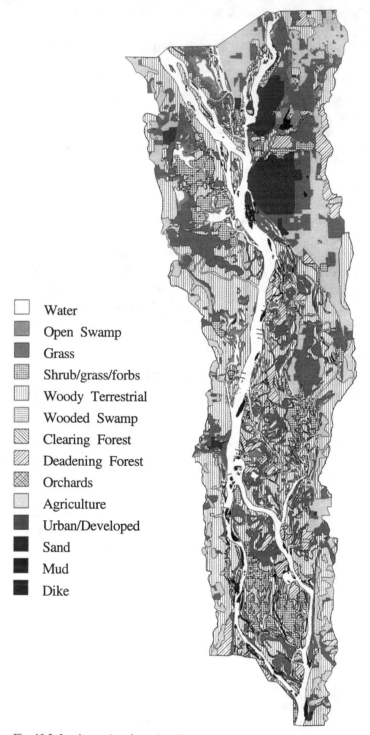

Water

Open Swamp

Grass

Shrub/grass/forbs

Woody Terrestrial

Wooded Swamp

Clearing Forest

Deadening Forest

Orchards

Agriculture

Urban/Developed

Sand

Mud

Dike

Fig. 10.2. Land cover/use from the 1891 Mississippi River Commission Survey.

Table 10.1 Comparison of land cover changes between 1891 and 1989 for Pool 8, Upper Mississippi River System

| | 1891 | | 1989 | |
Classification	Frequency (# of polygons)	Area (hectares)	Frequency (# of polygons)	Area (hectares)
Open water	541	3,261	751	6,494
Marsh	52	343	3,092	2,995
Grasses/forbs	292	4,441	1,646	1,479
Woody terrestrial	581	5,834	2,326	2,614
Sand/mud	174	355	97	31
Agriculture	37	587	14	84
Urban/developed	10	238	85	1,362

resource for fish, wildlife, and recreation. These projects focus on island construction by strategic placement of channel dredge material. A GIS application conducted at the Environmental Management Technical Center used existing land cover/use data in combination with terrain elevation to analyze the effects of levee placement (McConville 1996). An interface was developed to allow the novice GIS user to locate levees by on-screen digitizing. The model analyzes effects on the habitat based on levee placement and elevation and the rise of floodwaters. Information provided through this research and analysis effort will help guide construction projects to maximize benefits to riverine habitats.

Once analytical models and interfaces are developed for demonstration pools, they can be extrapolated and applied to any pool where similar databases exist. Efforts are under way to complete automation of detailed land cover/use GIS coverages for the entire study area from the aerial photography collected in 1989. This coverage will provide the database for a landscape analysis of the entire system based on the poolwide models currently under development at the center.

In addition to pool-specific studies, systemic analyses at a lower resolution are being conducted using satellite imagery. Landsat TM imagery has been used to classify the entire study area, and research on systemic change detection was conducted in 1994 (Laustrup and Lowenberg 1994). While databases derived from satellite imagery are not as detailed as those developed from low-level aerial photography, they are adequate for systemic landscape analysis (e.g., broad categories within the floodplain can be quantified). Detailed land cover classes from the interpreted aerial photography can be

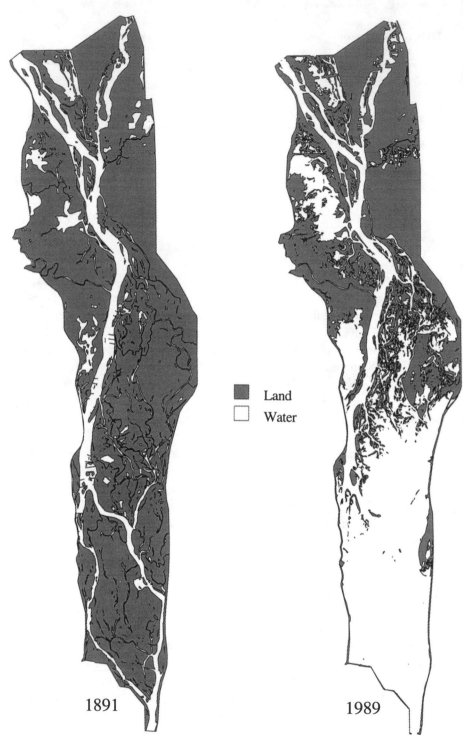

1891   1989

Fig. 10.3. Comparison of land cover/use for Pool 8, Upper Mississippi River System, between 1891 and 1989.

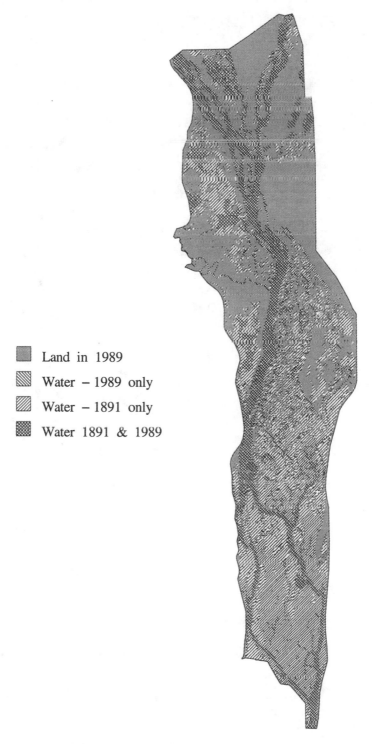

Fig. 10.4. Change in open water for Pool 8, Upper Mississippi River System, from 1891 to 1989.

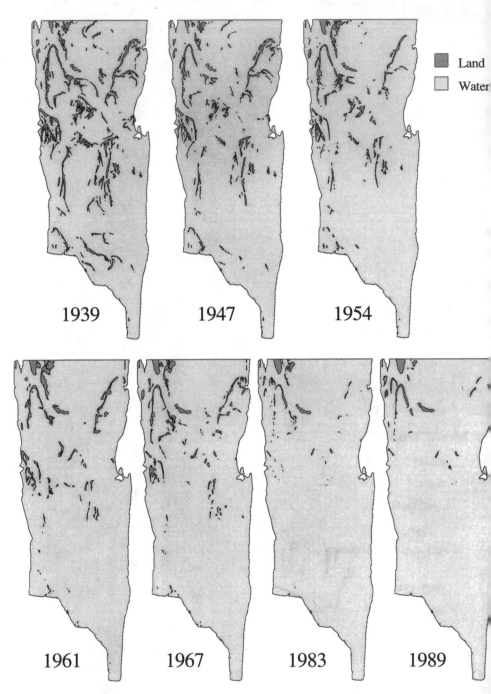

Fig. 10.5. Geomorphology of lower Pool 8, Upper Mississippi River System, between 1939 and 1989.

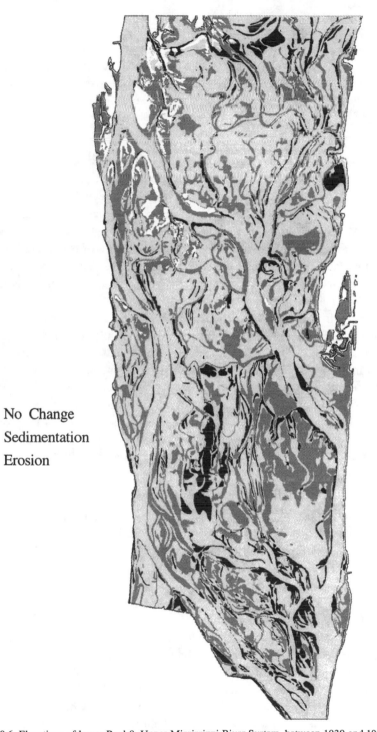

No Change
Sedimentation
Erosion

Fig. 10.6. Elevations of lower Pool 8, Upper Mississippi River System, between 1939 and 1989.

generalized to match the satellite imagery-derived classes, and comparisons then can be made to evaluate change over time. Satellite imagery provides a means to discern broad-based ecosystem changes over time in such generalized landscape characteristics as wetlands and forested areas.

CASE STUDY 2: MIGRATORY BIRD ANALYSIS

Three national wildlife refuges are located within the Upper Mississippi River basin: the Upper Mississippi River National Wildlife and Fish Refuge, Mark Twain National Wildlife Refuge, and Trempealeau National Wildlife Refuge. The Mississippi River corridor is a major waterfowl flyway during spring and fall migration. In addition, the river environment of backwaters, wooded bottom lands, open water, and wooded bluffs serves as a migration corridor to more than 290 species of raptors, shorebirds, and songbirds (NBS and USFWS 1995).

In an effort to take a proactive approach to managing the habitats used by migratory bird populations in the Upper Mississippi River basin, a pilot study was undertaken to analyze migratory bird habitats using GIS and remote sensing technologies. Agency representatives of the states of Illinois, Iowa, Minnesota, Missouri, and Wisconsin, the U.S. Army Corps of Engineers, the National Biological Service, and the U.S. Fish and Wildlife Service have agreed to implement a migratory bird management strategy for the river corridor. The Environmental Management Technical Center was selected to conduct the spatial analysis for the pilot project (Lowenberg 1996).

Study Region and Methods

The study area for the migratory bird project includes the Upper Mississippi River corridor from Wabasha, Minnesota, to St. Louis, Missouri. Pool 8 was selected for the pilot study. This river reach includes the impounded water above Lock and Dam 8 near Genoa, Wisconsin, to Lock and Dam 7 at Dresbach, Minnesota (fig. 10.7). The spatial analysis for this project used land cover/use maps developed from 1:15,000-scale aerial photography of the floodplain. Some models included satellite imagery to extend the boundaries into the bluffs above the floodplain and to evaluate the possible substitution of satellite imagery for modeling where detailed data are unavailable.

First, a literature search was conducted to determine life-cycle habitat requirements of a selected sample of migratory birds. This search identified parameters for habitats critical to the selected migratory bird species (Jacobson 1993). Next, matrices linking these habitat descriptions with land cover/

Fig. 10.7. Migratory bird study area: Pool 8, Upper Mississippi River System.

use spatial data coverages were developed. Individual range maps could then be generated for each migratory bird species.

Information from this search and other sources provided the means to categorize habitat use based on several life cycle variables (spring migration, prebreeding, nesting, brood rearing, postbreeding, fall migration, and wintering). The habitats identified were linked with the land cover/use classes of the GIS database and were used to generate GIS-predicted range maps. Some models were derived directly from matrices that cross-referenced vegetation types with bird species. Other models were more complex, requiring distance and neighborhood analyses.

A simple nesting habitat matrix association model for the pileated woodpecker *(Dryocopus pileatus)* is displayed in figure 10.8. The literature search revealed that the pileated woodpecker prefers mature dense forest stands with high snag densities (Bull 1975). The land cover/use database included tree height and percent vegetation cover. By selecting for the tallest trees with the greatest percentage of cover, it is assumed that mature dense forest stands are selected and that they contain a large quantity of snags. This prediction of snag density is crucial, because snags provide nest sites for the target species. Field verification is necessary to confirm these assumptions.

Another migratory bird model in the pilot study demonstrates the use of GIS to analyze habitat using neighborhood analysis techniques. The American bittern *(Botaurus lentiginosus)* was found to nest in thick marsh grass, sometimes adjacent to stands of willow and tamarack, within 6.1 m of water (Bohlen 1989). Habitat analysis was carried out using GIS procedures to create a 6.1-m buffer around all aquatic areas. Then, terrestrial habitats that met nesting requirements were selected from within this buffer (fig. 10.9). Although this type of buffer analysis is not complex using GIS, an analysis of preferred habitats with this requirement would be very difficult, if not impossible, to perform manually. Identification of critical habitats delineates areas to be considered for protection to help maintain populations of target species.

Use of Satellite Data

We have thus far discussed migratory bird habitats within a reach of the floodplain of the Mississippi River. Because an ecosystem approach was desired, spatial data from Landsat TM satellite imagery were combined with the floodplain data, allowing comparison of the high-resolution floodplain data and the coarser satellite imagery.

For this exercise, a 32 km swath of TM imagery on either side of the floodplain was classified for forested areas, grasslands, wetlands, urban, and agricultural areas. The floodplain data cover bluff to bluff, and the satellite imagery covers land cover on the bluffs. Due to the coarser TM resolution, small wetland areas could be misclassified; therefore, digital wetlands data from the National Wetlands Inventory (Cowardin et al. 1979) were incorporated into the TM imagery. Digital wetlands data were available for all but the southwest portion of the study area.

Two migratory bird species were chosen for this application, the cerulean warbler *(Dendroica cerulea)* and the canvasback *(Aythya valisineria)*. Cerulean warbler habitat includes forested areas larger than 16.2 ha (Bond 1957), while canvasback require large open bodies of water (Korschgen 1989). Cerulean warbler habitat within the floodplain is somewhat restricted, which could lead to the conclusion that this species' habitat needs protection, but when the search for the bird's habitat requirements is extended above the bluffs, we find that abundant habitat exists (fig. 10.10). Conversely, canvasback habitat would be restricted to the large body of open water at the lower

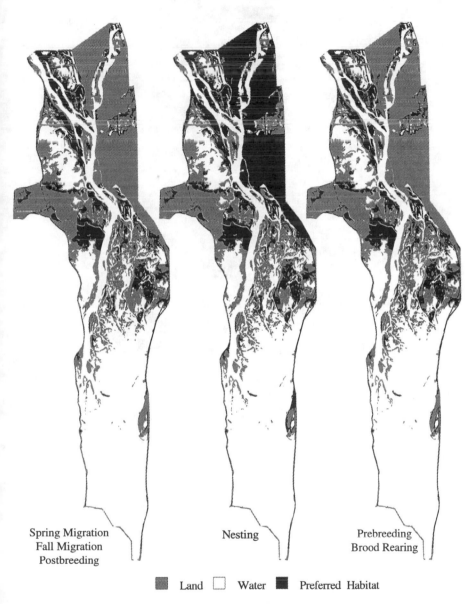

Spring Migration
Fall Migration
Postbreeding

Nesting

Prebreeding
Brood Rearing

■ Land □ Water ■ Preferred Habitat

Fig. 10.8. Pileated woodpecker life cycle habitat preferences.

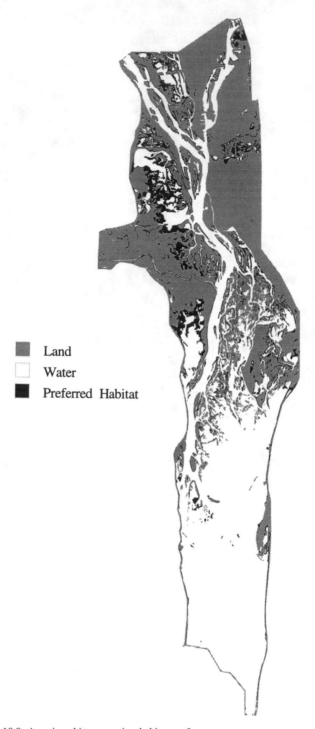

Land
Water
Preferred Habitat

Fig. 10.9. American bittern nesting habitat preferences.

end of Pool 8, with no adequate habitat available in the bluffs above the floodplain. In this example, a resource manager could make a more informed decision regarding habitat protection with the extended knowledge of the conditions in the area immediately surrounding the floodplain. By studying the landscape, it becomes apparent that, in this case, sufficient habitat for the cerulean warbler exists in the adjacent blufflands. Therefore, management objectives in the floodplain may be better served by focusing on canvasback habitat.

The pilot study combined all subject species' habitat ranges to develop a simulated map of species richness. Selected species data on preferred nesting habitats within the floodplain were combined and polygons were ranked according to number of species present (fig. 10.11). By overlaying the habitat requirements of multiple species, critical areas can be identified. This information is provided to resource managers and others responsible for wildlife habitat management and acquisition, who can then make decisions based on the conglomerate map, taking into account the status of individual species. Areas of importance to large numbers of species of concern (e.g., listed or declining) could be given a higher priority in proactive management decisions and habitat mitigation efforts.

The next step in the pilot study was to verify the models, which was accomplished by comparing migratory bird sightings during field studies with GIS-predicted habitats. Great blue heron *(Ardea herodias)* rookeries were located and plotted on a base map that displays the GIS-predicted preferred habitat for the species (fig. 10.12). The association between known sightings and GIS-predicted habitat preferences is needed to refine the model and verify the results of the analysis. Field sightings and other field information helped to fine-tune the models for each migratory bird species in the pilot study. The verification process provides the information necessary to validate and/ or update models.

Management of critical habitats for the benefit of groups of wildlife species, particularly migratory birds, was the focus of this analysis, rather than taking a single-species approach. If the methodologies applied to a sample area prove useful, the protocols could be applied over a larger area. Once the models are verified, a GIS interface will be developed to provide managers with an easy-to-use tool for analyzing habitats.

CASE STUDY 3: UPPER MIDWEST GAP ANALYSIS PROGRAM

The Gap Analysis Program is a nationally implemented National Biological Service effort which seeks to identify the degree to which plant and animal

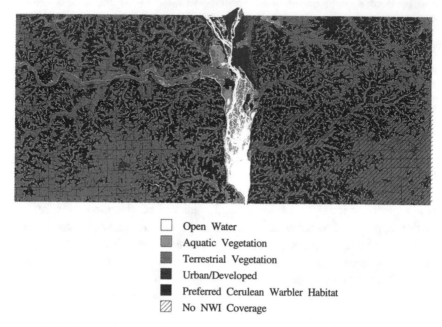

☐ Open Water
▨ Aquatic Vegetation
▨ Terrestrial Vegetation
■ Urban/Developed
■ Preferred Cerulean Warbler Habitat
▧ No NWI Coverage

Fig. 10.10. Cerulean warbler habitat preferences derived from aerial photographs and satellite imagery.

communities are or are not represented in areas being managed for the long-term protection of biological resources. Natural communities not adequately represented in such areas constitute "gaps" in biodiversity conservation (Scott et al. 1993). Cooperating organizations include private business corporations, nonprofit groups, and state and federal government agencies.

Gap Analysis projects typically are completed on a state-by-state basis. However, because of the contiguous biologically diverse habitats in the upper Midwest, the Environmental Management Technical Center proposed initiation of a multistate approach in the region to ensure consistency and continuity of land cover and other spatial databases across political boundaries. In cooperation with the state, federal, and nongovernment agencies in the region, the center has begun development of a land cover database for a three-state area (D'Erchia et al. 1993).

Study Region and Methods

The study area covers the states of Michigan, Minnesota, and Wisconsin. This area includes a portion of the Upper Mississippi River ecosystem and

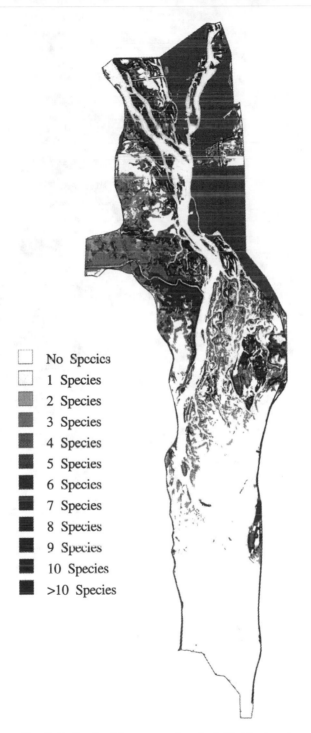

Legend:
- No Species
- 1 Species
- 2 Species
- 3 Species
- 4 Species
- 5 Species
- 6 Species
- 7 Species
- 8 Species
- 9 Species
- 10 Species
- >10 Species

Fig. 10.11. Species richness map of nesting habitat for a combination of several migratory bird species.

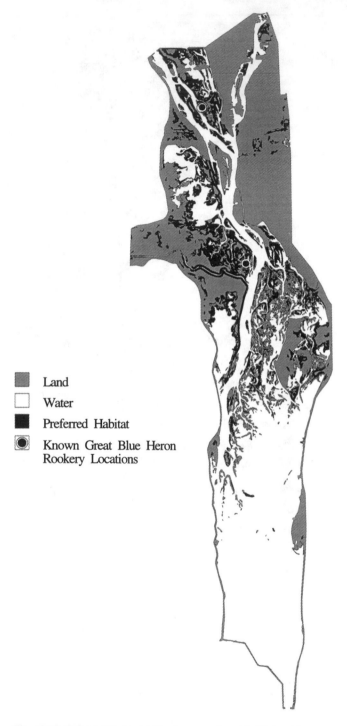

Land

Water

Preferred Habitat

Known Great Blue Heron Rookery Locations

Fig. 10.12. GIS-predicted great blue heron preferred habitat and known rookery locations.

extends into the watershed beyond the floodplain. The Gap Analysis Program in the upper Midwest is currently in its initial stages. Land cover is being mapped using Landsat TM satellite imagery. To accomplish the goals of this three-state effort, Landsat TM scenes will be acquired for full coverage of the study area. To assist in classification discrimination, especially of tree species, a second date for each scene will be compared to incorporate the temporal variations between scene dates. Scene pairs, with dates that coincide with peak biomass and leaf senescence, will be selected from the available cloud-free inventory of images not more than three years old. Several methods are possible for integrating the two-date scenes, including separate scene classification, combining all bands from both images, or some combination of principal components. Such ancillary information as aerial photography and existing digital land cover data will be integrated with the satellite imagery where it is available consistently for the three states (Lillesand 1994).

The standard minimum mapping unit for the Gap Analysis Program is 100 ha; however, the upper Midwest effort will maintain the 30 m pixel resolution of the TM imagery for land cover mapping. Uniquely classified individual pixels will be grouped with surrounding pixels, maintaining a minimum mapping unit of less than 1 ha. This system will result in a higher resolution and uniform classification across the three states. In addition, mapping of presettlement vegetation will be completed for the three states. Using a habitat association approach, wildlife species range maps will be generated to link potential habitat for a variety of wildlife species with the habitat types in the land cover maps. Land ownership maps will also be developed for the three-state area.

Wildlife-habitat relations are developed by linking vegetation classes with preferred habitats of individual wildlife species. A species-habitat matrix has been developed for the forested areas of the Upper Great Lakes region, which includes portions of the states of Michigan, Minnesota, and Wisconsin (Benyus et al. 1992). This habitat matrix displays 389 species of birds, mammals, reptiles, and amphibians with their associated habitats. Additional work will be required to develop habitat associations with land cover types for the lower portions of each state, as well as additions and enhancements to the existing information.

Maps of species-rich areas, individual species of concern, and overall vegetation types will be generated. Several species range maps will then be combined to identify the species-rich areas for the subject wildlife group. Using GIS, this information can be overlaid on maps of land stewardship to display those areas not adequately protected, identifying where efforts need to be focused to most efficiently achieve conservation of biodiversity.

The Long Term Resource Monitoring Program is an ongoing effort to evaluate and understand the dynamics of a changing floodplain environment. It is the goal of the monitoring program to conduct research and analysis in an effort to link environmental effects with the physical parameters operating within the floodplain ecosystem. The results of monitoring program studies are used by state and federal agencies in implementing management strategies.

Successful application of GIS technology in the migratory bird pilot study has demonstrated how this tool can provide valuable information to resource managers. A methodology was needed to assist resource managers in implementing a landscape approach to proactive habitat management of the migratory bird flyway in the Upper Mississippi River corridor. By evaluating habitat abundance for individual species and combining all species to predict areas of critical importance, GIS technology will provide resource managers with the information needed to make informed management decisions. Additional information, such as identification of keystone or indicator species' habitats and determination of their role in the ecosystem, can also be displayed in GIS maps.

The migratory bird pilot study utilizes a Gap Analysis Program approach in identifying species-rich areas; in addition to the migratory bird component, the Gap Analysis Program analyzes habitats for native birds, mammals, and other vertebrate species. The Upper Midwest Gap Analysis effort represents a broad landscape approach to the Upper Mississippi River and Great Lakes Basin ecosystem and will yield information critical to resource managers in this region.

Many federal land management agencies are moving toward ecosystem management objectives. The U.S. Fish and Wildlife Service is reorganizing its divisions into ecosystems based on watershed boundaries (James Fisher, USFWS, Winona, Minnesota, personal communication). The U.S. Forest Service is also shifting toward an ecosystem management approach, focusing on large-scale ecological regions that go beyond national forest boundaries (Thomas 1995).

Successful migration from the historical systems of monitoring and managing biological resources within agency boundaries to implementing large-scale, multitemporal assessments will require access to GIS and remote sensing technologies. Data collected via satellite, aerial photography, or videography will provide a historical record, and GIS can be used to automate, manage, and analyze the information collected. The Global Positioning System provides biologists and managers with the ability to collect accurate location information in the field which can be related and integrated with other spatial data using a GIS.

In an effort to define ecosystem management through an in-depth literature review, Grumbine (1993) found that one of the specific goals frequently endorsed was the incorporation of data on human use and occupancy within an ecosystem. GIS provides ample opportunity to integrate social and economic factors into the analysis of ecosystems. The technology is recognized as a viable and valid scientific tool and is being increasingly used to enhance the broad landscape-scale analysis required for ecosystem management.

LITERATURE CITED

Benyus, J. M., R. R. Buech, and M. D. Nelson. 1992. Wildlife in the Upper Great Lakes region: a community profile. USDA, North Central Forest Experiment Station. Research Paper NC-301.

Bohlen, H. D. 1989. The birds of Illinois. Indiana University Press, Indianapolis.

Bond, R. R. 1957. Ecological distribution of the upland forests of southern Wisconsin. Ecological Monographs 27:351-384.

Bull, E. L. 1975. Habitat utilization of the pileated woodpecker: Blue Mountains, Oregon. M.S. thesis, Oregon State University, Corvallis.

Cowardin, L. M., V. Carter, F. C. Golet, and E. T. LaRoe. 1979. Classification of wetlands and deepwater habitats of the United States. FWS/OBS-79.31.

D'Erchia, F. 1993. Metadata catalog of spatial data for the Upper Mississippi River System Long Term Resource Monitoring Program. U.S. Fish and Wildlife Service, Environmental Management Technical Center, Onalaska, Wisconsin, November 1993. EMTC 93-P009 (NTIS #PB94–130002).

D'Erchia, F., B. Drazkowski, and T. Owens (editors). 1993. Proceedings of the Upper Midwest Gap Analysis Workshop, January 12–14, 1993. Hosted by the National Biological Survey, Environmental Management Technical Center, Onalaska, Wisconsin. EMTC 93-P013, December 1993 (NTIS #PB94–138880).

Grumbine, R. E. 1993. What is ecosystem management? Conservation Biology 8:27–38.

Jacobson, T. R. 1993. Annotated bibliography of birds using the Upper Mississippi River. 55th Midwest Fish and Wildlife Conference, December 11–15, 1993, St. Louis, Missouri.

Korschgen, C. E. 1989. Riverine and deepwater habitats for diving ducks. Pages 157–180 in L. M. Smith, R. L. Pederson, and R. M. Kaminski, editors. Habitat management for migrating and wintering waterfowl in North America. Texas Technical University Press, Lubbock.

Laustrup, M. S., and C. D. Lowenberg. 1994. Development of a systemic land cover/land use database for the Upper Mississippi River System derived from Landsat Thematic Mapper satellite data. National Biological Survey, Environmental Management Technical Center, Onalaska, Wisconsin, May 1994. LTRMP 94-T001 (NTIS #PB94–186889).

Lillesand, T. M. 1994. Suggested strategies for satellite-assisted statewide land cover mapping in Wisconsin. *In* Proceedings of the American Society for Photogrammetry and Remote Sensing annual meeting, February 14–19, 1993, New Orleans, Louisiana.

Lowenberg, C. D. 1996. Geospatial application: geographic information system modeling procedures for the Upper Mississippi River System migratory bird pilot project. National Biological Service, Environmental Management Technical Center, Onalaska, Wisconsin.

McConville, D. R., T. W. Owens, and A. S. Redmond. 1996. Geospatial application: a geographic information system interface designed for use in river management. National Biological Service, Environmental Management Technical Center, Onalaska, Wisconsin.

Moorehouse, S. 1985. ARC/INFO: a geo-relational model for spatial information. Publication No. 640. Environmental Systems Research Institute, Redlands, California.

National Biological Service (NBS) and U.S. Fish and Wildlife Service (USFWS). 1996. The great river flyway: the management strategy for migratory birds on the Upper Mississippi River System. T. D'Erchia and S. Dreiband, editors. Upper Mississippi Science Center, La Crosse, Wisconsin; Environmental Management Technical Center, Onalaska, Wisconsin; and USFWS Region 3, Fort Snelling, Minnesota.

Owens, T. W., and Hop, K. 1995. Long Term Resource Monitoring Program standard operating procedures: photointerpretation. National Biological Service, Environmental Management Technical Center, Onalaska, Wisconsin, July 1995. LTRMP 95-P008-1.

Rogers, S. J. 1994. Preliminary evaluation of submersed macrophyte changes in the Upper Mississippi River. Lake and Reservoir Management 10(1):35–38.

Scott, J. M., F. Davis, B. Csuti, R. Noss, B. Butterfield, C. Groves, H. Anderson, S. Caicco, F. D'Erchia, T. C. Edwards, Jr., J. Ulliman, and R. G. Wright. 1993. Gap analysis: a geographic approach to protection of biological diversity. Wildlife Monograph 123:1–41.

Sparks, R. E., P. B. Bayley, S. L. Kohler, and L. L. Osborne. 1990. Disturbance and recovery of large floodplain rivers. Environmental Management 14(5):699–709.

Stow, D. A. 1993. The role of geographic information systems for landscape ecological studies. Chapter 2 *in* R. Haines-Young, D. R. Green, and S. H. Cousins, editors. Landscape ecology and GIS. Taylor and Francis, Bristol, Pennsylvania.

Thomas, J. W. 1995. Forest Service land management planning process. Presentation before the Subcommittee on Forests and Public Land Management Committee on Energy and Natural Resources, U.S. Senate. April 5.

U.S. Fish and Wildlife Service (USFWS). 1992. Operating plan for the Upper Mississippi River System Long Term Resource Monitoring Program. Environmental

Management Technical Center, Onalaska, Wisconsin, Revised September 1993. EMTC 91-P002. (NTIS #PB94-160199).

Upper Mississippi River Basin Association (UMRBA). 1982. Comprehensive master plan for the Upper Mississippi River System. 2 volumes. Upper Mississippi River Basin Commission, Bloomington, Minnesota.

White, B. M., and T. W. Owens. 1991. Geographic information system pilot project for the Upper Mississippi River System. U.S. Fish and Wildlife Service, National Ecology Research Center, Fort Collins, Colorado.

Chapter 11 **Population Viability Analysis: Adaptive Management for Threatened and Endangered Species**
Mark S. Boyce

Maintaining viable populations of all native animal and plant species is a central theme of ecosystem management.
—*Merrill Kaufmann and colleagues*

Population viability analysis (PVA) is the process of estimating the probability of persistence of a population for some arbitrary time into the future (Soulé 1987, Boyce 1992). PVA has its origins in the conservation biology movement; indeed, it is one of the keystone ideas of conservation biology (Wagner 1989). Performing a PVA entails compiling available biological data on a species and using these data as the basis for a simulation model for the population. The model then can be used to project future population trajectories from which one may estimate the probability that it will persist, for, say, one hundred years, or other related estimates, such as the probability of extinction or expected time extinction (Dennis et al. 1991).

The probability of extinction emerging from PVA would appear fundamental to establishing priorities for conservation based on guidelines that have been proposed for the categorization of species by International Union for Conservation of Nature and Natural Resources (IUCN; Mace and Lande 1991). In other applications, attempts are made to determine the minimum viable population (MVP) necessary to meet conservation objectives. Unfortunately, such applications are premature because we cannot reliably estimate the extinction probability for any species (Lebreton and Clobert 1991, Boyce 1992).

Yet, I believe that PVA can be enormously valuable if viewed in the context of adaptive management. The process of pulling together all available data and building a simulation model constitutes a synthesis of our current understanding of the population. Simulation models can be used to generate hypotheses of how we expect the system to respond to perturbations or management manipulations (Boyce 1991b). If this is followed by monitoring the consequence of management actions, PVA clearly is within the framework of adaptive management (Walters 1986).

LIMITATIONS

We do not know how many individuals are necessary to prevent population extinction, and there is insufficient empirical and theoretical basis on which to make such extrapolations. Small populations may remain viable over quite long periods of time. For example, the Socorro Island red tailed hawk *(Buteo jamaicensis socorroensis)* has persisted for well over forty years with a population of 20 ± 5 (Walter 1990). Although small populations gradually lose genetic variability due to drift, these populations may be important because geographic isolates often are genetically distinct (Lesica and Allendorf 1992). Small populations clearly are much more prone to extinction due to chance events, inbreeding depression, or an Allee effect (Soulé 1987, Dennis 1989). But we do not have sufficient knowledge of any of these processes to make defensible proclamations of a minimum viable population for any species.

Lack of Genetic Basis for Assigning MVP

It is common to place a target of an effective population size (N_e) of fifty for a short-term MVP, presumably based on the assumption that a 1 percent loss of heterozygosity is acceptable (Franklin 1980, Lacava and Hughes 1984). Then what often follows are calculations to estimate N_e based on data on sex ratio and mating system (Harris and Allendorf 1989).

Although N_e may give insight into the consequences of drift to loss of genetic diversity, there are numerous measures of effective population size depending upon the mechanisms affecting drift. Ewens (1990) reviews calculation of N_{ei} relative to inbreeding, N_{ev} for the variance in gene frequencies among subpopulations, N_{ee} to target the rate of loss of genetic variation, and N_{em} for mutation effective population size. Yet another measure, $N_e^{(meta)}$, defines the effective population size in a metapopulation experiencing repeated extinction-recolonization events (Gilpin and Hanski 1991). Each of these basic measures of N_e, then, is subject to adjustment for unequal sex ratio, age structure, and variable population size (Harris and Allendorf 1989). There is no sound basis for selecting one of these basic measures of N_e over another, yet, as Ewens (1990) shows, they can lead to radically different estimates of MVP.

Likewise, there is no solid basis for the often-cited rule of thumb that five hundred individuals may be sufficient to maintain long-term viability of a species. Unfortunately, the 50/500 rule does not have a sound genetic or demographic basis (Lande and Barrowclough 1987, Ewens 1990). And there is no theoretical or empirical justification for basing MVP on an estimate of N_e.

Yet the 50/500 rule is very popular. Clearly such simple guidelines would be very useful as we confront the global extinction crisis. It simply is not feasible to postpone conservation programs while we conduct a detailed PVA for each population of concern. Happily, there is some evidence that we may be able to come up with empirical justification for such rules of thumb. For example, studies of bighorn sheep *(Ovis canadensis;* Berger 1990) and birds on oceanic or habitat islands (Jones and Diamond 1976, Pimm et al. 1988, Soulé et al. 1988) consistently show that populations smaller than fifty are insufficient, and the probability of extinction is high for such small populations. Persistence of populations between fifty and two hundred is highly variable, whereas populations more than two hundred are unlikely to go extinct over the time frames of these studies.

Inferences from these few studies should be restricted to particular taxa, and we may require larger numbers for populations that vary more, for example, insect and small mammal populations (Thomas 1990, Tscharntke 1992). Also wise is Soulé's (1987) rule of thumb that one should always attempt to maintain three or more replicate populations. Further empirical evidence is urgently needed to justify the use of rules of thumb for MVP. But until such evidence becomes available, reliance on rules of thumb, such as the 50/500 rule, is arbitrary and capricious.

PVA Lacks Statistical Reliability

Performing a PVA almost always is severely constrained by the availability of data. Securing precise population estimates usually is difficult at best (Seber 1982, Richter and Söndgerath 1990), and for some populations it may not be possible to obtain estimates for many demographic parameters. Furthermore, any realistic population projection model requires knowledge of the population-regulating mechanism (Sinclair 1989), thus requiring estimates of a density-dependent function (McCullough 1990). But absolutely essential is that the model structure be defensible (Grant 1986, contra Ginzburg et al. 1990).

Assigning a hard number to a MVP is not possible (Thomas 1990). If the model is sufficiently complex to be realistic, we typically do not have enough data to do a conscientious job of estimating all of the population parameters. When these sampling errors are propagated by stochastic population projection, the confidence intervals surrounding some future probability of extinction are so large that the entire process becomes questionable (Lebreton and Clobert 1991). These problems are particularly severe for threatened and endangered species where the entire living population may be insufficient to yield acceptable levels of precision in estimates of such demographic parameters as survival.

SIMULATION APPROACHES

Problems with parameter estimation are indeed serious. But to my mind, the greatest value in PVA is not in the numbers generated by the models but in the identification of a model that formalizes our current understanding of the ecology of a particular population or species. Results from this model constitute testable hypotheses about the behavior of the system.

Software packages for PVA should be used cautiously because each case must be modeled uniquely. Models should be developed that capture the essential ecology of the system yet are as simple as possible to reduce the number of parameters that must be estimated. To illustrate the diversity of approaches that may be taken, I will review examples that use a variety of structures and modeling approaches.

The first PVA was Shaffer's (1983) model for grizzly bears *(Ursus arctos horribilis)* in the Greater Yellowstone Ecosystem. This was a stochastic simulation model that emphasized demographic structure. One approach is to explore the sensitivity of various variables in the model. By so doing, it became clear that adult survival was among the most sensitive elements in the model. PVA thereby offered valuable insight into the management of grizzly bears and contributed to the development of programs to enhance adult bear survival by minimizing conflicts with humans.

In contrast to the demographic approach used for grizzly bears, Foin and Brenchley-Jackson (1991) modeled critical habitat for the endangered light-footed clapper rail *(Rallus longirostris)* in southern California. Reliable demographic details for the rail were unavailable, and the only well-documented connection between the bird and its habitat was a linear relation between the biomass of Pacific cordgrass *(Spartina foliosa)* and the number of rails. But the salinity, transpiration, and soil moisture of salt marshes are essential to the development and maintenance of cordgrass stands used by rails.

For many species, focus on habitat in a PVA model is the correct focus, and I have chosen the light-footed clapper rail example because it does not dwell on the demographic structure of the population. Indeed, such details often are not known and may be best left out of the models. Eberhardt (1987) reviewed data from a number of large mammal populations to show that simple models without age structure could offer quite sufficient descriptions of population dynamics. For many threatened and endangered species, the most fundamental management programs will entail habitat management. Details of demographic structure for these species may be of little value.

The most extensive PVA program has been on the northern spotted owl *(Strix occidentalis caurina),* stimulated by the severe economic consequences of habitat protection for the subspecies (Boyce and Irwin 1990). The first effort included simple Leslie matrix projections with random elements (USDA Forest Service 1986, Marcot and Holthausen 1987). Use of an exponential growth model clearly was inadequate, and the prognosis for the owls was grim irrespective of future habitat management. A more realistic model by Lande (1988) includes density dependence via dispersal of young owls. This was subsequently expanded into a dynamic model (Lamberson et al. 1992) and then interfaced with explicit landscapes imported on a geographic information system (McKelvey et al. 1992). Lande's hypothesis regarding population regulation via juvenile dispersal remains untested, but it forms the basis for many of the Interagency Scientific Committee's management recommendations for the northern spotted owl (Thomas et al. 1990).

ADAPTIVE MANAGEMENT

PVA models by themselves usually are weak and cannot be counted on to provide reliable population projections. But when combined with an iterative process of model improvement and validation, the model can provide a progressively more robust understanding of the dynamics of a species and its habitat; and a model developed in such a way can be a powerful tool for management.

How can PVA be incorporated into adaptive management protocols? Active adaptive management proposes application of different management tactics in time and space, essentially as experiments, to develop a better understanding of the behavior of the system (Walters and Holling 1990). For endangered species applications, it may be possible to implement various management strategies in spatially separated subpopulations. Active management must be part of such a program, and may encompass a variety of activities, such as habitat manipulation, predator or disease control, manipulation of potential competitors, winter provision of food, transplanting individuals from other subpopulations to sustain genetic variation, and supplementation of population with releases of captive stock. Monitoring of the genetic and population consequences of such manipulations then provides data to validate and/or refine the PVA model.

Management of grizzly bears in the Greater Yellowstone Ecosystem has proceeded according to an adaptive management protocol. High sensitivity of population growth rate to adult survival suggested the importance of minimizing adult mortality factors. Aggressive programs to eliminate bear/human conflicts focused on areas identified as mortality sinks—localities where repeated bear mortalities had been documented. As prescribed by an adaptive management program, after the recovery program had been implemented and additional data were obtained, Shaffer's model was updated (Suchy et al. 1985). Preliminary evidence suggests that the program was highly successful. Indeed, federal officials recently have entertained the possibility of delisting grizzly bears and returning management to respective state and federal agencies (Boyce 1991a). However, extensive wildfires during the summer of 1988 altered habitat for the bears, and further updates to the bear model will need to be incorporated once the demographic response to the fires has been documented.

Another adaptive management program has been proposed for the management of endangered populations of *Banksia cumeata* in Western Australia. Based upon their PVA modeling, Burgman and Lamont (1992) recommended watering seedlings in several subpopulations to enhance seedling survival. Such programs require careful monitoring because watering or other forms of "enrichment" can have community-level effects that could be counter productive (Rosenzweig 1971). For example, it is conceivable that competing species or herbivores might respond more vigorously to watering than the target species.

For the northern spotted owl, the Interagency Scientific Committee (ISC) explicitly acknowledged the importance of adaptive management approaches for evaluating and updating their conservation strategy, posed as an appendix in the ISC report (Thomas et al. 1990). Adaptive management would require implementation of various timber harvest programs and associated landscape manipulations and then documentation of the consequences for spotted owl populations. Thus far, no such programs have been implemented because litigation has interfered with the ability of management agencies to develop timber harvests.

For several years, the Captive Breeding Specialists Group (CBSG) of the IUCN has been organizing Population and Habitat Viability Analysis (PHVA) workshops for various threatened and endangered species. These have been enormously successful at bringing together available data on a species, identifying possible structures for a PVA model, and stimulating agency coordination for conservation programs. One cannot place much stock in MVP estimates that emerge from these exercises, but if they help provide a structure that will encourage adaptive management approaches, they perform an exceedingly valuable function.

"Adaptive management is learning by doing" (Lee and Lawrence 1986). But agency restrictions may severely limit our ability to actually do management with threatened and endangered species. Naturally, any programs that might pose a risk to a threatened or endangered species will meet strong resistance from agencies charged with protecting the species. Yet creative manipulations may be allowed if only they could be viewed as enhancing conditions for the species of concern.

In a legal context, PVA probably will face many challenges because of omnipresent biological uncertainty. Given the statistical weakness of population parameter estimates and our inability to generate robust population projections, any PVA will be open to question even though the PVA constitutes our best statement of the expected behavior of a population. Such uncertainty recently was used in court to challenge the proposed adoption of the Interagency Scientific Committee's conservation strategy for the northern spotted owl by the USDA Forest Service. Although Lee and Lawrence (1986) suggest that biological uncertainty may often frustrate attempts to manage by adaptive management, it is through adaptive management that we can hope to resolve some of the uncertainty associated with PVA. It is the best we can do, and we know of no better way to gain "reliable knowledge" about managing our natural resources (Romesburg 1981).

Population viability analysis (PVA) entails evaluation of data and models for a population to anticipate the likelihood that a population will persist for some arbitrarily chosen time into the future. Models vary depending upon the availability of data and the particular ecology and life history of the organism. Unfortunately, we have insufficient data to validate PVA models for most endangered species. Seldom, if ever, do replications exist, and small sample sizes typically result in projections bearing large confidence intervals.

A great danger exists that resource managers may lend too much credence to a model when they may not fully understand its limitations.

There is much to be gained by developing a stronger understanding of the system by modeling—and much to be lost by shirking modeling for fear of its being misinterpreted. PVA as a process can be an indispensable tool in conservation, and it involves much more than attempts to calculate statistically feeble estimates of minimum viable populations or probabilities of extinction. PVA entails the process of synthesizing information about a species or population and developing the best possible model for the species given the information available. When done properly, this involves working closely with natural resource managers to develop a long-term iterative process of modeling and research that can reveal more about how best to manage a species. Done properly, PVA is a variation on Walter and Holling's notion of adaptive management.

Adaptive management proposes application of different management tactics in time and space to develop a better understanding of the behavior of the system. For application to endangered species problems, implementation of various management strategies may be attempted in spatially separated subpopulations. Active manipulation must be part of such a program. Monitoring of the genetic and population consequences of such manipulations then provides data to validate and/or refine the PVA model.

I thank R. A. Lancia, E. H. Merrill, C. Nations, and M. L. Shaffer for comments and discussion. I received support from the National Council for Air and Stream Improvement and from the Wyoming Water Research Center. This chapter is adapted from a paper previously published in the Transactions of the North American Wildlife and Natural Resources Conference.

LITERATURE CITED

Berger, J. 1990. Persistence of different-sized populations: an empirical assessment of rapid extinctions in bighorn sheep. Conservation Biology 4:91–96.

Boyce, M. S. 1991a. Natural regulation or the control of nature? Pages 183–208 *in* R. B. Keiter and M. S. Boyce, editors. The Greater Yellowstone Ecosystem: redefining America's wilderness heritage. Yale University Press, New Haven, Connecticut.

———. 1991b. Simulation modelling and mathematics in wildlife management and conservation. Pages 116–119 *in* N. Maruyama, editor. Wildlife conservation: present trends and perspectives for the 21st century. Japan Wildlife Research Center, Tokyo.

―――. 1992. Population viability analysis. Annual Reviews of Ecology and Systematics 23:481–506.

Boyce, M. S., and L. L. Irwin. 1990. Viable populations of spotted owls for management of old growth forests in the Pacific Northwest. Pages 133–135 *in* R. S. Mitchell, C. J. Sheviak, and D. J. Leopold, editors. Ecosystem management: rare species and significant habitats. Bulletin 471. New York State Museum, Albany.

Burgman, M. A., and B. B. Lamont. 1992. A stochastic model for the viability of *Banksia cumeata* populations: environmental, demographic and genetic effect. Journal of Applied Ecology 29:719–727.

Dennis, B., P. L. Munholland, and J. M. Scott. 1991. Estimation of growth and extinction parameters for endangered species. Ecological Monographs 61:115–143.

Eberhardt, L. L. 1987. Population projections from simple models. Journal of Applied Ecology 24:103–118.

Ewens, W. J. 1990. The minimum viable population size as a genetic and a demographic concept. Pages 307–316 *in* J. Adams, D. A. Lam, A. I. Hermalin, and P. E. Smouse, editors. Convergent issues in genetics and demography. Oxford University Press, Oxford, England.

Foin, T. C., and J. L. Brenchley-Jackson. 1991. Simulation model evaluation of potential recovery of endangered light-footed clapper rail populations. Biological Conservation 58:123–148.

Franklin, I. R. 1980. Evolutionary change in small populations. Pages 135–149 *in* M. E. Soulé and B. A. Wilcox, editors. Conservation biology. Sinauer, Sunderland, Massachusetts.

Gilpin, M. E., and I. Hanski. 1991. Metapopulation dynamics. Academic Press, London.

Ginzburg, L. R., S. Ferson, and H. R. Akcakaya. 1990. Reconstructibility of density dependence and the conservative assessment of extinction risks. Conservation Biology 4:63–70.

Grant, W. E. 1986. Systems analysis and simulation in wildlife and fisheries science. John Wiley, New York, New York.

Harris, R. B., and F. W. Allendorf. 1989. Genetically effective population size of large mammals: an assessment of estimators. Conservation Biology 3:181–191.

Jones, H. L., and J. M. Diamond. 1976. Short-time-base studies of turnover in breeding bird populations on the California Channel Islands. Condor 78:526–549.

Lacava, J., and J. Hughes. 1984. Determining minimum viable population levels. Wildlife Society Bulletin 12:370–376.

Lamberson, R. H., R. McKelvey, B. R. Noon, and C. Voss. 1992. A dynamic analysis of northern spotted owl viability in a fragmented forest landscape. Conservation Biology 6:505–512.

Lande, R. 1988. Demographic models of the northern spotted owl *(Strix occidentalis caurina)*. Oecologia 75:601–607.

Lande, R., and G. F. Barrowclough. 1987. Effective population size, genetic variation, and their use in population management. Pages 87–123 *in* M. E. Soulé, editor.

Viable populations for conservation. Cambridge University Press, Cambridge, England.

Lebreton, J.-D., and J. Clobert. 1991. Bird population dynamics, management, and conservation: the role of mathematical modelling. Pages 105 125 in C. M. Perrins, J.-D. Lebreton, and G. J. M. Hirons, editors. Bird population studies: relevance to conservation and management. Oxford University Press, Oxford, England.

Lee, K. N., and J. Lawrence. 1986. Adaptive management: learning from the Columbia River Basin Fish and Wildlife Program. Environmental Law 16:431–460.

Lesica, P., and F. W. Allendorf. 1992. Are small populations of plants worth preserving? Conservation Biology 6:135–139.

Mace, G. M., and R. Lande. 1991. Assessing extinction threats: toward a reevaluation of IUCN threatened species categories. Conservation Biology 5:148–157.

Marcot, B. G., and R. Holthausen. 1987. Analyzing population viability of the spotted owl in the Pacific Northwest. Transactions of the North American Wildlife and Natural Resources Conference 52:333–347.

McCullough, D. A. 1990. Detecting density dependence: filtering the baby from the bathwater. Transactions of the North American Wildlife and Natural Resources Conference 55:534–543.

McKelvey, K., B. R. Noon, and R. Lamberson. 1992. Conservation planning for species occupying fragmented landscapes: the case of the northern spotted owl. Pages 424–450 in P. M. Kareiva, J. G. Kingsolver, and R. B. Huey, editors. Biotic interactions and global change. Sinauer, Sunderland, Massachusetts.

Pimm, S. L., H. L. Jones, and J. Diamond. 1988. On the risk of extinction. American Naturalist 132:757–785.

Richter, O., and D. Söndgerath. 1990. Parameter estimation in ecology: the link between data and models. VCH Verlagsgesellschaft mbH, Weinheim, Germany.

Romesburg, H. C. 1981. Wildlife science: gaining reliable knowledge. Journal of Wildlife Management 45:293–313.

Rosenzweig, M. L. 1971. Paradox of enrichment: destabilization of exploitation ecosystems in ecological time. Science 171:385–387.

Seber, G. A. F. 1982. The estimation of animal abundance. 2d edition. Griffen, London.

Shaffer, M. L. 1983. Determining minimum viable population sizes for the grizzly bear. International Conference on Bear Research and Management 5:133–139.

Sinclair, A. R. E. 1989. The regulation of animal populations. Pages 197–241 in J. M. Cherrett, editor. Ecological concepts. Blackwell, Oxford, England.

Soulé, M. E., D. T. Bolger, A. C. Alberts, J. Wright, M. Sorice, and S. Hill. 1988. Reconstructed dynamics of rapid extinctions of chaparral-requiring birds in urban habitat islands. Conservation Biology 2:75–92.

Suchy, W., L. L. McDonald, M. D. Strickland, and S. H. Anderson. 1985. New estimates of minimum viable population size for the grizzly bears of the Yellowstone ecosystem. Wildlife Society Bulletin 13:223–228.

Thomas, C. D. 1990. What do real population dynamics tell us about minimum viable population sizes? Conservation Biology 4:324–327.

Thomas, J. W., E. D. Forsman, J. B. Lint, E. C. Meslow, B. R. Noon, and J. Verner. 1990. A conservation strategy for the northern spotted owl. U.S. Government Printing Office, Portland, Oregon.

Tscharntke, T. 1992. Fragmentation of *Phragmites* habitats, minimum viable population size, habitat suitability, and local extinction of moths, midges, flies, aphids, and birds. Conservation Biology 6:530–536.

USDA Forest Service. 1986. Draft supplement to the environmental impact statement for an amendment to the Pacific Northwest regional guide. Vols. 1, 2. USDA Forest Service, Portland, Oregon.

Wagner, F. H. 1989. American wildlife management at the crossroads. Wildlife Society Bulletin 17:354–360.

Walter, H. S. 1990. Small viable population: The red-tailed hawk of Socorro Island. Conservation Biology 4:441–443.

Walters, C. J. 1986. Adaptive management of renewable resources. Macmillan, New York, New York.

Walters, C. J., and C. S. Holling. 1990. Large-scale management experiments and learning by doing. Ecology 71:2060–2068.

MAKING IT HAPPEN

1911

1984

Ecosystem management is more about people than anything else.
—Hal Salwasser

Overleaf: Most of the Chequamegon and Nicolet National Forests in northern Wisconsin were acquired subsequent to extensive clearcutting, fires, and futile attempts at agriculture in the late 1800s and early 1900s. This pair of photographs illustrates the reforestation of an area near Trout Lake that was clearcut in 1911 (courtesy of Wisconsin Department of Natural Resources).

Chapter 12 Ecosystem Protection and Restoration: The Core of Ecosystem Management

Reed F. Noss and J. Michael Scott

*Ecosystem management defines a paradigm that weaves
biophysical and social threads into a tapestry of beauty,
health, and sustainability. It embraces both social and
ecological dynamics in a flexible and adaptive process.
Ecosystem management celebrates the wisdom of both our
minds and hearts, and lights our path to the future. Why
ecosystem management now? Why wait?!*
—*Zane J. Cornett*

Ecosystem management promises much. As our epigraph from a Forest Service report suggests, at least some land management professionals seem to see a light at the end of the tunnel through which they have been crawling for years in search of better land stewardship. The seemingly insoluble conflicts between endangered species protection and commodity extraction, the court battles, the political turmoil, communities pitted against one another, the lack of unified leadership and direction, all showed some signs by early 1994 of ameliorating, though polarization has since increased. Maybe humans and nature *can* live together. Perhaps we can manage the land in a way that is kinder and gentler, and that better mimics the ebbs and flows of ecological processes. Maybe we can find the middle ground between preservation and exploitation in a landscape without lines, a landscape where nature reserves are no longer necessary because our management is no longer destructive.

Or can we? Have we learned so well from our past mistakes that we can now be confident about our abilities to manage the land wisely? An ecosystem perspective in land management is sorely needed, and recent developments in agency policies and practices suggest that such a perspective is slowly gaining ground. Piecemeal management, which considers each resource, each site, each species, and each project in isolation from its ecological context, is still with us but is giving way to strategies that emphasize maintenance of biodiversity and ecological integrity as essential to future productivity. The spatial scale of planning is shifting from the site to the landscape. Piecemeal management is no longer defensible scientifically or politically. The sciences of ecology and conservation biology have had a profound impact on the way land managers think about and execute their responsibilities. People *are* thinking bigger.

Yet despite these auspicious trends, we believe it is premature to claim that our new ways of managing landscapes will ultimately prove successful. Ecosystem management is an experiment and in many cases a poorly controlled experiment. The words of ecologist Frank Egler (1977)—"ecosystems are not only more complex than we think, but more complex than we *can* think"—ring just as true today as when Egler first uttered them. Or as a Koyukon elder repeatedly lectured anthropologist Richard Nelson (1993), "Each animal knows way more than you do." In the face of our scientific uncertainty and profound ignorance about the effects of our activities on the land and its inhabitants, we believe the key principle is prudence. It is too early to throw away the traditional conservation practices of protecting species and ecosystems in reserves. Instead, those practices must be augmented by better management of the seminatural matrix, where, as Brown (1988) noted, the vast bulk of biodiversity is located.

In this chapter, we develop the notion that the core of ecosystem management is ecosystem protection and restoration. After describing ecosystem diversity as a higher level of biodiversity, we review the tremendous losses of ecosystem diversity that have occurred in the United States as agriculture, resource extraction, urban development, and other human activities have transformed natural ecosystems into highly altered systems. Then we discuss some ecosystem protection approaches, beginning with a general justification for the "coarse filter" (ecosystem level) strategy, followed by a brief review of current and proposed approaches to ecosystem protection in the United States. We conclude with some specific recommendations for conserving ecosystems, including protection and restoration of endangered ecosystems and management to maintain ecological integrity in ecosystem types that are still common.

ECOSYSTEM DIVERSITY

The biodiversity concept has revolutionized the way conservationists and land managers think about their duties and about the world in which they work. Only recently has it become widely accepted that the 1.5 million species formally described by scientists represent only a tiny fraction of the 10 million to 100 million species that occupy the earth (Wilson 1988, 1992). And while destruction of natural areas and extinction of species have concerned scientists and some laypersons since the eighteenth century (Grove 1992), the magnitude of the losses was, until recently, thought to be small compared with the great mass extinctions of the geologic past. We now know that we are losing at least 10,000 species each year worldwide and that the current extinction event is likely to rival all but the largest of the past (Wilson 1992).

But the biodiversity crisis involves more than the loss of species. Almost all definitions of biodiversity recognize the importance of natural variety at several levels of biological organization—for example, genetic, population-species, community-ecosystem, and landscape levels (Noss 1990a). An eco-system is a biotic community (an assemblage of organisms from many taxonomic groups living together) plus the abiotic environment. All ecosystems are open systems, exchanging energy and materials with other systems around them. Furthermore, we have learned that species usually respond in an individualistic way to environmental gradients in space and time (Gleason 1926, Whittaker 1960, Davis 1981). These facts have led some ecologists to conclude that the boundaries we draw around ecosystems are wholly arbitrary. However, discontinuities do occur in environmental gradients, and these discontinuities are reflected in the distributions of plants and animals. Boundaries between ecosystems, although they are often diffuse or ecotonal, can be recognized at one scale or another and ecosystems can be classified by their distinctive vegetation or other prominent characteristics. Hence, ecosystems can be mapped.

Ecosystem diversity, then, is the variety of ecosystem types in a defined area. This diversity in large part subsumes the diversity of species. Representing all species in protected areas is one of the best-accepted of all conservation goals worldwide. As noted by Margules et al. (1988), ''a prerequisite for preserving maximum biological diversity in a given biological domain is to identify a reserve network which includes every possible species.'' However, the distributions of the vast majority of species, partic ularly invertebrates, fungi, bacteria, and other lesser known taxonomic groups, in most regions are poorly known. Thus, vegetation, physical habitats, or some other ecosystem typology is used to provide a surrogate for species diversity—the so-called coarse filter (Noss 1987, Hunter et al. 1988, Hunter 1991, Scott et al. 1993).

By protecting examples of ecosystem types, we presumably protect the bulk of species. This is not to suggest that an ecosystem approach will meet all our conservation goals; a comprehensive conservation strategy must be hierarchical and address multiple levels of organization, from genes to landscapes (Noss 1990a). But the most efficient strategy is an iterative one that begins with the coarsest filter by protecting multiple examples of all major vegetation or landscape types, then moves progressively to finer and finer scales by identifying species, populations, and genetic variants that are not protected by the coarse filters. Although logical, this iterative approach will not be easy or cheap. Indeed, only recently have we begun a systematic effort to map the actual vegetation, as opposed to "potential natural vegetation" (Kuchler 1966), of the United States at even a coarse scale (Scott et al. 1993).

Conservation at the ecosystem level involves more than setting aside examples of defined ecosystem types—we must also consider ecological processes. Ecosystems remain viable only when their processes—nutrient cycling, energy flow, hydrology, disturbance-recovery regimes, predator-prey dynamics, and so on—continue to operate within an acceptable range of variation. In fragmented landscapes and other areas with a long history of human occupation, human management is often necessary to substitute for natural processes that have been disrupted. Prescribed burning, for example, is a useful management tool for vegetation types adapted to frequent fire. In fragmented landscapes with abundant firebreaks, lightning fires may not occur often enough to maintain the characteristic vegetation. In other cases, burning by aboriginal humans maintained pyrogenic vegetation over thousands of years (Pyne 1982). But before we worry about the specific management prescriptions necessary to keep ecosystems healthy, we must first assure that all types of ecosystems and the species they contain are represented adequately in a network of protected areas. We have to be sure we've saved all the pieces before we begin playing with those pieces. As we will point out later, the current level of ecosystem representation in the United States is quite poor.

ECOSYSTEM DECLINES

Loss of biodiversity at the ecosystem level occurs when distinct habitats, species assemblages, and natural processes are diminished or degraded in quality. The destruction of tropical forests is probably the first ecosystem-level conservation issue to achieve worldwide attention. However, tropical forests are hardly the only ecosystems in trouble. Most of the temperate zone has already been altered much more than the tropics. Some temperate habitats, such as freshwaters in California (Moyle and Williams 1990) and old-growth forests in the Pacific Northwest (Norse 1990), are being destroyed faster than most tropical rainforests and stand to lose as great a proportion of their species. Areas with a long history of intense use by humans and their domestic animals have suffered profound but largely unknowable losses of biodiversity at all levels of organization (Gore 1992).

Losses of biodiversity at the ecosystem level may be either quantitative or qualitative. Quantitative losses, as when a prairie is plowed to create a corn-field, are the most obvious and can be measured by a decline in areal extent of a mapped ecosystem type. On the other hand, qualitative losses involve a change or degradation in the structure, function, or composition of an eco-system (see Franklin et al. 1981, Noss 1990a). For example, a marsh may be invaded by purple loosestrife *(Lythrum salicaria)*, or a longleaf pine *(Pinus palustris)* forest may be invaded by oaks and other hardwoods in the absence of fire. Changes in the relative abundances of plant species in a community may in turn lead to changes in physiognomy, animal populations, and such ecological processes as nutrient cycling. Invasions by exotics have led to major disruptions of ecological processes and losses of native biodiversity (Mooney and Drake 1986, Vitousek 1988). Qualitative changes may be ex-pressed quantitatively—for instance, by reporting that 99% of the sagebrush steppe has been affected by livestock grazing (West 1996). In many cases, qualitative changes in the structure, function, or composition of an ecosys-tem—though they may be more subtle to the untrained eye and may take place over many years—are every bit as severe as outright habitat conversion.

A recent review of ecosystem declines in the United States (Noss et al. 1995) confirmed that considerable biotic impoverishment at the ecosystem level has occurred since European settlement. The study was based on a literature review and survey of conservation agencies (particularly the state natural heritage programs) and professionals. One limitation to assessing the status of ecosystems nationally is that no standardized and well-accepted national classification or map of ecosystems currently exists. Therefore, the review considered ecosystems defined at various spatial scales, including veg-etation types, plant associations, natural communities, and other types defined by floristics, structure, age, geography, condition, and other ecologically rel-evant factors. For each ecosystem type the most reliable and recent estimates of loss or degradation were cited. If or when a national classification is completed, ecosystem types reviewed in this study can be cross-walked within a consistent framework.

The results of our review of ecosystem declines (Noss et al. 1995; see also Noss and Peters 1995) are presented in table 12.1 by listing ecosystems as critically endangered (a greater than 98% decline), endangered (85–98% decline), and threatened (70–84% decline). Grasslands and savannas stand out in the critically endangered category, but in the larger list of threatened and endangered ecosystems wetlands and forests appear just as imperiled. Aquatic and near-shore marine systems, though under-represented in terms of number of "types" imperiled, largely because of sampling problems and other biases (for example, state heritage programs rarely employ aquatic ecologists, and most people have a hard time recognizing degradation in aquatic systems so long as those systems remain wet), have been seriously degraded throughout the nation. For instance, 81% of fish communities nationwide are adversely affected by human activities (Judy et al. 1984), and 98% of streams are degraded enough to be unworthy of federal designation as wild or scenic rivers (Benke 1990). Losses of all kinds of ecosystems have been most pronounced in the South, Northeast, Midwest, and California, but no region of the U.S. has escaped damage (Noss et al. 1995, Noss and Peters 1995).

The most significant finding of this study was that much more biodiversity at the ecosystem level has been lost than is generally recognized in environmental policy debates. The public and Congress, for example, are generally familiar with endangered species issues (at least with regard to charismatic fauna) but seem to have little awareness that entire assemblages of species and their habitats have already been destroyed over much of the country. We suggest that discussions of ecosystem management options include a historical perspective and begin with an explicit recognition of what has already been lost. Otherwise, a false impression may be given that we are starting with a clean slate and have complete, healthy ecosystems to begin tinkering with. The old-growth debate in the Pacific Northwest often seems to reflect this illusion. When 90% of an ecosystem has been eliminated (Norse 1990), there is little rational basis for continued exploitation.

The consequences of ecosystem conversion and degradation extend beyond the loss of biodiversity at the ecosystem level. Indeed, habitat destruction is widely agreed to be the major cause of extinction and endangerment of species today (Ehrlich and Ehrlich 1981, Diamond 1984, Ehrlich and Wilson 1991, Soulé 1991). A list of more than two hundred extinct and possibly extinct plant species in the U.S. and Canada (Russell and Morse 1992) is dominated by narrow endemics. Presumably, these species went extinct when their localized habitats were modified by human activities. Eighty-one percent

Table 12.1 Critically endangered, endangered, and threatened ecosystems of the United States

Critically endangered (> 98% decline)

Old-growth and other virgin stands in the eastern deciduous forest biome

Spruce-fir forest in the southern Appalachians

Red and white pine forests (mature and old-growth) in Michigan

Longleaf pine forests and savannas in the Southeastern Coastal Plain

Pine rockland habitat in south Florida

Loblolly/shortleaf pine–hardwood forests in the West Gulf Coastal Plain

Arundinaria gigantea canebrakes in the Southeast

Tallgrass prairie east of the Missouri River and on mesic sites across the range

Bluegrass savanna–woodland and prairies in Kentucky

Black belt prairies in Alabama and Mississippi and Jackson prairie in Mississippi

Ungrazed dry prairie in Florida

Oak savanna in the Midwest

Wet and mesic coastal prairies in Louisiana

Lakeplain wet prairie in Michigan

Sedge meadows in Wisconsin

Hempstead plains grasslands on Long Island, New York

Serpentine barrens, maritime heathland, and pitch pine–heath barrens in New York

Coastal rocky headlands in New Hampshire

Prairies (all types) and oak savannas in Willamette Valley and the foothills of Coast Range, Oregon

Palouse prairie (Washington, Oregon, and Idaho, plus similar communities in Montana)

Native grasslands (all types) in California

Alkali sink scrub in southern California

Coastal strand in southern California

Needlegrass steppe in California

Ungrazed sagebrush steppe in the intermountain West

(continued)

Table 12.1 (cont.)

Critically endangered (cont.)

Basin big sagebrush in the Snake River Plain of Idaho

Atlantic white cedar stands in the Great Dismal Swamp of Virginia and North Carolina, and possibly across the entire range

Streams in the Mississippi Alluvial Plain

Endangered (85–98% decline)

Old-growth and other virgin forests in regions and states other than those listed above, except Alaska

Mesic limestone forest and barrier island beaches in Maryland

Coastal plain Atlantic white cedar swamp, maritime oak–holly forest, maritime red cedar forest, marl fen, marl pond shore, and oak openings in New York

Coastal heathland in southern New England and Long Island

Pine-oak-heath sandplain woods and lake sand beach in Vermont

Floodplain forests in New Hampshire

Red spruce forests in central Appalachians (West Virginia)

Upland hardwoods in the Coastal Plain of Tennessee

Lowland forest in southeastern Missouri

High-quality oak-hickory forest on the Cumberland Plateau and Highland Rim of Tennessee

Limestone cedar glades in Tennessee

Wet longleaf pine savanna and eastern upland longleaf pine forest in Louisiana

Calcareous prairie, Fleming glade, shortleaf pine/oak-hickory forest, mixed hardwood–loblolly pine forest, eastern xeric sandhill woodland, and stream terrace sandy woodland/savanna in Louisiana

Slash pine forests in southwestern Florida

Red and white pine forests in Minnesota

Coastal redwood forests (California)

Old-growth ponderosa pine forests in northern Rocky Mountains, the intermountain West, and the eastern Cascades Mountains

Riparian forests in California, Arizona, New Mexico

Coastal sage scrub (especially maritime) and coastal mixed chaparral in southern California

Table 12.1 (cont.)

Endangered (cont.)

Dry forest on main islands of Hawaii

Native habitats of all kinds in lower delta of Rio Grande River, Texas

Tallgrass prairie (all types combined)

Native shrub and grassland steppe in Oregon and Washington

Low-elevation grasslands in Montana

Gulf coast pitcher plant bogs

Pocosins (evergreen shrub bogs) and ultramafic soligenous wetlands in Virginia

Mountain bogs (southern Appalachian bogs and swamp forest–bog complex) in Tennessee and North Carolina

Upland wetlands on the Highland Rim of Tennessee

Saline wetlands in eastern Nebraska

Wetlands (all types combined) in south-central Nebraska, Missouri, Iowa, Illinois, Indiana, Ohio, California

Marshes in the Carson-Truckee area of western Nevada

Low-elevation wetlands in Idaho

Woody hardwood draws, glacial pothole ponds, and peatlands in Montana

Vernal pools in Central Valley and southern California

Marshes in Coos Bay area of Oregon

Freshwater marsh and coastal salt marsh in Southern California

Seasonal wetlands of San Francisco Bay, California

Large streams and rivers in all major regions

Aquatic mussel beds in Tennessee

Submersed aquatic vegetation in Chesapeake Bay, Maryland and Virginia

Mangrove swamps and salt marsh along the Indian River Lagoon, Florida

Seagrass meadows in Galveston Bay, Texas

(continued)

Table 12.1 (cont.)

Threatened (70–84% decline)

Riparian forests nationwide (other than in regions listed above), including southern bottomland hardwood forests

Xeric habitats (scrub, scrubby flatwoods, sandhills) on Lake Wales Ridge, Florida

Tropical hardwood hammocks on central Florida keys

Northern hardwood forest, aspen parkland, and jack pine forests in Minnesota

Saline prairie, western upland longleaf pine forest, live oak–pine–magnolia forest, western xeric sandhill woodland, slash pine–pond cypress/hardwood forest, wet and mesic spruce pine–hardwood flatwoods, wet mixed hardwood–loblolly pine flatwoods, and flatwoods ponds in Louisiana

Alvar grassland, calcareous pavement barrens, dwarf pine ridges, mountain spruce–fir forest, inland Atlantic white cedar swamp, freshwater tidal swamp, inland salt marsh, patterned peatland, perched bog, pitch pine–blueberry peat swamp, coastal plain poor fens, rich graminoid fen, rich sloping fen, and riverside ice meadow in New York

Maritime-like forests in Clearwater Basin of Idaho

Woodland and chaparral on Santa Catalina Island

Southern tamarack swamp in Michigan

Wetlands (all kinds) in Connecticut, Maryland, Arkansas, Kentucky

Marshes in Puget Sound region, Washington

Cienegas (marshes) in Arizona

Coastal wetlands in California

Note: Decline refers to outright destruction, *conversion* to other land uses or significant degradation of ecological structure, function, or composition since European settlement. Estimates (see references in Noss et al. 1994) based on quantitative studies and qualitative assessments. Because coverage is uneven among states, the regional status of ecosystems listed should be evaluated. Ecosystems at high risk of further decline should be the highest priorities for conservation.

of the federally listed threatened and endangered plant species in the U.S. are threatened by agriculture, mining, urban and suburban development, introduction of exotic species, and other human activities (Cook and Dixon 1989). A review of factors threatening biodiversity in rivers and streams concluded that habitat degradation and species introductions are the leading causes of species declines (Allan and Flecker 1993). Habitat degradation was a contributing factor in the extinction of at least 73% of the twenty-seven species and thirteen subspecies of freshwater fishes lost from North America over the last century (Miller et al. 1989) and is the major factor threatening freshwater fishes today, followed closely by introductions of alien species (Williams et al. 1989). The decline of neotropical migrant birds is closely linked to habitat destruction, particularly of forests, but is a complex problem because forests are being lost and fragmented on the wintering grounds in the neotropics, on the breeding grounds in temperate regions, and in migratory stopover habitat in between (Robbins et al. 1989).

Why have we allowed our native ecosystems to become so degraded? In most cases, of course, people did not destroy or degrade ecosystems out of spite. The ecosystems most endangered today are typically those at low elevations, near the coast or major rivers, with fertile soils, benign climate, easy terrain, or abundant natural resources—in other words, the places where people settled in greatest numbers. The farmers who originally cleared the virgin forests, plowed the prairies, and drained the wetlands were simply trying to feed their families. Their exploitation was not particularly efficient or sustainable, but usually only because they lacked the knowledge or technology for better practices. Later, market forces fueled by increases in the human population and consumption of resources led to greater and greater demands for natural resources. Logging, livestock grazing, mining, and other commodity developments have been responsible for the degradation of many areas that were not converted to agriculture. The political lobbies associated with these industries often demand levels of exploitation that greatly exceed the real needs of society (Noss and Cooperrider 1994). Recently, residential and commercial development has assumed a greater role in habitat destruction, particularly in the Sun Belt and other areas with rapid human population growth.

Yet in all these cases where legitimate human needs have driven habitat destruction, much could have been done to reduce the damage. Although a steadily increasing demand for natural resources is usually taken for granted, this belief assumes continued waste, inefficiency, and consumption of resources far in excess of our needs. In opposition to this assumption is the idea that conservation, recycling, and reuse of resources can reduce demand for raw materials dramatically, thus easing the pressures on natural ecosystems. For example, credible estimates suggest that use of wood products in the United States can be reduced by 50% through aggressive recycling (Postel and Ryan 1991). By comparison, the national forests produce only 14% of the timber cut in the U.S. (Waddell et al. 1989). Arguments that public lands must be exploited to meet society's need for resources are increasingly being questioned. Management options that do not pit jobs versus the environment and that allow for constructive partnerships between public and private sectors are greatly in need. In the next section we review some strategies for conserving ecosystems, but note here that none of the goals of these strategies can be attained without fundamental changes in the way we exploit and consume natural resources.

ECOSYSTEM PROTECTION AND RESTORATION STRATEGIES

We have already said that the coarse-filter strategy seeks to protect the vast majority of species efficiently by maintaining viable examples of the ecosystems in which they live (Noss 1987, Hunter et al. 1988, Hunter 1991). This approach is entirely consistent with the first stated goal of the Endangered Species Act, "to provide a means whereby the ecosystems upon which endangered species and threatened species depend may be conserved." However, Congress provided no explicit directives to federal agencies on how to implement this goal. Today the need for conservation at the ecosystem level is being increasingly emphasized as conflicts over individual endangered species have escalated.

Representation

Ecosystem management is experimental, and all good experiments are controlled. Thus, it makes sense to set aside examples of ecosystems to serve as benchmarks or control areas for comparison to manipulated sites which are the experimental treatments. Only then can effects of management practices be separated from natural processes. Thus, the conservation goal of representing all ecosystem types in reserves serves a new purpose.

Representation is one of the best-established goals of conservation. But most strategies for representing ecosystems have had to be content with small natural areas that harbor incomplete biological communities. For example, research natural areas (RNAs) are established on federal lands primarily to represent examples of natural vegetation types. However, an analysis of all 213 USDA Forest Service RNAs as of 1990 determined that 93% were smaller than 1000 ha, and all were smaller than 5000 ha (Noss 1990b). Obviously, reserves this small cannot represent many vertebrate species associated with natural vegetation types, particularly if surrounding lands become unsuitable due to intensive commodity production.

This coarse-filter idea of protecting species by setting aside examples of natural ecosystems has not been tested empirically. Without complete species lists for a region, it is impossible to know whether a large majority of species will be included within a reserve network established to represent ecosystems.

Thus, a comprehensive biological survey is ultimately needed to test the coarse-filter hypothesis. Nevertheless, the hypothesis is reasonable and, as noted earlier, diversity of ecosystem types is one legitimate measure of biodiversity, though no single measure of biodiversity is adequate for all purposes.

A problematic limitation of any representation strategy based on vegetation is that natural communities are not stable; they change as species respond more or less independently to environmental gradients in space and time. For instance, when climate changes, species track shifting habitat conditions at different rates determined by their dispersal capacities and other aspects of autecology (Davis 1981, Graham 1986). Given the instability of natural communities, one way to represent biodiversity at the ecosystem level is to maintain the full array of physical environments and gradients in an interconnected network of reserves (Noss 1987, 1992, Hunter et al. 1988). World Wildlife Fund Canada, in consultation with the Canadian Council on Ecological Areas, is conducting an "enduring features" gap analysis (Iacobelli et al. 1995) to assess representation of environments defined by slope position, topography, soil type, bedrock geology, and other abiotic variables within each natural region of the country. Gaps in representation of habitats defined in this way would become priorities for protection.

Practical advantages of a coarse filter include efficiency and cost-effectiveness (it is easier to deal with dozens or even hundreds of ecosystem types than thousands or millions of species) and the assumed ability to protect species we know nothing about and could not begin to inventory individually. Furthermore, a coarse filter can be applied at any level of classification hierarchy. Unfortunately, no accepted classification of communities or ecosystems exists for the United States (Orians 1993). At which level of classification hierarchy should natural communities be recognized? In California, for example, should we recognize 52 wildlife habitat types (Mayer and Laudenslayer 1988), 375 natural communities (Holland 1987), or something in between? The level chosen will dramatically affect assessments of how well we are succeeding in protecting natural diversity. In most cases, the coarser the classification, the less overall area needed to be protected; thus, the level of classification hierarchy chosen has political as well as ecological ramifications (Orians 1993).

Gap Analysis

A significant national effort in the United States to determine the protection status of ecosystems is the gap analysis project of the National Biological Service. Gap analyses are being conducted state by state, carried out through the cooperative research units and cooperating state and federal agencies and universities (Scott et al. 1991, 1993). Gap analysis is basically a coarse-filter assessment of representation of vegetation types and species in protected areas, using satellite imagery, ancillary data on vegetation, wildlife-habitat association models, and GIS mapping. Gaps in the representation of species, ecosystems, and hot spots of species richness are selected as priorities for conservation.

The major elements of a gap analysis project are:

1. *Vegetation mapping.* Vegetation polygons (areas of relatively homogeneous cover at least 100 ha in the West, 40 ha in the East) are delineated and labeled through a combination of visual photointerpretation of satellite photographic images (LANDSAT TM), digital classification of satellite data, and reference to aerial photographs, existing vegetation maps, and field surveys. The vegetation map corresponds to a classification of plant communities to the series (dominant plant species) level. The vegetation map and all subsequent gap analysis maps are produced at a scale of 1:100,000 to 1:500,000.

2. *Species range maps.* Existing range maps for species of vertebrates and other well-inventoried species are refined by identifying polygons of suitable vegetation within overall range boundaries. Species are then associated with particular vegetation types on the basis of simple habitat relation models. A GIS overlay of each species' known distribution (from various sources such as range maps and museum specimens) and the vegetation-derived map of suitable habitat provides a map of predicted current distribution.

3. *Species richness maps.* Distribution maps for individual species are overlaid to produce maps of species richness. Such maps can be constructed to show the number of species expected for each vegetation polygon, or a grid can be overlaid across polygons to show expected species richness across environmental gradients. Hot spots of high species richness usually show up clearly in such analyses. Species-rich areas are important because they represent opportunities to protect large numbers of species efficiently.

4. *Aquatic, wetland, and rare species.* The minimum mapping unit of some state gap analyses is too large to portray small, patchy habitats, such as many wetlands. Many states have National Wetland Inventory maps, which can be used to map wetlands and wetland-associated species when incorporated as a gap analysis data layer. For streams, riparian areas, lakes, and associated species, digital line graphs can be used (Scott et al. 1993). Rare species, whether terrestrial or aquatic, are generally localized and cannot be mapped on the basis of habitat relations. Data from natural heritage programs on occurrences of rare plants and animals can be en-

tered into the GIS as point locations. Clusters or concentrations of points show hot spots of rare species richness.

5. *Land ownership and management status.* How habitats of high conservation value are being managed is of obvious concern. If biodiversity hot spots are already located within national parks or other reserves, we may have some confidence that they are being protected (albeit most reserves were not established to maintain biodiversity and many are not managed toward this goal). If, on the other hand, biodiversity hot spots are on multiple-use public lands or private lands, their protection is not assured. Four classes of management status are recognized: (a) most national parks, Nature Conservancy and Audubon Society preserves, some wilderness areas and national wildlife refuges, research natural areas, and other areas managed for their natural values have management status 1; (b) most wilderness areas, national wildlife refuges, BLM areas of critical environmental concern, and other areas generally managed for natural values but subjected to some uses that degrade natural qualities have management status 2; (c) most nondesignated public lands, such as national forests, BLM lands, state parks, and other lands with some legal mandates for conservation but many potentially damaging uses have management status 3; (d) private or public lands without legal mandates to protect natural qualities and managed primarily for intensive human uses have management status 4.

6. *Finding the gaps.* A comparison of maps showing biodiversity hot spots and other priority sites with the management status overlay shows gaps in protection. Unprotected and underrepresented vegetation types and hot spots are areas that warrant the most immediate conservation action. Iterative computer algorithms can be used to determine efficient ways of capturing all species and/or vegetation types in a representative reserve network (Margules et al. 1988, Pressey et al. 1993).

Gap analysis is a procedure for identifying sites of high conservation value and assuring that all inventoried elements of biodiversity are represented. Issues of reserve design and land management are outside the scope of gap analysis per se. An obvious limitation of gap analysis is its coarse level of resolution. To provide a national or regional picture of the status of biodiversity cost-effectively, maps cannot contain too much detail (Scott et al. 1993). The best way to view gap analysis is as a first-cut, coarse-filter assessment of biodiversity representation. It is intended to provide land managers with focus and direction and researchers with a set of testable biogeographic hypotheses. Areas that stand out as important on the basis of gap analysis can then be subjected to more detailed study. A comprehensive biodiversity analysis of any region will require a plurality of databases and evaluation criteria (Noss and Cooperrider 1994).

The Wildlands Project

An interesting trend in the environmental movement in North America recently has been the emergence of grassroots regional groups concerned about the biodiversity and ecological integrity of their respective bioregions (Foreman 1991). An effort called the Wildlands Project has been organized to provide technical guidance and support to regional groups across North and Central America by linking conservation biologists with activists (Foreman et al. 1992). These alliances seek to produce scientifically credible conservation plans that have the involvement and support of at least a significant minority of the local populace. The land conservation strategy for the Wildlands Project was summarized by Noss (1992), and can be described as long-range (planning over decades and centuries), biocentric, and optimistic. A conservation plan for the Oregon Coast Range (Noss 1993) was the first case study for the Wildlands Project published in the technical literature.

The Wildlands Project has four fundamental biological conservation objectives (Noss 1992):

1. Represent, in a system of protected areas, all native ecosystem types and seral stages across their natural range of variation. This is the coarse-filter strategy as described earlier.
2. Maintain viable populations of all native species in natural patterns of abundance and distribution. This complements the representation goal and concentrates on species most vulnerable to human activities, such as large carnivores.
3. Maintain ecological and evolutionary processes, including natural disturbance-recovery regimes, hydrological processes, nutrient cycles, such biotic interactions as predator-prey and mutualistic relations, and genetic differentiation of populations.
4. Manage the land conservation system to be responsive to short-term and long-term environmental change in an unpredictable world.

The Wildlands Project seeks to fulfill these goals, recognizing that some may take decades to attain and will ultimately require reduction of human population and resource consumption regionally and globally. A basic assumption of the strategy is the ''blip theory,'' where the present ''binge of [human] reproduction will be a transient blip on the graph rather than a surge to a plateau of permanent planetary obesity'' (Soulé 1992) and that the peak in human population in the twentieth or early twenty-first centuries will be followed by a slow implosion in the twenty-first and twenty-second centuries. This assumption gives conservationists hope for the future; hope is a better motivation than despair. Aside from long-term sociopolitical adjustments, the process for fulfilling the goals of the Wildlands Project is primarily one of establishing a series of regional reserve networks that include zones with varying types and intensities of human use, as in the familiar biosphere reserve model (UNESCO 1974) and the multiple-use module (Harris 1984, Noss and Harris 1986, Noss 1987). Protected areas managed primarily to maintain their natural values form the backbone of each regional reserve system. Core reserves should collectively encompass the full range of communities, ecosystems, physical habitats, environmental gradients, and natural seral stages in each region.

Legislation

Interest is increasing, at least among conservation biologists, for legislation that assures adequate representation of ecosystem diversity in the United States. To this end, scientists and environmentalists have called for an endangered ecosystems act (Noss and Harris 1986, Hunt 1989, M. Liverman unpublished), an endangered habitats act (Ehrlich and Ehrlich 1986), a native ecosystems act (Noss 1991a, 1991b), a sustainable ecosystems act (Jontz 1993), and other legislation designed to protect ecosystems.

One comprehensive approach to conservation and restoration of ecosystems is the Native Ecosystems Act (Noss 1991a, 1991b). This act would serve to protect and restore the entire spectrum of native plant and animal communities, ecosystems, and landscapes across the United States. The Native Ecosystems Act would have three primary concerns: (1) endangered ecosystems, (2) representative ecosystems (including attention to conservation of ecosystem types that are still common), and (3) ecosystem inventory, research, and monitoring.

The endangered ecosystems section of the Native Ecosystems Act would be modeled after the Endangered Species Act. Ecosystems would be defined according to a hierarchical classification, such as that being developed jointly by the Nature Conservancy, the U.S. Environmental Protection Agency, and the National Biological Service. A draft hierarchical classification to the series (dominant plant species) level has been produced for the western United States (Bourgeron and Engelking 1992) and is consistent with a modified international UNESCO classification (Driscoll et al. 1984, Jennings 1993). This classification system, when complete, will be used to standardize results of the National Biological Service gap analysis nationwide. Other classifications, including those used by state heritage programs, can be cross-walked to the national system. Natural community types defined by heritage programs are usually at a very high level of resolution, for example, the 375 natural community types recognized for California (Holland 1987). To supplement classifications based primarily on vegetation or floristics, ecosystems defined by seral stage, structure, functional relations, condition, and other ecologically relevant factors should be recognized (Noss et al. 1995). Thus, virgin and old-growth forests, ungrazed sagebrush steppe, seepage bogs, vernal pools, free-flowing rivers, and seagrass meadows can all be recognized as endangered ecosystems.

Endangerment of ecosystems would be first assessed by extent of decline since European settlement, with decline including outright destruction, conversion to other land uses, or significant degradation of ecological structure, function, or composition (Noss et al. 1994). Ecosystem types at any level of classification hierarchy that have declined by 98% or more would be considered critically endangered; those that have declined by between 85 and 98% would be considered endangered; and those that have declined by 70–84% would be considered threatened (table 12.1). Within these categories, ecosystem types or geographic areas that are at greatest risk of further decline, perhaps due to development pressures, would be the highest priorities for immediate protective action (analogous to emergency listing and other actions for critically endangered species).

Endangered and critically endangered ecosystems, like species listed under the Endangered Species Act, would be protected from further "taking" that would degrade them in any way. Thus, roading, logging, livestock grazing, mining, development, or other habitat alteration would be prohibited, unless necessary for restoration (for example, thinning of trees that have invaded due to fire suppression or grazing to take the place of extirpated native herbivores that for some reason cannot quickly be reintroduced). Recovery goals would be set and recovery (restoration) plans developed, to reestablish large, viable examples of these ecosystem types in their native landscapes. The natural distribution of vegetation along environmental gradients in the landscape mosaic would be restored, as would natural disturbance regimes and populations of extirpated species. Threatened ecosystems would be listed, monitored, and managed so as to prevent further degradation. Recovery goals would be established as appropriate.

The second major section of the Native Ecosystems Act, on representative ecosystems, would seek to represent viable examples of all native ecosystem types in a network of protected areas, regardless of their current rarity and across their full range of natural variation (Noss 1991a, 1991b). The emphasis in this section is to assure that ecosystems that are still relatively secure and healthy will remain that way. The same hierarchical classification system would be used as for endangered and threatened ecosystems. Recent analyses have shown that existing reserves do a poor job of representing ecosystem diversity (Crumpacker et al. 1988, Noss and Cooperrider 1994). A representative ecosystems section would complement the recovery process for endangered and critically endangered ecosystems and prevent further degradation of threatened and nonthreatened ecosystems. The gap analysis project would be used to determine where reserves would be located in order to capture centers of native species richness and endemism within each major ecosystem type. General reserve design and management guidelines also would be provided in this section of the act. Because most ecosystem types are already heavily modified, ecological restoration (including species reintroductions) would be proposed as needed.

The final major section of the Native Ecosystems Act would provide guidelines for inventory, research, and monitoring of ecosystems. These functions would be coordinated by the National Biological Service and would use existing gap analysis and heritage program resources. However, these efforts would be augmented by more specific studies of status and trends, for example, analyses of extent of decline since settlement for each ecosystem type, ecological consequences of ecosystem loss and degradation, and current and potential threats to each ecosystem. A nationwide and global monitoring system (using remote sensing and ground samples) would track trends in ecosystem distribution, quality, management, and protection status. Information from research and monitoring would be used adaptively to revise lists of threatened and endangered ecosystems, modify recovery goals and methods, and eventually to delist ecosystems that have been restored adequately. Degenerative trends in nonendangered ecosystems would be noted and mitigated to prevent the need for the emergency responses required for endangered and threatened ecosystems.

Management

We emphasize ecosystem protection and restoration as fundamental to ecosystem management. This view considers protection of biodiversity preeminent and recognizes reserves as critical and essential components of land-use strategy. Thus, we differ from some of our colleagues who downplay the importance of reserves and emphasize refinements of multiple-use management as key to protecting biodiversity. But we agree with our colleagues (for example, Franklin 1993) that selecting and designing a system of reserves takes us only so far toward maintaining viable populations and meeting other basic conservation goals. How reserves and the lands surrounding them are managed remains a central issue and brings us full circle—just as ecosystem protection and restoration form the core of ecosystem management, management is essential to ecosystem protection and restoration.

Reserves must be managed as surely as multiple-use lands. As noted by Noss and Cooperrider (1994):

> All land management is biodiversity management, whether intended or not. All land-use decisions—including a decision to designate a reserve, put a fence around it, and leave it alone—are land management decisions with significant consequences for biodiversity. It is much better to manage biodiversity by design rather than by default. As the most powerful species on Earth, we can alter biodiversity worldwide and are expressing that capability in a frightening way. Accepting responsibility for our actions means not only that we carefully consider the effects of management on biodiversity, but also that our management programs be designed explicitly to protect and restore native biodiversity.

Ecosystem management runs the full gamut from intensive exploitation to rigorous protection. Guidelines for managing reserves and multiple-use areas in forests, rangelands, and aquatic ecosystems should be based on the same fundamental ecological principles (see Noss and Cooperrider 1994). Discussions of ecosystem management should explicitly recognize the breadth of the land-use spectrum and not be limited by the assumption that a landscape must either be exploited for resources or set aside in a hands-off preserve. A sensible approach is to consider each landscape in a broad context and determine where in the land-use spectrum it properly lies. For a landscape rich in endemic species and wilderness values, or one containing populations of species sensitive to human access (for example, large carnivores), the best management plan would arguably be one that minimizes potentially disruptive human activities.

But even if an area is designated as wilderness, active management may be needed to emulate or supplement natural processes. Few wilderness areas in today's world can manage themselves and maintain their biodiversity. For example, a reserve containing vegetation that depends on recurrent fire may be surrounded by anthropogenic habitats where fires are suppressed. Lightning ignitions may not occur often enough within the reserve to maintain natural vegetation structure; fires that once spread from single ignitions throughout a vast landscape are now constrained by roads and other artificial firebreaks. Therefore, prescribed burning is often a necessary management action and has the further advantage of reducing woody fuels within the reserve and limiting the potential for fires to spread from the reserve to adjacent private lands. In other cases, management of hydrological regimes, control of exotic species, reintroductions of extirpated native species, or control of visitor activities may be management necessities (Noss and Cooperrider 1994).

The important point is that management and protection should not be seen as inimical to each other. Increasingly, we need to view landscapes in a unified way and determine where and how various human uses and activities should occur. We do not believe, however, that a "landscape without lines" ideal, where all uses occur everywhere but more gently than in the past, is feasible with current human attitudes and technologies. Rather, zoning of the landscape appears necessary to protect and buffer those habitats and species that are most sensitive to human activities. How much land must occur within strictly protected zones to meet conservation goals will vary from landscape to landscape, but most scientifically credible estimates range from 25% to 75% of a given region—an entire order of magnitude above that currently protected (see review in Noss and Cooperrider 1994, Noss 1996).

Experiments in ecosystem management are encouraging in that they signal a new appreciation for the complexity of nature and an awareness that we cannot legitimately manage resources in isolation from their ecological context. However, ecosystem management experiments are also dangerous to the extent that they assume more knowledge of ecological phenomena than we actually possess, or that they simply substitute one way of attempting to achieve mastery over nature for another. We should have learned by now that nature is always capable of surprising us. Expanding our system of protected areas and restoration zones to conserve ecosystem diversity will help keep potential abuses of the ecosystem management concept in check.

Wild areas serve several functions that complement the ideal of ecological sustainability that underlies ecosystem management (Noss 1991c):

1. They provide a standard of relatively healthy and unmodified land; thus, they serve as benchmarks or control areas for our management experiments. Aldo Leopold (1941) stressed this value of wilderness more than fifty years ago.
2. They serve as habitat refugia for species sensitive to human persecution or disturbances, such as large carnivores and furbearers.
3. They provide a source of humility, a reminder that Nature remains more powerful than we are and is ultimately unconquerable.
4. They and the wild species they contain have intrinsic or existence values.

Although intrinsic value is not objective—it cannot be proved or disproved by science—there is also no objective support for the widespread belief that humans are fundamentally superior to other life-forms. If we have value, then all life has value.

Recognizing these values, we urge that ecosystem management be conservative. Placing limits on the scale and intensity of human activities is not only ethical behavior from a biocentric standpoint, it is also prudent from an anthropocentric standpoint. When our knowledge of natural ecosystems is incomplete—as it always will be—then saving all the cogs and wheels (Leopold 1953) makes abundant sense. Natural ecosystems have evolved their functional relations over thousands or millions of years; our experiments in manipulating ecosystems are comparatively brief. Future generations will be best served if those who experiment with today's landscape are cautious and willing to risk erring on the side of protecting too much. And increasingly, we need to devote effort to repairing damage already done.

To conclude, we make these summary recommendations for ecosystem conservation in the United States:

1. Efforts to develop a national classification of ecosystems in the United States should be expedited. This classification should be hierarchical and cross-walked to other classifications here and in other countries, particularly Canada and Mexico.
2. The gap analysis project should receive increased funding for improvement and timely completion.
3. Detailed studies of ecosystem status and trends should be conducted nationwide, with most attention given to those ecosystems already identified to be greatly reduced, degraded, or at imminent risk of loss (Noss et al. 1995, Noss and Peters 1995).
4. Ecosystem conservation plans (including guidelines for specific management and restoration projects) should be developed at landscape to regional scales.
5. Experimental (adaptive) ecosystem management should be implemented in a scientifically sound but cautious manner.
6. Creative partnerships of public and private entities should be formed to implement ecosystem conservation at landscape and regional scales.

LITERATURE CITED

Allan, J. D., and A. S. Flecker. 1993. Biodiversity conservation in running waters. BioScience 43:32–43.

Benke, A. C. 1990. A perspective on America's vanishing streams. Journal of the North American Benthological Society 91:77–88.

Bourgeron, P. S., and L. Engelking (editors). 1992. Preliminary compilation of a series level classification of the vegetation of the western United States using a physiognomic framework. Report to the Idaho Cooperative Fish and Wildlife Research Unit. Western Heritage Task Force, The Nature Conservancy, Boulder, Colorado.

Brown, J. H. 1988. Alternative conservation priorities and practices. Paper presented at 73d annual meeting, Ecological Society of America. Davis, California.

Cook, R. E., and P. Dixon. 1989. A review of recovery plans for threatened and endangered plant species. Unpublished report. World Wildlife Fund, Washington, D.C.

Cornett, Z. J. 1993. Ecosystem management: why now? Unpublished report. USDA Forest Service, n.p.

Crumpacker, D. W., S. W. Hodge, D. F. Friedley, and W. P. Gregg, Jr. 1988. A preliminary assessment of the status of major terrestrial and wetland ecosystems on federal and Indian lands in the United States. Conservation Biology 2:103–115.

Davis, M. B. 1981. Quaternary history and the stability of forest communities. Pages 132–153 in D. C. West, H. H. Shugart, and D. B. Botkin, editors. Forest succession. Springer-Verlag, New York, New York.

Diamond, J. M. 1984. Historic extinctions: A Rosetta stone for understanding prehistoric extinctions. Pages 824–862 in P. S. Martin and R. G. Klein, editors. Quaternary extinctions: a prehistoric revolution. University of Arizona Press, Tucson.

Driscoll, R. S., D. L. Merkel, D. L. Radloff, D. E. Snyder, and J. S. Hagihara. 1984. An ecological land classification framework for the United States. USDA Forest Service, Misc. Pub. 1439, Washington, D.C.

Egler, F. 1977. The nature of vegetation: its management and mismanagement. Aton Forest, Norfolk, Connecticut

Ehrlich, A. H., and P. R. Ehrlich. 1986. Needed: an endangered humanity act? Amicus Journal. Reprinted as pages 298–302 in K. A. Kohm, editor. 1991. Balancing on the brink of extinction: the Endangered Species Act and lessons for the future. Island Press, Washington, D.C.

Ehrlich, P. R., and A. H. Ehrlich. 1981. Extinction: the causes and consequences of the disappearance of species. Random House, New York, New York.

Ehrlich, P. R., and E. O. Wilson. 1991. Biodiversity studies: science and policy. Science 253:758–762.

Foreman, D. 1991. The new conservation movement. Wild Earth 1(2):6–12.

Foreman, D., J. Davis, D. Johns, R. Noss, and M. Soulé. 1992. The Wildlands Project mission statement. Wild Earth (special issue):3–4.

Franklin, J. F. 1993. Preserving biodiversity: species, ecosystems, or landscapes? Ecological Applications 3:202–205.

Franklin, J. F., K. Cromack, W. Denison, A. McKee, C. Maser, J. Sedell, F. Swanson, and G. Juday. 1981. Ecological characteristics of old-growth Douglas-fir forests.

Gen. Tech. Rep. PNW-118. USDA Forest Service, Pacific Northwest Forest and Range Experiment Station, Portland, Oregon.

Gleason, H. A. 1926. The individualistic concept of the plant association. Bulletin of the Torrey Botanical Club 43:463–481.

Gore, A. 1992. Earth in the balance: ecology and the human spirit. Houghton Mifflin, Boston, Massachusetts.

Graham, R. W. 1986. Response of mammalian communities to environmental changes during the Late Quaternary. Pages 300–313 *in* J. Diamond and T. J. Case, editors. Community ecology. Harper and Row, New York, New York.

Grove, R. H. 1992. Origins of western environmentalism. Scientific American, July: 42–47.

Harris, L. D. 1984. The fragmented forest: island biogeography theory and the preservation of biotic diversity. University of Chicago Press, Chicago, Illinois.

Holland, R. F. 1987. Is *Quercus lobata* a rare plant? Approaches to conservation of rare plant communities that lack rare plant species. Pages 129–132 *in* T. S. Elias, editor. Conservation and management of rare and endangered plants. California Native Plant Society, Sacramento.

Hunt, C. E. 1989. Creating an Endangered Ecosystems Act. Endangered Species Update 6(3–4):1–5.

Hunter, M. L. 1991. Coping with ignorance: the coarse-filter strategy for maintaining biodiversity. Pages 266–281 *in* K. A. Kohm, editor. Balancing on the brink of extinction: the Endangered Species Act and lessons for the future. Island Press, Washington, D.C.

Hunter, M. L., G. L. Jacobson, and T. Webb. 1988. Paleoecology and the coarse-filter approach to maintaining biological diversity. Conservation Biology 2:375–385.

Iacobelli, T., K. Kavanagh, and S. Rowe. 1995. A protected areas gap analysis methodology: planning for the conservation of biodiversity. World Wildlife Fund Canada, Toronto, Ontario.

Jontz, J. 1993. The Sustainable Ecosystems Act. Draft report. Silver Lake, Indiana.

Judy, R. D., Jr., T. M. Murray, S. C. Svirsky, M. R. Whitworth, and L. S. Ischinger. 1984. 1982 national fisheries survey. Volume 1. FWS/OBS-84/06. U.S. Fish and Wildlife Service, Washington, D.C.

Kuchler, A. W. 1966 (revised 1985). Potential natural vegetation (map). U.S. Geological Survey, Reston, Virginia.

Leopold, A. 1941. Wilderness as a land laboratory. Living Wilderness 6(July):3.

———. 1953. Round River. Oxford University Press, New York, New York.

Margules, C. R., A. O. Nicholls, and R. L. Pressey. 1988. Selecting networks of reserves to maximize biological diversity. Biological Conservation 43:63–76.

Mayer, K. E., and W. F. Laudenslayer. 1988. A guide to wildlife habitats of California. California Department of Forestry and Fire Protection, Sacramento.

Miller, R. R., J. D. Williams, and J. E. Williams. 1989. Extinctions of North American fishes during the past century. Fisheries 14(6):22–38.

Mooney, H. A., and J. Drake (editors). 1986. The ecology of biological invasions of North America and Hawaii. Springer-Verlag, New York, New York.

Moyle, P. B., and J. E. Williams. 1990. Biodiversity loss in the temperate zone: decline of the native fish fauna of California. Conservation Biology 4:4/5–484.

Nelson, R. 1993. Searching for the lost arrow: physical and spiritual ecology in the hunter's world. Pages 201–228 in S. R. Kellert and E. O. Wilson, editors. The biophilia hypothesis. Island Press, Washington, D.C.

Noss, E. A. 1990. Ancient forests of the Pacific Northwest. The Wilderness Society and Island Press, Washington, D.C.

Noss, R. F. 1987. From plant communities to landscapes in conservation inventories: A look at The Nature Conservancy (USA). Biological Conservation 41:11–37.

———. 1990a. Indicators for monitoring biodiversity: a hierarchical approach. Conservation Biology 4:355–364.

———. 1990b. What can wilderness do for biodiversity? Pages 49–61 in P. Reed, editor. Preparing to manage wilderness in the 21st century. General Technical Report SE-66. USDA Forest Service, Southeastern Forest Experiment Station, Asheville, North Carolina.

———. 1991a. From endangered species to biodiversity. Pages 227 246 in K. A. Kohm, editor. Balancing on the brink of extinction: the Endangered Species Act and lessons for the future. Island Press, Washington, D.C.

———. 1991b. A Native Ecosystems Act. Wild Earth 1(1):24.

———. 1991c. Sustainability and wilderness. Conservation Biology 5:120–121.

———. 1992. The Wildlands Project: land conservation strategy. Wild Earth (special issue):10–25.

———. 1993. A conservation plan for the Oregon Coast Range: some preliminary suggestions. Natural Areas Journal 13:276–290.

———. 1996. Protected areas: how much is enough? In R. G. Wright, editor. National parks and protected areas. Blackwell, Cambridge, Massachusetts.

Noss, R. F., and A. Y. Cooperrider. 1994. Saving nature's legacy: protecting and restoring biodiversity. Defenders of Wildlife and Island Press, Washington, D.C.

Noss, R. F., and L. D. Harris. 1986. Nodes, networks, and mums: preserving diversity at all scales. Environmental Management 10:299–309.

Noss, R. F., E. T. LaRoe, and J. M. Scott. 1994. Endangered ecosystems of the United States: a preliminary assessment of loss and degradation. National Biological Service. Report No. 28, Washington, D.C.

Noss, R. F., and R. L. Peters. 1995. Endangered ecosystems of the United States: a status report and plan for action. Defenders of Wildlife, Washington, D.C.

Orians, G. H. 1993. Endangered at what level? Ecological Applications 3:206–208.

Postel, S., and J. C. Ryan. 1991. Reforming forestry. Pages 74–92 in L. Starke, editor. State of the world 1991: A Worldwatch Institute report on progress toward a sustainable society. W. W. Norton, New York, New York.

Pressey, R. L., C. J. Humphries, C. R. Margules, R. I. Vane-Wright, and P. H. Williams. 1993. Beyond opportunism: Key principles for systematic reserve selection. Trends in Ecology and Evolution 8:124–128.

Pyne, S. J. 1982. Fire in America: a cultural history of wildland and rural fire. Princeton University Press, Princeton, New Jersey.

Robbins, C. S., J. R. Sauer, R. S. Greenberg, and S. Droege. 1989. Population declines in North American birds that migrate to the neotropics. Proceedings of the National Academy of Sciences 86:7658–7662.

Russell, C., and L. Morse. 1992. Extinct and possibly extinct plant species of the United States and Canada. Unpublished report. Review draft, March 13. The Nature Conservancy, Arlington, Virginia.

Scott, J. M., B. Csuti, and S. Caicco. 1991b. Gap analysis: assessing protection needs. Pages 15–26 in W. E. Hudson, editor. Landscape linkages and biodiversity. Defenders of Wildlife and Island Press, Washington, D.C.

Scott, J. M., F. Davis, B. Csuti, R. Noss, B. Butterfield, C. Groves, J. Anderson, S. Caicco, F. D'Erchia, T. C. Edwards, J. Ulliman, and R. G. Wright. 1993. Gap analysis: a geographical approach to protection of biological diversity. Wildlife Monographs 123:1–41.

Soulé, M. E. 1991. Conservation: tactics for a constant crisis. Science 253:744–750.

———. 1992. A vision for the meantime. Wild Earth (special issue):7–8.

UNESCO. 1974. Task force on criteria and guidelines for the choice and establishment of biosphere reserves. Man and the Biosphere Report No. 22. Paris.

Vitousek, P. M. 1988. Diversity and biological invasions of oceanic islands. Pages 81–89 in E. O. Wilson, editor. Biodiversity. National Academy Press, Washington, D.C.

Waddell, K., D. Oswald, and D. Powell. 1989. Forest statistics of the United States, 1987. USDA Forest Service, Portland, Oregon.

West, N. E. 1996. Strategies for maintenance and repair of biotic community diversity on rangelands. Pages 326–346 in R. C. Szaro and D. W. Johnston, editors. Biodiversity in managed landscapes: theory and practice. Oxford University Press, New York, New York.

Whittaker, R. H. 1960. Vegetation of the Siskiyou Mountains, Oregon and California. Ecological Monographs 30:279–338.

Williams, J. E., J. E. Johnson, D. A. Hendrickson, S. Contreras-Balderas, J. D. Williams, M. Navarro-Mendoza, D. E. McAllister, and J. E. Deacon. 1989. Fishes of North America endangered, threatened, or of special concern. Fisheries 14(6):2–20.

Wilson, E. O. 1988. The current state of biological diversity. Pages 3–18 in E. O. Wilson, ed. Biodiversity. National Academy Press, Washington, D.C.

———. 1992. The diversity of life. Belknap Press of Harvard University Press, Cambridge, Massachusetts.

Chapter 13 Silviculture and Ecosystem Management
John Kotar

Ecosystems comprise all living and nonliving components that interact on a particular segment of a landscape. For research or management purposes, boundaries between different terrestrial ecosystems can arbitrarily be established, but absolute boundaries between adjacent ecosystems cannot be defined. No one claims to understand functions of even the simplest of ecosystems enough to manipulate it with predictable results. Thus ecosystem management is more an approach to management than a cause-effect method of control. I believe that the goal of forest ecosystem management should be the development of methods of extracting human commodities from forest ecosystems in ways that do not greatly alter the processes that shape the development of natural forest communities. In other words, we are not attempting to take control of the ecosystem processes themselves, since we do not fully understand them, but we wish to interfere with them as little as possible. Presumably, with our current state of knowledge, this approach is the best way to assure sustainability of those ecosystem outputs upon which societies depend.

"Natural" or "native" forests are assemblages of those regional floristic and faunal components that can coexist, in a particular environment. This composition is in a constant state of flux as individual populations react to changes in their immediate environments brought about by internal as well as external forces. Natural forests can be simple or complex in terms of species composition and arrangement of size or age classes (structure), depending on the stage of their development and on physical limitations of the site. On a landscape scale, and over long periods of time, natural communities display the greatest amount of biological diversity possible under the prevailing climatic and edaphic conditions of a given region.

Managed forests are manipulated to produce specific commodities. Traditionally, wood production and consumptive wildlife have been the dominant commodities. Managed forests are generally less complex than natural forests because management attempts to optimize only a few species: those of high commercial value and fast growth, or those that can be grown in pure stands or in relatively simple mixtures. Economic considerations often lead to other structural and compositional limitations, among them economic rotation age, maximum diameter, fixed spacing, and elimination of competing vegetation.

These simplifications of the composition and structure of managed forests reduce the diversity of resources and number of possible niches in a forest ecosystem, resulting in lower biodiversity.

SILVICULTURE AND FOREST MANAGEMENT

Although *silviculture* and *forest management* have very different meanings within the forestry profession, the terms are often used interchangeably and contribute greatly to misunderstandings between foresters and other resource managers, as well as for the public. Furthermore, both terms are often interpreted as synonyms for timber production. Silviculture has been defined in various ways: the art of producing and tending a forest; the theory and practice of controlling forest establishment, composition, and growth; and the application of the knowledge of silvics (biological and ecological characteristics of trees) in the treatment of a forest (Smith 1986). The aim of the practice of silviculture has traditionally been production of a certain crop. However, silviculture can also be applied to manipulating forest composition for other purposes, such as to improve habitat for certain species of wildlife or desirable plants, or to enhance recreational or aesthetic values.

Forest management, as defined by the Society of American Foresters, is "The application of business methods and technical forestry principles to the operation of forest property" (Davis 1966). It is obvious that the owner's objectives and economic factors limit the range of silvicultural options for managing a given forest. However, because social and economic conditions, and therefore the owner's objectives, are always changing, it is appropriate from time to time to reexamine and reevaluate silvicultural techniques. This is essential because we know that under any management system some forest benefits are always being compromised. At the same time we must not assume that any adjustment in silvicultural techniques to better accommodate another forest benefit is necessarily uneconomical.

Silvicultural Methods and Silvicultural Systems

Silvicultural methods are procedures by which stand properties are controlled. A reproduction method, for example, is a procedure by which a stand is established or renewed. The process is accomplished during the regeneration period by means of natural or artificial reproduction. The more comprehensive term *silvicultural system* signifies a planned program of silvicultural treatments during the whole life of a stand; it includes not only the reproduction cutting but all intermediate manipulations. The reproduction method employed has such a significant influence on the subsequent handling of the stand that the name of the method is commonly applied to the silvicultural system as well. The shelterwood system, for example, leads to reproduction by means of the shelterwood method of cutting. Under this method a stand is cut in two or more stages, each stage creating conditions favorable for seed germination or establishment of young trees. The method is generally applied to species that do not reproduce under dense shade of normal stands.

The distinction between these two terms is important for several reasons. First, because there are only a few standard reproduction methods (selection, shelterwood, seed-tree, and clearcutting), one may get a false impression that silvicultural systems themselves are equally limited. In fact, the four basic reproduction methods can be applied in various combinations, spatially and temporally, allowing for many customized silvicultural systems to serve a variety of needs. Second, in light of considerable public opposition to clearcutting in general, it does not help to label as a clearcutting method a silvicultural system that uses clearcutting only for regeneration and uses partial cuttings during most of the life span of a stand. Third, as forest management changes to equalize weight given to other forest values, silvicultural systems as practiced in the past are likely to lose their meaning, while forest reproduction methods will always be needed.

The Stand as a Silvicultural Unit

The basic unit of silvicultural manipulation is a *stand*. A stand may loosely be defined as a contiguous group of trees sufficiently uniform in species composition, arrangement of age classes and general condition to be considered a homogeneous and distinguishable unit. Although the stand has been a unit of silvicultural manipulation historically, it does not necessarily represent a homogeneous ecological unit. The properties by which it is recognized (for example, species composition, age structure, diameter distribution) are more likely to be a result of past treatment than the expression of community development under the control of a homogeneous site. This is especially so in the United States, where so-called "extensive" rather than "intensive" forestry is practiced. In extensive forestry there is a tendency to treat large groups of dissimilar stands as if they conformed to some hypothetical average. For the most effective ecological approach to management, one should recognize the smallest stands that can be conveniently delineated on the forest type maps. Where site classification systems have been developed, an attempt also should be made to place stand boundaries so that they do not encompass more than one ecological site unit.

Silviculture and the Concept of Forest Cover Type

Because so much of silvicultural experience and literature is organized by *forest cover type,* it is essential that we understand to what extent ecological factors are considered in defining cover types and how the cover type approach to management relates to the proposed ecosystem concept of management.

The English-language version of *Terminology of Forest Science, Technology, Practice, and Products* (Ford-Robertson 1971) gives two definitions of forest type: (1) forest site type—generally a category of forest or forest land, actual or potential; and (2) forest cover type, crop type, stand type—more particularly a category of forest defined by its vegetation or locality factors. The *Forest Cover Types of the United States and Canada* (Eyre 1980) follows the second definition with some further restrictions. In short, forest cover type is a descriptive classification of forest land based on present occupancy of an area by particular tree species. Forest cover types are named after predominant tree species as determined by basal area. Because an almost unlimited number of combinations of tree species occur in nature, only those of relatively wide distribution were formally recognized and described by Eyre (1980). Ninety-five cover types were described for the Eastern United States.

Forest cover types are useful for describing and mapping composition and distribution of forest vegetation as it appears at various periods of time and, to some extent, for summarizing the known ecological characteristics and silvicultural information. However, there are severe limitations to the use of generalized cover types as ecological units of vegetation management. Most tree species have a wide ecological amplitude (they occur over a wide range of environmental conditions) and therefore do not grow at the same rate nor respond uniformly to competition or disturbance. For the same reasons they also do not respond uniformly to given silvicultural techniques. Thus, for management purposes, models of ecological behavior of cover types across a range of sites must be developed locally.

To approach forest management on an ecosystem basis, foresters need to overcome the common practice of treating most cover types as natural, more or less permanent vegetation units. This practice was borrowed directly from the European experience of growing successive crops of economically desirable species on the same site for many generations. In fact, all silvicultural systems were originally developed for this type of crop-oriented management. However, the sustainability of such forest practices is now coming into question in many parts of the world (Oliver 1992, Leslie 1993, Smith 1993).

While ecologists have long been aware that many forests do not return to their original state in a one-step process following a disturbance, the mechanisms driving succession were not well enough understood to be incorporated into silvicultural systems. This is especially true for central European forests, where human influence has been historically most intense. More importantly, until very recently, the need for management strategies that follow natural forest development processes has not been widely advocated by the scientific community, nor has it been demanded by society.

Silvicultural Systems Can Mimic Natural Community Dynamics

To mimic natural community dynamics, composition and structure of managed forests should change with time. That is, cover types should be recognized and treated as successional or developmental stages; each cover type should not necessarily be perpetuated in place but should occur on the landscape at different times and in different locations in a cyclic process. Opportunities, of course, always exist to adjust the extent and duration of an existing type if desired for specific reasons. This, in essence, is resource utilization through ecosystem management.

New silvicultural techniques are not required to begin management based on ecosystem concepts. What is needed is a reassessment of management objectives for individual stands, along with willingness to use a variety of available silvicultural techniques simultaneously. For example, shelterwood, clearcutting, and single-tree selection methods may all be appropriate to establish regeneration in a same large heterogeneous stand. Most importantly, this means that specific stand treatments will have to be developed on the site and not in the office.

If we model silvicultural systems after natural processes that shape forest community development in a given environment (that is, site type), a long-term regional master plan for desired future condition is not necessary, or even possible, since conditions will be changing continuously. It is assumed that by applying this method all future conditions of the forest as a whole will be ecologically acceptable because individual stands received ecological considerations.

Nevertheless, some qualification is necessary of a plan to manage by accepting and aiding natural successional trends. If carried out universally and rigidly, the practice may lead to underrepresentation of young stages. To avoid this trend, regionwide landscape monitoring systems should be developed. In some circumstances it will be justifiable to maintain a particular successional stage for more than one generation. Also, it must be accepted that maintenance of many intolerant groups will require aggressive silvicultural measures. Relying on natural disturbances to accomplish this is no longer an option in human-dominated environments.

THE LAKE STATES FOREST EXAMPLE

Conditions Prior to Euro-American settlement

In order to implement management strategies that accommodate a variety of natural forest development processes and community structures, an understanding of the forest history of specific regions is necessary.

The upland forests of the Lake States region in the pre-Euro-American settlement period could be grouped into three broad categories: mixed pine, mixed oak, and hemlock–white pine–northern hardwood. It must be understood that individual forest communities within these types exhibited considerable variation in composition and structure. The pine forests were primarily a function of the substratum. The drought-prone, coarse-textured soils promoted a forest type which was extremely flammable and subject to stand-initiating fires at 130–260 year intervals (Whitney 1987). Fire eliminated most competing hardwoods, and the few pines that escaped fire restocked the land. Relative proportions of the three pine species *(Pinus banksiana, P. resinosa, P. strobus)* varied as a function of fire frequency and intensity. Red pine was most prevalent in mixtures in the region as a whole. Jack pine dominated on sites most prone to frequent fires, while white pine was more common on somewhat less xeric soils or where fire cycles were longer.

Oaks (largely *Quercus rubra–velutina–ellipsoidalis* complex and *Q. alba*) were generally not well represented in the canopy of northern forests but were common as sprouts in the understory. However, oak forests were common in southern parts of the region, where fires were even more frequent than in the north, and pines as well as mesic hardwoods were largely eliminated.

Fires were much less frequent in the moist hemlock–white pine–northern hardwood forests, which occurred on finer-textured morainic soils. Sugar maple *(Acer saccharum)*, hemlock *(Tsuga canadensis)*, yellow birch *(Betula alleghaniensis)*, and, in eastern parts of the region, beech *(Fagus grandifolia)* were the dominant species. Many of the largest white pine trees also were found scattered within this type. Aspen *(Populus tremuloides* and *P. grandidentata)* and paper birch *(Betula papyrifera)* communities were relatively uncommon.

The Postlogging Condition of the Forest

The intensive logging that preceded or paralleled Euro-American settlement profoundly altered the nature of the Lake States' forests. Selective logging of white pine, and later hemlock and the better hardwoods, converted the hemlock–white pine–northern hardwood type to sugar maple–dominated stands. Waves of fires, which closely followed the logging, also upset the apparent equilibrium of the Great Lakes mixed pine forests. The ignition of the logging slash destroyed the remaining seed trees and the seedling pine. The result was a forest of oak sprouts and aspen root suckers. Many of these postlogging stands were poorly stocked. Red maple *(Acer rubrum)* is another species that was seldom mentioned in descriptions of presettlement forests but became common in the burned-over landscape.

Wild fires continued in the Lake States following the logging era, and reforestation was impossible until they were brought under control, mainly in the period of 1920–1950. Early reforestation was hampered by the lack of seed source of most species, particularly the pines. Artificial regeneration accounts for most of the recent increase in the areal extent of the pine type, with strong predominance of red pine.

The cessation of fires and maturation of aspen and jack pine set the stage for the development of the new pulp-oriented industry. Most of the aspen and jack pine forests are managed by clearcutting on a 30–60-year rotation (Chase, Pfeifer, and Spencer 1970). Gradually other species, particularly red pine but also various hardwoods, have been entering economic maturity for pulpwood production and are increasingly being managed for this purpose.

It is not the use of the forest resource that is causing conflicts with ecosystem management but rather the methods we have developed for utilizing it. Because the pulpwood industry is based on utilization of relatively small-dimension wood, it encourages perpetuation of short-lived species, usually in pure cultures, or young stages of species with potentially greater longevity. This strategy perpetuates forests that were compositionally and structurally simplified during the logging and settlement period.

Perhaps the first step toward the development of the ecosystem approach to management of the Lake States forests is to move away from the concept of maintaining existing forest cover types and to guide their development through any of a variety of pathways that are a part of natural forest development for a particular site. The intent is not to reestablish the presettlement pattern of forest vegetation but to include in our managed forests some of the compositional elements that have been lost in the initial wave of forest utilization. This approach will continue to yield wood products and other forest commodities, but in different mixtures of species and by different schedules. Some express fear that this type of management would make it too difficult for procurement planning. However, the rapid expansion in computer technology should make monitoring and projecting of the available forest resources less of a problem than we now perceive it to be.

Some Examples of Opportunities for Ecosystem Approach to Management

Very broadly speaking, three types of forest communities represent the vast majority of forest currently managed in the Lake States: (1) aspen, (2) red pine, and (3) "northern hardwood," which has an overwhelming dominance of sugar maple.

Aspen, in particular, dominates landscapes that are capable of supporting a much greater range of community composition. To be sure, some aspen stands at maturity contain little or no advance reproduction of other potential species simply because their seed sources have been eliminated. Such stands can, of course, continue to be harvested and regenerated back to aspen until recruitment of other species occurs naturally, but artificial introduction of such species as white pine, for example, is also an alternative.

On the other hand, many aspen stands already contain adequate reproduction of white pine, red oak, red maple, balsam fir, and other species. These species will not be able to reach canopy status under the clearcutting system aimed at perpetuation of aspen. This is because aspen-regeneration clearcuts remove all size classes regardless of species. Rapidly growing aspen root suckers subsequently outgrow all other species that may have become established in the clearcut. There are many silvicultural techniques available, however, to maintain, or at least prolong, the presence of aspen in the stand and to encourage other species. Perhaps white pine offers the best opportunities for comanagement with aspen. When white pine saplings in aspen stands are of sufficient height, they are capable of staying ahead of aspen sprouts as mature aspen is removed from the stand. Because densities of white pine saplings are usually relatively low, another aspen generation can still be established. In fact, more than one rotation of aspen can be produced during a single rotation of white pine. If enough white pine is retained on the landscape to maintain the seed source, pine and aspen can be managed together indefinitely. Similar strategies may be possible with red oak and perhaps other species on appropriate sites.

As described earlier, where pine has been reestablished by artificial regeneration, pure red pine plantations rather than mixed stands prevail. At the time these plantations were established, artificial regeneration was the only alternative. However, it does not follow that a system based on clearcutting, site preparation, and replanting remains the only alternative to managing future pine forests. Where white pine seed sources exist, thinned red pine plantations often exhibit prolific advance reproduction of white pine, which can be regenerated with the shelterwood system.

Red pine plantations are notorious for their lack of structure and for their typical lack of any understory development. However, there is ample evidence that, with proper silvicultural treatment, such plantations can be transformed into much more diverse and structured stands. Many thinned plantations have a significant component of red oak regeneration. This species is not only valuable for wildlife but also plays an important role, as do all deciduous species, in the nutrient cycling process that improves soil fertility. Prolonging the rotation of some red pine and carrying oak through the stand regeneration process will preserve more management options. Under favorable conditions mixed stands of all three native pines can at least temporarily occupy a site. On all former mixed-pine landscapes, a mosaic of stands of the three pine species of various ages can be maintained.

The "northern hardwood" complex is a much oversimplified term for mixed communities, usually dominated by shade-tolerant sugar maple or American beech, with various mixtures of other shade-tolerant and midtolerant species. Two other species, eastern hemlock and yellow birch, also shared dominance on many sites prior to the logging era but are now much less abundant and often entirely absent. Single-tree or group-selection management as commonly practiced today successfully maintains this type, but it leads toward increasing dominance of sugar maple at the expense of all other species. Stand structure is maintained by following stocking guides based on control of basal area, stem density, and average diameter. Furthermore, stocking guide norms are achieved by removing stems in the following order of priority: (1) those that may not survive to the next cutting cycle, (2) those with unacceptable amounts of cull, (3) those showing poor vigor, and finally, (4) those that satisfy species considerations. In order to improve species composition, stocking guides may have to be followed more loosely, and criteria for order of removal may have to be adjusted according to local conditions. In addition, if hemlock and yellow birch abundance is to be increased, silvicultural procedures to stimulate regeneration will have to be applied. In those stands where hemlock and yellow birch seed sources are lacking, artificial introduction of these species may be considered.

Forest ecosystem management can be carried out only in the context of resource utilization. Although forests in all parts of the United States have been altered to one degree or another since Euro-American settlement, they have for the most part not lost the capacity to support again those compositional and structural components that have been eliminated. Managed forests can function as natural and sustainable ecosystems if we move from the static, cover type–based system to a dynamic approach to management. Although much remains to be learned about the function of forest communities and ecosystems, the essential tools for embarking on this type of management are clearly in the hands of the forestry profession. Effective forest ecosystem management is not possible without the use of the scientific and empirical knowledge accumulated by the forestry profession.

The primary obstacle to implementation of this new management strategy will not be the lack of ecological or technical knowledge but rather the difficulty of reaching society's consensus on which forest values will prevail. To illustrate this difficulty we only need look at the example of aspen resource described earlier. Aspen is not only an important species for pulp and paper industry, it is also vigorously promoted by wildlife interests. For example, the aspen resource today supports much higher population levels of such popular game species as white-tailed deer *(Odocoileus virginianus)* and ruffed grouse *(Bonasa umbellus)* than ever before. In order to proceed with the ecosystem approach to management, some of these values will have to be compromised. We must accept the reality that science cannot provide an objective basis for prioritizing society's values.

LITERATURE CITED

Chase, C. D., R. E. Pfeifer, and J. S. Spencer, Jr. 1970. The growing timber resource of Michigan, 1968. Resource Bull., NC-9, USDA, Forest Service, Washington, D.C.

Davis, K. P. 1966. Forest management: regulation and valuation. McGraw-Hill, New York, New York.

Eyre, F. H. (editor). 1980. Forest cover types of the United States and Canada. Society of American Foresters, Washington, D.C.

Ford-Robertson, F. C. 1971. Terminology of forest science, technology, practice and products. English-language version. Society of American Foresters. Washington, D.C.

Leslie, A. 1993. The discipline of forestry. International Union of Societies of Foresters (IUSF) Newsletter No. 34, August 1993, 1–2.

Oliver, C. D. 1992. Achieving and maintaining biodiversity and economic productivity. Journal of Forestry 90(9):20–25.

Smith, D. M. 1986. The practice of silviculture. 8th edition. John Wiley and Sons, New York, New York.

———. 1993. Silvicultural perspectives: have we been there before? Pages 16–21 *in* Society of American Foresters, national convention, Indianapolis, Indiana.

Whitney, G. G. 1987. An ecological history of the Great Lakes forest of Michigan. Journal of Ecology 75:667–684.

Chapter 14 **Policies for Protecting Aquatic Diversity**
Douglas J. Norton and David G. Davis

The U.S. Environmental Protection Agency (EPA) has a dual mission that includes both the protection of the environment and the protection of human health and welfare. Throughout most of its existence, the agency has tended to emphasize human health issues, while its attention to ecosystems has remained a secondary priority. A gradual but significant change is taking place at EPA. The agency's policies and programs are taking into account ecosystems, their condition, and their long-term sustainability, along with human health considerations, as EPA moves toward improving its protection of the environment as a whole.

One of EPA's greatest opportunities to influence ecosystem condition as well as guard human health is through its existing authorities to protect water resources. The Clean Water Act goal, to "restore and maintain the chemical, physical and biological integrity of the Nation's waters," is clearly applicable to the management of aquatic ecosystems and protection of biodiversity. Consistent with EPA's changes, the Clean Water Act has evolved from its narrow origins in water pollution control through technology-based, end-of-the-pipe discharge limits to a watershed-oriented statute that can contribute significantly to ecosystem management within watershed boundaries. EPA is moving toward an integrated view of environmental protection that does not unnecessarily segregate protecting human health and welfare from protecting and sustaining ecosystem health, and the Clean Water Act is providing some of its most effective tools.

EPA'S EMERGING CONCERN FOR HABITATS AND ECOSYSTEMS

Given EPA's dual mission to protect the environment as well as people, why has it historically concentrated on human health and why is the agency's concern for ecosystems relatively recent? Some of the contributing factors include limited resources, changes in priorities of different administrations, EPA's brief and still formative history, and statutory authorities that are limited and not designed for protecting ecosystems.

Because limited resources affect staff and funding to implement programs, and thus restrict the flow of scientific information upon which to base regulatory action, these problems have constrained EPA many times in its twenty-five-year history. In difficult economic times, the agency has had to reevaluate priorities among its many program elements. In the late 1970s, for example, EPA's leadership faced limited resources, an expanding mission, and heightened public concern over pollution. The agency's decision was to place a significantly higher priority on protection of human health, and EPA's emphasis on human health over environmental condition grew even stronger throughout the 1980s. The actions of congressional committees that directly affect staff and funding for specific programs have also reshaped EPA priorities, and the agency is also responsive to the public's perceptions and to social as well as environmental concerns. Further, because the relatively young agency has been asked many times during the past two decades to implement new environmental laws along with exercising its existing authorities, it is not surprising that EPA's role has not completely stabilized.

Finally, although EPA's statutory authority includes several useful environmental protection provisions that extend beyond human health, the interpretation of these statutes has tended to address parts of ecosystems rather than their condition as a whole. EPA's statutes separately address air and water media and the environmental effects of hazardous waste disposal and pesticide use, just to name a few. Nevertheless, virtually every EPA statute contains sufficiently broad language to address ecological concerns in EPA programs (Fischman 1992), and as a result, EPA has opportunities to consider the effects of its actions on whole ecosystems even when the individual action may be relatively narrow in scope. The Clean Water Act represents one of EPA's best opportunities to protect ecosystems, especially if the act is interpreted and implemented with a more distinctly ecological emphasis than it was in the 1970s and 1980s.

A turning point in EPA's perspective on the importance of ecosystem protection occurred in 1990, through the efforts of an independent panel of scientists responsible for advising the EPA administrator. The EPA's Science Advisory Board ranked loss of biodiversity, habitat degradation, and species extinction among the highest environmental risk problems facing EPA, alongside stratospheric ozone depletion and global climate change. In its September 1990 report entitled *Reducing Risk: Setting Priorities and Strategies for Environmental Protection,* the board advised the EPA administrator that "human health and welfare ultimately rely on the life support systems and natural resources provided by healthy ecosystems" and that "the Agency should communicate to the general public a clear message that it considers ecological risks to be just as serious as human health and welfare risks, because of the inherent value of ecological systems and their link to human health" (USEPA Science Advisory Board 1990). Furthermore, the board insisted it was well within EPA's responsibilities to take action on ecosystem protection: EPA should "attach as much importance to reducing ecological risk as it does to reducing human health risk." This now widely known report has moved EPA leadership toward recognizing the protection of human and environmental health as one.

In the fall of 1991, an EPA interoffice working group called the habitat cluster was established to formulate an EPA strategy for the problems facing the nation's ecosystems in response to the Science Advisory Board's report. The cluster researched the ecosystem protection-related elements in EPA legislation and programs and analyzed the agency's habitat information needs. The group used its findings to develop an agency ecosystem protection goal, describe EPA's role, and identify the central concepts of an agency strategic approach. The cluster also identified ecosystem and habitat protection programs of other federal, state and private organizations, realizing the unprecedented level of teamwork and coordination that would be necessary to protect and manage major ecosystems in their entirety.

In the habitat cluster's discussions on an ecosystem protection strategy, six basic themes repeatedly emerged as critically important (USEPA 1992a):

- improving the use of existing regulatory authorities;
- focusing EPA's nonregulatory programs;
- improving the habitat science base;
- providing better habitat information management;

- forming effective public and private partnerships;
- using a risk-based approach for setting priorities and making decisions.

The habitat cluster developed its strategy because the EPA's past approach to environmental management and protection had not been sufficient to manage whole natural systems and retain their values and functions, such as pollution filtering and processing, climate moderation, outdoor recreation, and biodiversity. The strategy later proved to be a valuable foundation for the EPA Ecological Protection Subcommittee report to the vice president's 1993 national performance review, as well as influencing many ongoing EPA programs that are beginning to bring about changes at the grassroots level.

Ecosystem protection has begun to enter into EPA decisions more frequently in the 1990s as the agency focuses more of its strategies on the whole environment instead of its isolated parts. Many existing programs within EPA, like the Clean Water Act's watershed-oriented programs and several of EPA's geographic initiatives, are helping to counter the degradation of valuable natural systems by identifying the main ecosystem-level functions and managing them. The agency has increased its use of ecological risk assessment, geographic targeting and basinwide planning, and monitoring status and trends in ecological resource extent and condition (USEPA 1991b, USEPA 1991c, Messer et al. 1991). These activities exemplify EPA's progress toward a more comprehensive, ecosystem-oriented approach to environmental problems and their remedies.

By 1994, EPA was taking decisive steps to institute a new ecosystem management approach that would affect virtually all its programs. EPA Administrator Carol Browner listed ecosystem protection among her top priorities and stated, "By protecting our environment, we are protecting human health. . . . EPA's new agenda means *comprehensive approaches*, not piece by piece, site-specific thinking. We need ecosystem protection, watershed protection, cross media protection" (USEPA 1993f). Although major statutory change was unlikely, EPA was capable of making changes in its methods of doing business under the current statutes. The challenge was to overcome the disjoint, media-specific limitations of EPA's statutes and lay a new foundation for environmental protection that would be consistent with ecosystem management principles as well as protection of human health and welfare. The solution appeared in the idea that environmental protection should be place-based, rather than environmental media- or national program-based, in order to contribute to ecosystem management on appropriate scales while accommodating locally variable settings.

EPA has instituted this new approach, called community-based environmental protection, to improve the effectiveness of its actions by making its national-scale programs more relevant and responsive to people and ecosystems on the local scale. The approach has several key elements. One goal is to assess and manage environmental quality as a whole, treating all the people and natural resources in any given place as interconnected parts of a system. Focusing on a clearly definable geographic unit, such as a watershed, enables nonscientists to more easily understand the concept of managing a natural system that needs care but provides benefits. The geographic focus speeds additional progress in identifying a system's condition, environmental problems and risks, involved parties, and desired environmental results. Community-based environmental protection also takes into account striking differences in environmental conditions, problems and solutions from place to place, counteracting the problems arising from inflexible, nationally designed regulations. Moreover, the approach capitalizes on working more effectively with many partners in environmental protection, public and private. Here, too, EPA's role would vary from place to place and issue to issue. EPA might lead an effort of national or interstate concern in one place while serving as an active partner or providing background assistance in other places.

The long-term effort to incorporate ecosystem protection principles as part of the EPA mission has also been strengthened by many career EPA managers and staff who have maintained the commitment to protect ecosystems through wide swings in the issue's popularity during past administrations. In spite of recent progress, however, phasing a community-based approach into the EPA, program by program, will continue to require substantial effort. Informing EPA's staff, cooperators, and the public of the agency's intent and approach to protect ecosystems is merely a start. Transforming some of the ways EPA's programs operate under statutory authorities—for example, using the Clean Water Act to protect and sustain biologically sound aquatic ecosystems and their watersheds—is the next, critically important step. The ultimate measure of success is the degree to which these new policies can transform not just programs, but places.

THE CLEAN WATER ACT: PROTECTING AQUATIC ECOSYSTEMS

Within what the courts have called the "waters of the United States" (NRDC 1975) is a significant share of our nation's most biologically diverse ecosystems—its lakes, rivers, and streams, its wetlands, and its estuaries. These same aquatic systems are priceless resources for such human uses as recreation, commercial and sport fisheries, and domestic and industrial water consumption. Balancing the human and nonhuman ecological dependency upon our nation's waters will be among the greatest challenges of the coming century. Aquatic ecosystems contain a large proportion of not only total species numbers but also endangered and threatened species, suggesting that these ecosystems may also be the most critical future battleground for the protection of biodiversity.

Among EPA's major statutes, the Clean Water Act provides EPA with its most straightforward mandate to protect habitat and biodiversity via its stated goal, to "restore and maintain the chemical, physical and biological integrity of the Nation's waters" (Fischman 1992). The act also refers to the goal of attaining sufficient water quality to provide for the protection and propagation of balanced indigenous populations of fish and wildlife. The act's applicability to policies governing water quantity or flow as an attribute of water quality has gained recent attention as a factor critical to the recovery of some of the country's most outstanding regional ecosystems, including salmon rivers of the Pacific Northwest and the Florida Everglades (Jefferson City 1994, Walters and Gunderson 1994). Despite the ecosystem orientation of Clean Water Act goals, however, the current regulations underlying the act do have limitations. Whereas some of the regulatory tools are well suited for applicability to ecosystem management, they do not cover all the major threats to aquatic ecosystems. Approaches for controlling nonpoint source pollution, for example, are very limited, and the Clean Water Act's reach does not extend to certain types of wetland losses, introduction of exotic species, many types of physical habitat alteration, and direct or indirect harvest (Wilcove and Bean 1994). Nevertheless, this statute offers significant opportunities to contribute to aquatic ecosystem management even if it alone cannot address all possible concerns.

Under the Clean Water Act, EPA administers several different programs dedicated to maintaining and improving the integrity of the nation's waters. Most of the act's programs and activities are implemented in concert with states, American Indian tribes, other federal agencies, local governments, private entities, and citizens. The Clean Water Act therefore not only is among the federal government's most powerful statutory tools for protecting, in the act's own words, "biological integrity" but also provides structure for forming multiple governmental and private partnerships to pursue Clean Water Act goals. The act is predominantly state implemented, with federal oversight.

The broadly defined goal of the act, with the periodic rewriting of key provisions during reauthorization, has allowed the Clean Water Act to evolve over the years from a narrowly interpreted, point-source pollution control law to a versatile statute that can support aquatic ecosystem management and watershed protection. Since 1972, EPA and the states have concentrated on controlling municipal and industrial discharges to achieve the "fishable, swimmable goals of the Act." However, success in controlling these point sources of pollution has revealed the significance of nonpoint-source pollution as a source of impairment to the nation's waters. In addition, the need to focus on the physical and biological components of water quality has become increasingly apparent.

To address the multiple, cumulative problems that threaten the nation's waters and yet remain focused on the goal of ecosystem integrity, the Clean Water Act operates at several interrelated levels (see fig. 14.1). The top level represents the ultimate intent of the law, which is to attain and maintain a level of ecosystem integrity that will sustain both human uses and ecosystem condition. The physical/chemical/biological goal statement emphasizes that the act is not limited to seeking acceptable water quality based on water chemistry alone. At the next level, water quality standards play the key role of defining the environmental goals in tangible, specific terms representing the overall conceptual goals at the top of the pyramid. Beneath this level, the watershed approach provides a unifying philosophy of comprehensive, integrated analysis and management that guides each individual program. The many EPA and state programs in water quality–based management constitute the bottom tier of the pyramid and represent the watershed approach's "tools in the toolbox." This diverse assortment of programs and activities range from the specific to the comprehensive. At this level, the Clean Water Act provides for:

- development and implementation of water quality standards;
- state water quality monitoring and assessment;
- identification of impaired waters that fail to support water quality standards and threatened waters that may have future problems;
- development of national effluent guidelines and implementation of technology-based controls for municipal and industrial point sources;
- development of stringent water quality-based controls when technology-based controls are insufficient to maintain water quality standards;
- wastewater discharge permits and enforcement;
- state development and implementation of nonpoint-source control programs;
- federal grants and loan programs that support treatment plant construction

Fig. 14.1. Conceptual diagram of how the Clean Water Act's programs, watershed protection approach, and state water quality standards support the ultimate goal of ecosystem integrity.

and upgrading, state nonpoint-source pollution programs, coastal programs, wetlands protection, and other activities.

Within the framework of these programs are latent opportunities to innovate under the broad and flexible Clean Water Act banner. Although there has been noticeable progress in broader and more creative interpretations of the act during the 1990s, the full range of opportunities is still underexploited and just beginning to be explored. For example, one can use the Clean Water Act to:

- expand EPA regional work with states, tribes, and other agencies to address ecosystem protection through basin management plans;
- expand efforts to develop and accelerate state adoption of better narrative and numeric biological criteria, criteria for physical habitat parameters, and wetland criteria;
- continue to develop chemical criteria to protect wildlife, sediment quality, and trophic condition;
- improve biological assessment and monitoring techniques and incorporate their use in the biennial national water quality inventory and report to Congress;
- expand use of total maximum daily loads (TMDLs) for nonchemical stressors; increase consideration of habitat degradation in each state's decisions on where to develop TMDLs;
- develop more rigorous design guidance, funding criteria, and technical tools

for ecological restoration; place highest priority on techniques that use natural ecological processes to attain clean water and stable ecosystems;

• increase the application of ecological risk assessment concepts to watersheds; develop and issue watershed ecorisk guidelines;

• use and increase the emphasis on the outstanding national resource waters provision to better define critical areas for aquatic biodiversity conservation in each state and nationwide;

• provide technical assistance to states implementing best management practices that directly enhance biodiversity; support education and outreach efforts concerning management practices for areas like riparian zones that are critically important to protecting aquatic biodiversity.

THE PHILOSOPHY OF WATERSHED PROTECTION

In EPA's Office of Water, watershed protection has become a conceptual foundation for its many Clean Water Act programs. The watershed approach, not in itself a government program, was derived from many years of collective experience in state and federal agencies. It is a flexible framework of guiding principles that supports and unifies existing water programs and helps them operate in a more comprehensive and coordinated manner. This underlying philosophy of ecosystem-oriented water quality management focuses on the whole watershed instead of only its component parts and the impacts of individual stressors.

Several main principles characterize the watershed approach. First, protection efforts focus on hydrologically defined natural systems including watersheds, bodies of water, aquifers, or composites of all three. Second, the watershed approach emphasizes public and private partnerships among the watershed's communities and involved stakeholders in the analysis of problems and the creation of solutions. Their involvement adds keen insight, heightens the sense of community stewardship of the watershed ecosystem, and provides a focus for public participation. Watershed management can then draw from the broad range of abilities and authorities that are available among the participants and which often exceed EPA's own contributions, and it can then use these tools to generate integrated solutions to the watershed's problems (USEPA 1991c). Third, the watershed approach employs sound environmental management techniques based on good data and good science. Watershed protection, focusing on the whole watershed as an ecosystem or mosaic of ecosystems and using the full range of technical abilities and authorities, is necessary to integrate all components of water quality–

based management and reach the Clean Water Act's highest goal of physical, chemical, and biological integrity.

A watershed approach to managing aquatic systems can also support protection and enhancement of biological diversity, at scales ranging from genetic diversity of an individual population to habitat diversity across whole regions. On the local scale, for example, Clean Water Act–supported clean lakes grants or best management practices for nonpoint-source pollution control can be tailor-made to reduce impacts on species diversity in resident aquatic communities (Bennett et al. 1993) or may conserve genetic diversity by prescribing target levels for smolt survival in a threatened salmon population (Baker et al. 1993). On broader scales, EPA is working with the states, other federal agencies, and American Indian tribes to develop statewide whole-basin management plans to meet statewide and regional goals (USEPA 1993c). Some of these plans explicitly address parameters crucial to the quantity and quality of aquatic habitat. Flow, volume, temperature, bank and riparian vegetation condition, sedimentation, large woody debris, and riffle/pool ratios are examples of such plan elements, and all can help determine habitat quality and affect biodiversity at several scales.

Watershed protection, as a term open to very different interpretations, does not automatically imply attention to ecological concerns. EPA's usage of this term, however, has clearly demonstrated its intentions to take an ecosystem management approach to its activities in watersheds. In a May 13, 1994, memorandum, Office of Water Assistant Administrator Robert Perciasepe wrote to regional and national directors in this EPA office, "The watershed approach is ecosystem management within watershed boundaries."

INNOVATORS: WATER PROGRAMS THAT SUPPORT ECOSYSTEM MANAGEMENT

Virtually dozens of federal and state programs exist under the Clean Water Act's expansive mission. Although some are narrowly defined activities linked to end-of-the-pipe pollution control, others have shown particular promise in addressing much broader ecological concerns, like restoration and protection of aquatic and riparian zone habitat and biodiversity.

Monitoring, Identification, and Listing of Impaired Waters

The Clean Water Act requires states to compile monitoring data, periodic reports, and other databases that can be valuable sources of ecosystem condition–related information over large areas. Although these activities have not yet involved a thorough characterization of the most valuable or pristine aquatic ecosystems, they do provide information on the stressors that may threaten them.* Clean Water Act section 303(d) requires states, or EPA in the absence of an acceptable state submission, to identify biennially bodies of water that do not meet or maintain water quality standards. High-priority waters from the 303(d) list are selected for development of total maximum daily loads. Section 305(b) water quality inventory reports to Congress identify bodies of water that do not fully support designated uses and are a source of information about stressors that may impact ecosystems. Lists submitted under these statutory requirements can indicate where biodiversity is declining. States are also encouraged under 303(d) to identify threatened bodies of water on these lists, although these may not be degraded at the time. Designation as a threatened body of water may be used as an early warning.

Water Quality Standards and Criteria

Setting standards for permissible levels of pollution in the environment is among EPA's core mandates, and thus the agency can use this authority to protect biodiversity and other aspects of ecosystem condition. Water quality standards, and particularly their numeric criteria, are among the most powerful Clean Water Act provisions in that they can provide reference points for supporting the aggressive use of many other Clean Water Act provisions to protect ecosystems. The regulatory muscle of the Clean Water Act is exercised, for example, in decisions to deny a national pollutant discharge elimination system (NPDES) permit that will violate standards, to fund a restoration project to improve water and meet unattained standards, or to implement pollutant loading reductions in order to meet standards.

Under the Clean Water Act's 1987 amendments, all states must adopt EPA-approved standards for their waters that define specific water quality goals for individual bodies of water.† Standards consist of three parts: designated beneficial uses to be made of the water, criteria to protect those uses, and antidegradation provisions to protect existing water quality. At a minimum, state beneficial uses must provide for "the protection and propagation of fish, shellfish, and wildlife" and must support "recreation in and on the water"

---

* The Clean Water Act at 40 CFR 131.12(a)(3) does provide for identifying outstanding national resource waters. Designation affords the highest degree of antidegradation protection. The provision, however, has not generally been used to provide comprehensive inventories or priority-ranked lists of such waters in most states.

† The term *states* as used here in reference to standards is meant to include tribes, which are also authorized to develop water quality standards.

where attainable (USEPA 1990). Whether these can address the ecosystem as well as chemical water quality depends on the extent that the specific standards directly address valued ecological features and functions. Much less can be accomplished, however, where standards and criteria are not sensitive to biological types of impairment in the first place.

Ironically, water quality standards can be among the act's weakest as well as its strongest elements for protecting ecosystems because of differences in and deference to state implementation. Although there are national EPA guidelines and mandated state adoption of standards for 126 toxic pollutants, water quality standards still vary considerably from state to state and from one body of water to another. Some states' standards are much more strict than others' in addressing biotic components, and this disparity aggravates competitive efforts between states and related conflicts between economic growth and environmental protection (Wilcove and Bean 1994).

Standards and particularly criteria can be tailored to provide ecosystem integrity targets and a system for measuring progress achieved by water quality programs. EPA is directing states to develop and adopt biological criteria into state water quality standards over the next few years. Biological criteria are numerical values or narrative expressions that describe reference levels for the biota expected to be present in waters, with standards specifying a given aquatic life support use (USEPA 1990). Whereas most chemical or physical criteria can reasonably be expected to help protect the ecosystem indirectly, biological criteria have the potential for a more direct, positive effect. Biological criteria are also valuable because they indicate the condition of the biological community at risk and may detect problems that other (for example, chemical) methods may underestimate or miss entirely (USEPA 1990).

Ohio provides an example of a state that has effectively developed and implemented biologically oriented standards. The Ohio EPA's surface water monitoring and assessment program have been progressive in their use of biological criteria. Since the late 1970s, Ohio has operated biological surveys to provide a baseline set of biological and chemical data to support monitoring and development of water quality standards. The surveys have led to the development of biological criteria as an ambient aquatic life assessment tool. The many uses of their biological criteria include numeric standards development, point-source discharge permitting, enforcement/litigation, and storm water management.

Ohio's criteria were developed with reference to common indexes of biological and community integrity and were derived from results of sampling conducted at more than three hundred marginally impacted reference sites.

Fish and macroinvertebrate data were used to establish attainable, baseline expectations within different habitats and ecoregions across the state, and were arranged within the framework of a tiered system of aquatic life use designations. The state is also attempting to develop ways to recognize impact signatures from specific stressors that may alter the ecosystem in a reasonably consistent and recognizable pattern (Ohio EPA 1990).

As Ohio implemented its use of biological criteria in statewide biennial monitoring, the perception of relative impairment of the state's bodies of water changed significantly. Although many factors undoubtedly contribute to the perceived differences in results, differential sensitivity of biotic and chemical measurements in detecting impairment was believed to be an important part of the explanation. Between the 1986 and 1988 305(b) reports, for example, failure to attain designated aquatic life uses rose from 9% (using chemical water quality criteria) to 44% (using numeric biological criteria linked to expected aquatic community composition in the ecoregion). Further comparing chemical and biological criteria, the state found that biological criteria detected an impairment in 49.8% of the situations where no impairment was evident with chemical criteria alone. Agreement between chemical and biological criteria was evident in 47.3% of the cases, while chemical criteria detected an impairment in only 2.8% of the cases where biological criteria indicated attainment (Ohio EPA 1990). Based on their experiences, the state's surface water monitoring and assessment program has recognized that biological criteria must play a key role in defining water quality standards and in evaluating and monitoring standards attainment if the goal to restore and maintain the chemical, physical, and biological integrity of their waters is to be met.

The other two components of standards—designated uses and antidegradation—have not received the degree of attention focused on criteria. But a 1994 U.S. Supreme Court ruling (Jefferson City 1994) endorsed for the first time the antidegradation regulation and reinforced the recognition of water quantity as integral to water quality and related designated uses. The decision upheld the State of Washington's right to require a minimum water flow necessary to protect salmon and steelhead and to disapprove a hydroelectric plant application that would have diminished the existing flow. Writing for the Court, Justice Sandra Day O'Connor labeled the separation of quality and quantity of water an artificial distinction that had no place in a law intended to give broad protection to the physical and biological integrity of water. Justice O'Connor added that a lowering of water quantity or flow was capable of destroying all designated uses, and that the Clean Water Act's definition of pollution was broad enough to encompass the effects of reduced water flow. In short, the decision upholds states' ability to ensure compliance with the narrative as well as numeric criteria components of their water quality standards and reinforces the broad interpretation of the Clean Water Act in the critically important area of water flow.

Total Maximum Daily Loads

When technology-based controls on point-source discharges such as discharge permits are insufficient to maintain water quality standards, more stringent, water quality–based controls must be developed. This process involves determining for a given pollutant or stressor the total maximum daily load (TMDL) that a given waterbody can assimilate and still meet water quality standards. TMDLs thereby establish a key link between water quality standards and point- *and* nonpoint-source pollution control actions designed to achieve and maintain a desired condition in the waterbody. States develop and implement TMDLs based on a priority ranking of impaired or threatened bodies of water (USEPA 1991a).

A TMDL can theoretically be developed for any type of stressor. The basic TMDL process models the loading capacity of a body of water for a given stressor and ultimately provides a plan for allocating loadings (or external inputs) among pollutant sources, many of which are then requested to reduce their loading contribution. In doing so, the TMDL quantifies the relations among sources, stressors, water quality conditions, ecological effects, and recommended controls. Pollutant loadings can be expressed in a number of ways, using terms such as mass per unit of time, energy (heat, for example), toxicity, or other appropriate measure. In addition, controls identified relative to implementing the TMDL may range from discharge permits for point sources to incentives or voluntary best-management practices. For example, a TMDL may mathematically show how using a variety of control measures is necessary to reduce pollutant loading from sources throughout the water shed and to reach the pollutant concentration that will meet a state water quality standard.

The term *total maximum daily load* does not immediately convey the full meaning of the TMDL concept. Historically, there has been confusion concerning the applicability of TMDLs, particularly with respect to nonpoint-source pollution, nonattainment of narrative water quality standards, or impairments such as physical degradation of aquatic habitat. In fact, the TMDL process has the flexibility needed for developing comprehensive, watershed-based solutions for many types of problems that impact aquatic ecosystems. As larger numbers of permits incorporate water quality-based effluent limits, TMDLs are becoming an increasingly important component of state point-source control programs. The TMDL process is also appropriate for addressing such nonpoint and cross-media problems as aerial deposition of pollutants, pollutant transfer through contaminated sediments, inflow of contaminated ground water, and pollutant migration from waste sites. TMDLs are applicable to any kind of body of water impacted by point sources only, nonpoint sources only, or a combination of both point and nonpoint sources, and can be used to address the effects of loading on a literally unlimited variety of ecological endpoints of concern.

EPA regional offices and state water programs across the country are exploring the application of the TMDL process in a series of recent case studies (USEPA 1992–1994). Many of these were directed at ecological rather than strictly water quality problems. The development of a TMDL for fine sediment on Idaho's Salmon River South Fork, for example, addressed sedimentation of salmon and steelhead spawning areas caused by silviculture (USEPA 1992d). EPA, the state, and the USDA Forest Service set allowable levels for depth of fine sediments and cobble embeddedness based directly on the designated use by cold water biota implicit in the Salmon River's water quality standards. The controls agreed upon by the participants prescribed forestry best management practices to minimize sedimentation and selected rehabilitation projects such as sand removal from pools, closing and reclaiming logging roads, and stabilizing slopes. As with any TMDL that is undergoing phased implementation due to limited data, this TMDL provides for monitoring to track progress and revise plans if necessary.

Other case studies also show how the TMDL process can help protect ecosystems. Some, although targeting pollutants and standards not directly involving ecological features, have led to measurable benefits in biodiversity; for example, implementation of the TMDLs developed for metals and instream flow at Clear Creek, Colorado, was followed by a dramatic increase in macroinvertebrates and the resident trout population (USEPA 1992e). Other studies, while not yet producing actual TMDLs, followed creative versions of the basic TMDL process to achieve ecosystem protection goals. In Boulder Creek, Colorado, ammonia and nutrients from numerous point and nonpoint sources were quantified through monitoring and modeling, and then the city addressed the aquatic life impairment problem through a combination of wastewater treatment plant upgrade, best management practices, and ecological restoration projects (USEPA 1993g). In the Tar-Pamlico Basin of North Carolina, the state used the TMDL requirement to consider all sources of a pollutant to arrange a trading program between municipal dischargers and farmers in the same watershed as an innovative and cost-effective way to meet nutrient reduction goals (USEPA 1993h).

Developing TMDLs for several of an area's most valued aquatic ecosystems, whether threatened or already impaired, would be a highly logical, quantitative, stress-response basis for identifying these ecosystems' management needs and planning their protection. Such stressors as physical alteration of habitat may not clearly fit traditional concepts such as chemical loading, and this may make it challenging to formulate the problem in TMDL terms; however, this does not signify in any way that it is inappropriate to develop a TMDL of this type. The fact that TMDLs have not been more widely used to implement ecosystem management is a reflection of historical state water quality management priorities and the limited data and resources available to state agencies, not an indication of limited applicability of TMDLs to ecological concerns.

Nonpoint-Source Control

As directed by Section 319 of the Clean Water Act, enacted in the 1987 amendments, all states and territories have developed and are presently implementing EPA-approved statewide nonpoint-source management programs. The state programs focus on preventing and controlling nonpoint-source pollution to waters previously identified by the state as either impaired or threatened by nonpoint sources. As of 1994, Congress had appropriated more than $270 million to states to assist them in implementing approved nonpoint-source management programs. Funded activities were statewide or watershed-specific. These activities have ranged from education and outreach, technical assistance, and effectiveness monitoring, to projects that demonstrate nonpoint-source control technology and achieve on-the-ground implementation of effective best-management practices. Whenever possible, Section 319 activities are integrated with the resources and programs of such other federal agencies as the U.S. Department of Agriculture and the U.S. Department of the Interior to amplify the resources being used to resolve nonpoint-source water quality problems. Many states also provide additional funding for nonpoint-source management activities.

Restoration of aquatic ecosystems is gaining momentum as an effective contributor to pollution control and processing, as well ecosystem protection. In 1994 EPA directed that 10 percent of each state's Section 319 funds target watershed resource restoration activities. Restoration activities were defined broadly and included activities that restore wetlands, lakes, rivers, streams, coastal zones, estuaries, shorelines, riparian areas, seagrass beds, coral reefs, and other aquatic habitats. EPA's guidance for awarding these funds cautioned that restoration planning should consider the ecosystem's natural processes as well as root causes of deterioration or decline.

The emphasis on nonpoint pollution in coastal zone protection is also increasing. As required under the Coastal Zone Act Reauthorization, EPA has issued national guidance on the best available management measures to control nonpoint pollution. EPA's guidance goes beyond describing typical structural controls for preventing physical and chemical water quality problems to recommending land use and other management practices for maintaining riparian zones and instream habitats.

Although most remedies for nonpoint-source pollution secondarily improve habitats and aquatic communities, the nonpoint-source program under the Clean Water Act can be intentionally focused on ecological features of concern through such approaches as funding certain types of best-management practices, development of nonpoint-source habitat-oriented TMDLs, and ecological restoration projects. For example, a State of Virginia Section 319 project on the Holston River (Bennett et al. 1993) used advanced monitoring techniques to evaluate best-management practice options and recommend loading reductions to meet standards. This study attempted to formulate a procedure to use biological indicators to estimate nonpoint-source loading allocations. The investigators used the index of biotic integrity, or IBI (Karr et al. 1986), to evaluate the biological conditions in the subwatersheds of their study area. They then used geographic information systems and airphoto analysis to document subwatershed land uses in fine spatial and categorical detail to identify and quantify all the stressors.

The index of biotic integrity and individual land use/land cover types in the Holston subwatersheds were related through statistical correlation analysis. Land cover parameters with the highest correlations with the IBI included sediment yield and the ratio of good pasture to overgrazed pasture. Based on the results, the geographic information system was then used to develop a "what if" methodology to evaluate best-management practice scenarios. Strong points of this geographic information system application included the ability to demonstrate the scenarios publicly at meetings and the capability to target specific properties and land cover types that were making the most critical contribution of pollutants. The resulting data thus provide the state with an opportunity to consider funding incentives for best-management practices that are most likely to reduce risks to the Holston River's biota, as well as the needed quantitative data on stressors, loadings, and necessary loading reductions to develop and implement the TMDL process.

In some locations, nonpoint-source control and the TMDL process are being merged to address ecological problems. In the Upper Grande Ronde River watershed of northeastern Oregon, logging, grazing and other land use

practices have removed shade along the river and changed channel morphology. These practices have elevated water temperatures to levels that are
harmful or lethal to juvenile and adult salmonids (including trout and
salmon). More than 80% by length of the basin's tributaries are affected.
With virtually no point sources involved, the controls recommended by the
TMDL will need to reduce heat loads from nonpoint sources. The controls
will reduce heat loading by reestablishing stream shading, through forestry
best-management practices, riparian zone fencing, and a range of riparian
zone restoration techniques. Many federal agencies, tribes, local citizens, and
nongovernmental organizations have become involved in the Grande Ronde
effort as landowners, scientists, or funding partners. Cooperation among them
in implementing the voluntary controls will be crucial to achieving the
TMDL's target levels for temperature and saving the salmonid populations
(USEPA 1994a).

Wetlands Protection

As wetlands are included within "waters of the U.S.," they are afforded
protection under Clean Water Act. The wide variety of wetland flora and
fauna, including many endangered species, attests to the critical importance
of wetlands protection as a component of any effort to sustain biodiversity
(CEQ 1991). For this and other reasons, the protection of wetlands under
Clean Water Act has had a deeper relation to ecosystem protection than most
other Clean Water Act programs that have been traditionally associated with
pollution control and the protection of human uses.

Most federal wetlands protection is accomplished through the Clean Water
Act Section 404 program. In addition to wetlands, the 404 program covers rivers, lakes, streams, and other "special aquatic sites," such as mud flats or riffle
and pool complexes, that may possess special ecological characteristics of productivity, habitat, wildlife protection, or other important and easily disrupted
ecological values. Under this program, EPA oversees and holds veto authority
over U.S. Army Corps of Engineers decisions on permit applications to discharge dredged or fill material into the waters of the United States, including
wetlands.* Such discharges are commonly associated with such activities as
filling to create development sites, water resources development projects such
as dams or impoundments, and conversions of wetlands for farming and forestry (USEPA 1993b). Applications are evaluated through a process called se-

* EPA's veto authority is described at CWA Section 404(c). An as-yet unused provision of
this authority allows EPA to protect intact, high-quality aquatic ecosystems by restricting in
advance any permit application that might otherwise destroy or degrade the ecosystem.

quencing; the sequencing options, from most to least favorable, include impact avoidance, minimization, and compensatory action. The permit review is based on the premise that no discharge of dredged or fill material is permitted if there is a practicable alternative that would be less damaging to the aquatic environment. In addition, the applicant must show that all unavoidable adverse impacts will be minimized to the extent practicable. Finally, the applicant must mitigate or compensate (for example, though restoration or creation of similar aquatic habitats) for the losses incurred.

Protection of wetlands biodiversity and habitat is a consideration at any step in the sequencing process. Army Corps of Engineers permit review guidelines, for example, specifically call for the consideration of loss of biodiversity through fragmentation during review. In addition, consideration is given to cumulative effects of proposed discharges. Should an application continue to the compensation stage, a mitigation project involving wetlands creation and/or restoration may be proposed. If a wetland can be restored or created that is of equal or greater value (as reflected by its capacity to provide similar physical and biological functions) than the wetland being destroyed, mitigation plans may be approvable. Mitigation banking has been endorsed in the Clinton administration's wetlands plan and in EPA and Army Corps of Engineers guidance documents (USEPA and USACOE 1993). This market-oriented approach creates, restores, enhances, or preserves wetlands in advance of permit applications that are expected to require mitigation. Credits accrued in a mitigation bank may be used by bank sponsors to compensate for their own impacts or sold to others who are required to provide mitigation in the same watershed or ecoregion in which the bank is located. Although wetlands creation has thus far exhibited only a mixed record of success, mitigation banking's proponents cite several potential advantages that banking holds over individual mitigation projects. For example, banked wetland sites should be completed and functional before credits may be used, and mitigation sites are expected to be monitored to assure they are functioning as expected. Moreover, the possibility exists under banking to address specific resource needs (for example, habitat for a species of particular concern), or to replace fragmented wetlands in developing areas with contiguous, larger wetlands in better landscape positions.

The advance identification process is yet another Section 404 provision that is highly applicable to ecosystem protection (USEPA 1993a). Advance identification affords EPA and the Army Corps of Engineers the opportunity to issue preemptive notice of likelihood of permit approval or disapproval for activities in wetlands in a given area. As the process may involve substantial data gathering and analysis in order to form advance opinions on permit suitability, advance identifications are performed on a limited number of sites. Generally, the chosen sites have both exceptional ecological values and a significant degree of threat to those values. While not in itself a final, regulatory decision, the product of the advance identification process is a planning and communication tool that indicates to the public and local officials the likely response to permit applications in special areas. Fischman (1992) argues for combining advance identification designation of exceptional wetland ecosystems with EPA's use of advance 404(c) veto as a powerful, two-part use of the agency's resources for wetlands protection.

Grants and Loans

The EPA provides states with financial assistance under several different parts of Clean Water Act to help support many of their pollution control programs. These programs include the state revolving loan program for construction and upgrading of municipal wastewater treatment plants; water quality monitoring, permitting, and enforcement; and the development and implementation of nonpoint-source pollution controls, combined sewer and storm water controls, groundwater strategies, estuary and near-coastal management programs, and wetlands protection activities.

Many of these funding programs can support ecosystem management (USEPA 1993i, USEPA 1994b) in ways that regulatory actions lack the flexibility to accomplish. There are also opportunities for states to influence the evaluation of funding applications to favor ecosystem protection. For example, a state may assign priority to restoration projects aimed at biotic communities over projects targeting recreational uses that may conflict with some aquatic species. Or a state may select to place the highest priority on applications consistent with a multiple-use philosophy where the proposed uses are sustainable and do not compete or disrupt ecosystem stability. The flexibility in state-delegated Clean Water Act funding authorities allows for considerable variety in the ways these funds can be applied.

Addressing the Future: Efforts to Reauthorize the Clean Water Act

Efforts to pass a reauthorized Clean Water Act drew significant attention and debate in the 103d and 104th Congresses, but no consensus has yet been reached. The bill considered by the 103d Congress was supported by the Clinton administration and contained many progressive themes consistent with ecosystem management and watershed protection.

In January 1994, the president publicly stated his position on many of the issues fundamental to Clean Water Act reauthorization in *President Clinton's*

Clean Water Initiative, commonly called the *Green Book* (USEPA 1994b). This document asked Congress, in reauthorizing the Clean Water Act, to "enter a new era of environmental protection. Instead of simply controlling the end of the discharge pipe, we propose to protect and conserve our water, aquatic habitats and the living resources within, through an integrated, holistic approach, based on natural watersheds, and aimed at reducing pollutants from all sources that impair water quality" (USEPA 1994b, v–vi). Many of the initiative's main points remain relevant to strengthening ecosystem management under a reauthorized act.

Increasingly Making Watersheds the Basis for Analysis and Action

Whereas watershed protection has thus far been an emerging philosophy, the initiative proposed a new provision in the Clean Water Act to establish state-wide programs for comprehensive watershed management. The state programs would focus attention on hydrologically defined watershed units, involve stakeholders throughout the public and private sectors, and act on priorities to address the problems unique to each watershed.* Along with this emphasis on watershed management, the initiative also proposed guidelines for market-based approaches to point- and nonpoint-source pollution controls in the watershed and the development of wetlands management plans and permitting procedures on a watershed basis. Watershed management plans would include a thorough characterization of the watershed, including analyses of pollution sources, wetland and groundwater resources, and valuable aquatic habitats. Enforcement actions formerly conducted through separate Clean Water Act programs would consider their broader implications in the whole watershed context.

Reducing Nonpoint-Source Pollution Through Enforceable Minimum Controls in Selected Waters

The initiative proposed that EPA, in conjunction with other federal agencies and states, should establish national guidance for the best available management measures to control nonpoint pollution. States should apply their nonpoint programs to existing nonpoint sources in targeted waters whose quality is impaired, threatened, or deserving of special attention, and to new sources. Federal agencies must carry out state programs on federal lands as nonfederal entities do elsewhere.

---

* The initiative includes all states, territories, and eligible tribes in its usage of the term *state watershed programs.*

Establishing More Current and Comprehensive Water Quality Standards

Since the mid-1980s, research and criteria development have expanded beyond the focus on chemicals to incorporate the full range of ecological problems. Scientific advancements have improved the ability to set ecological criteria based on sound science, but more progress is still needed. The initiative stated that EPA should develop criteria and implementation guidance based on current science, with development priorities based on risk and effectiveness. The initiative recommended that states be required to adopt and apply biological, nutrient, habitat, and other forms of ecological criteria as expeditiously as possible, but not later than four years from their publication in impaired or threatened waters. Water body use designations were also recommended for review and improvement, to fill gaps where designated uses are incomplete. The Clean Water Act, the initiative stated, should also require protection against unacceptable degradation of special-value waters. The initiative recommended several enhancements to the outstanding national resource waters provision, aimed at encouraging states to increase its designation of important waters with additional tiers of protection and less rigid restrictions for outstanding national resource waters designation. EPA would gain authority to designate outstanding national resource waters or challenge inappropriate uses of designated waters.

Improving Comprehensive Inventory and Monitoring of Bodies of Water

Currently, the biennial 305(b) reports cover less than 20% of the river miles in the nation and an even lower percentage of the nation's lakes, estuaries, wetlands, and coastal shores. Clean Water Act Sections 303 (TMDLs), 304 (problem waters), 314 (lakes grants), and 319 (waters needing nonpoint controls) require separate lists, also not comprehensive in coverage. Resources sufficient to fund comprehensive coverage in monitoring and assessment remain improbable. Inventories of state waters, however, can be improved, according to the initiative, within two years of enactment and required every five years thereafter. Each inventory would record (1) all types of waters for which pollution controls will not be likely to help achieve standards; (2) waters threatened by any pollutant or stressor; (3) waters requiring protection of drinking water; (4) waters requiring protection of endangered species habitat, special protection waters, or designated outstanding national resource waters; (5) sources contributing to the failure to meet standards for each body of water; and (6) delineation of watersheds for all bodies of water listed, including urban waters, wetlands, and estuaries. EPA would be responsible for review and approval.

Establishing Watershed Management Grants and Expanding State Revolving Fund Applicability to a Wider Range of Activities

The several separate grant authorizations currently supporting Clean Water Act water programs would be modified to reduce the administrative and tracking burden they represent and to assure their accordance with the watershed management provisions of the act. The more flexible, consolidated grants and funding programs would allow states to manage federal funds through a single plan and enable states to shift funds to high priorities. Project funding eligibility for municipalities would also be expanded to include riparian habitat restoration in priority watersheds, control of all municipal pollution sources, and programs for water conservation and pollution prevention.

The federal government is still laboring over the reauthorization of the Clean Water Act. Many federal, state, and local agencies and many citizens and private organizations await the new legislation with great anticipation: will the act continue to evolve, and in what ways? How might our experiences since the 1987 reauthorization be used to improve Clean Water Act? In any event, the Clean Water Act's history has shown that an overriding goal to "restore and maintain the chemical, physical and biological integrity of the Nation's waters" leads inevitably to a focus on the watershed as the critical management unit and aquatic biodiversity as the primary indicator of our degree of success in aquatic ecosystem management.

The authors gratefully acknowledge Ann Beier, Don Brady, John Ettinger, Geoffrey Grubbs, Susan Jackson, Tom Kelsh, Theresa Tuano, and anonymous reviewers for providing information and/or reviewing this chapter.

This chapter has been reviewed in accordance with U.S. Environmental Protection Agency policy and approved for publication.

LITERATURE CITED

Baker, R., R. Shanks, R. Rains, and L. Walker. 1993. Watershed management approach to toxicity control: California style. Water Environment Federation 66th annual conference and exposition, Anaheim, California. October.

CEQ. 1991. Twenty-first annual report of the Council on Environmental Quality. U.S. Government Printing Office, Washington, D.C.

Federal Water Pollution Control Act of 1972. Public Law 92-500, 86 Stat. 816. Amended in 1977 and 1987, referred to as the Clean Water Act, codified at 33 U.S.C. 1251-1387 (1988).

Fischman, R. L. 1992. Biological diversity and environmental protection: authorities to reduce risk. Environmental Law 22(435):435–502.

Jefferson City. 1994. Jefferson City Public Utility District v. Ecology Dept. of Washington, No. 92-1911. U.S. Supreme Court Opinion.

Karr, J. R., K. D. Fausch, P. L. Angermeier, P. R. Yant, and I. J. Schlosser. 1986. Assessing biological integrity in running waters: a method and its rationale. Illinois Natural History Survey special publication no. 5.

Messer, J. J., R. A. Linthurst, and W. S. Overton. 1991. An EPA program for monitoring ecological status and trends. Environmental Monitoring and Assessment 17: 67–78.

NRDC. 1975. Natural Resources Defense Council v. Callaway, 392 F. Supp. 685 (D.D.C. 1975).

Ohio EPA. 1990. Use of biocriteria in the Ohio EPA surface water monitoring and assessment program. Division of Water Quality Planning and Assessment, Columbus, Ohio.

USEPA. 1990. Biological criteria: national program guidance for surface waters. Criteria and Standards Division, Office of Water, EPA-440/5-90-004. U.S. Environmental Protection Agency, Washington, D.C.

———. 1991a. Guidance for water quality-based decisions: the TMDL process. EPA 440/4-91-001, USEPA Office of Water, Washington, D.C.

———. 1991b. Framework for Ecological Risk Assessment. EPA/625/R-92/001, USEPA Risk Assessment Forum, Washington, D.C.

———. 1991c. The watershed protection approach: an overview. EPA 503/9-92/002, USEPA Office of Water, Washington, DC.

———. 1992a. Protecting habitats and ecosystems: an EPA strategy. USEPA Habitat Cluster, Washington, D.C. (Unpublished draft).

———. 1992b. Securing our legacy: an EPA progress report, 1988–1991. Report no. 175R-92-001, EPA Office of Communication, Education and Public Affairs. Washington, D.C.

———. 1992c. National water quality inventory: 1990 report to Congress. EPA503/9-92/006, USEPA Office of Water, Washington, D.C.

———. 1992d. TMDL case study: South Fork of the Salmon River. EPA841-F-93-002, USEPA Office of Water, Washington, D.C.

———. 1992e. TMDL case study: West Fork of Clear Creek. EPA841-F-93-003, USEPA Office of Water, Washington, D.C.

———. 1992–1994. TMDL case studies 1–10. EPA841-F-93-001–010, USEPA Office of Water, Washington, D.C.

————. 1993a. Advance identification (ADID). Wetlands fact sheet #28; 843-F-93-001bb, USEPA Office of Water, Washington, D.C.

————. 1993b. Clean Water Act section 404: an overview. Wetlands fact sheet #7; 843-F-93-001g, USEPA Office of Water, Washington, D.C.

————. 1993c. The watershed protection approach: annual report 1992. EPA840-S-93-001, USEPA Office of Water, Washington, D.C.

————. 1993d. Clean Water Act reauthorization: special edition. OWOW quarterly highlights, summer.

————. 1993e. Oregon developing TMDLs that address habitat alteration. OWOW quarterly highlights, summer.

————. 1993f. New directions at EPA. EPA InSight policy paper EPA-175-N-93-026, December.

————. 1993g. TMDL case study: Boulder Creek, Colorado. EPA841-F-93-006, USEPA Office of Water, Washington, D.C.

————. 1993h. TMDL case study: Tar-Pamlico Basin, North Carolina. EPA841-F-93-010, USEPA Office of Water, Washington, D.C.

————. 1993i. A commitment to watershed protection: a review of the Clean Lakes Program. EPA841-R-93-001, EPA Office of Water, Washington, DC.

————. 1994a. Upper Grande Ronde River riparian characterization and temperature modeling projects. Project summary, January. EPA841-S-94-001, USEPA Office of Water, Washington, D.C.

————. 1994b. President Clinton's Clean Water Initiative. EPA800-R-94-001, USEPA Office of Water, Washington, D.C.

USEPA and USACOE. 1993. Memorandum: establishment and use of wetland mitigation banks in the Clean Water Act section 404 regulatory program. August 23.

USEPA Science Advisory Board. 1990. Reducing risk: setting priorities and strategies for environmental protection. Science Advisory Board SAB-EC-90-021, Washington, D.C.

Walters, C. J., and L. H. Gunderson. 1994. A screening of water policy alternatives for ecological restoration in the Everglades. Pages 757–767 in S. M. Davis and J. C. Ogden, editors. Everglades: the ecosystem and its restoration. St. Lucie Press, Delray Beach, Florida.

Wilcove, D. S., and M. J. Bean (editors). 1994. The big kill: declining biodiversity in America's lakes and rivers. Environmental Defense Fund, Washington, D.C.

Chapter 15 Ecological Restoration: A Practical Approach

Steven I. Apfelbaum and Kim Alan Chapman

Traditional management of ecological systems focuses on products or services that people want. Commodities that are sold and consumed are managed most intensively. Resource managers learn just enough about ecosystems to maximize the production of these commodities. As a result, ecosystems are overused and misunderstood in an ecological sense.

A different perspective and approach to management is required to avoid damage to ecosystems that people use. Only recently have resource managers begun to appreciate the relation between an ecosystem's condition and the sustainability of human use. Some studies suggest that regional ecosystem degradation can lead to a decline in the production of natural resource commodities (e.g., Yonzon and Hunter 1991). Evidence of widespread ecosystem decline is seen in the growing number of threatened plant and animal species worldwide. Development not only causes declines in species due to habitat loss but also disrupts the disturbances that sustain species composition and structure, thus producing further changes in ecosystems (Haney and Apfelbaum 1990).

Species and ecosystems decline and disappear at local and regional scales, an event that takes only a few years or several decades to complete (Wilson 1988). Ideally, ecological restoration must work on mosaics of ecosystems and ecotones over large landscapes, while at the same time paying close attention to localized species populations, isolated habitats, and the smaller levels of ecological organization.

We define ecological restoration as a practical management strategy that uses ecological processes in order to maintain ecosystem composition, structure, and function with minimal human intervention. A successful ecological restoration requires a full understanding of the ecological problems, a defined course of scientific study through experimental management, and the development of a program for carrying out restoration.

Our definition of restoration is human-centered because restorations come alive to the degree that they satisfy people, whether The Nature Conservancy or a housing developer. The restorationist will be peripheral in America unless he or she acknowledges the dominance of human values in setting goals for restorations.

We are nevertheless enthusiastic and unflagging champions of historical conditions as a *tool* or source of reference material for understanding the composition, structure, and function of modern ecosystems. Other information also serves restoration—including the experiences of rotational grazers (Allan Savory, the Land Stewardship Project), creators of wetlands (the U.S. Fish and Wildlife Service, state agencies), alternative agriculturalists, foresters, and mine site reclaimers. Compared with ecologists, these practitioners refer only minimally if at all to historical systems. Ultimately, the experience of the restorationist will far outweigh the historical information, because the latter reveals the past, while the former discovers the present and future.

Experience through experimentation, in the long run, will tell us much more than tree-core data, peat stratigraphy, General Land Office survey notes, and the anecdotes of settlers. We are patient and ardent students of these sources and the researchers who delve into them. However, while engaged in restoration, hypothesis testing, and recalibration of approaches, our own—and your own—experiences with modern ecosystems will "save all the pieces" and set ecosystems on a trajectory that demands the least amount of input from people to continue it.

If people want restorationists to produce ecosystems that as far as humanly possible re-create historical conditions, then re-creating historical conditions becomes the goal. Such a goal is possible only if people value historical conditions.

In our mind, ecological restoration's goal is the establishment of sustainable, productive ecosystems that humans subsequently benefit from. Analyzing the cost of a restoration and its immediate benefits is very different from short-term cost-benefit analyses for commodity production because the latter does not factor in long-term loss resulting from ecosystem deterioration. However, both commodity production and ecological restoration seek to maximize future human enjoyment and quality of life.

INTRODUCTION TO PRESETTLEMENT ECOLOGICAL SYSTEMS

We will discuss some recent changes in Midwestern ecosystems. Although we describe historical ecosystem conditions, in practice we begin with current conditions because we believe that these better define the limits and possibilities of a restoration. In many cases we find that restoration of all elements of the former composition and structure is not practical or economically possible. We and other restorationists also understand that small-scale restorations will be more successful and less costly when we consider processes at work over an entire region, not just in the neighborhood. The following examples identify some regional issues that should be considered in planning restorations.

Prairie Ecosystems

As late as 1860, prairie ecosystems occupied millions of hectares of North America. Elk *(Cervus elaphus),* bison, and carnivores with large home ranges (for example, the wolf *[Canis lupus]*), together with wildfire, structured the plant and animal communities (Nature Conservancy 1990). Patches of different successional stages were created by the interaction of these regional forces, while small disturbances such as animal burrowing maintained local biodiversity. Settlement reduced the size and complexity of prairies and eliminated animals that needed large acreages (fig. 15.1). The frequency and pattern of fire, grazing, and soil disturbance consequently was changed. The prairie ecosystem was and remains dependent on disturbances to maintain its diversity. However, in the modern landscape, disturbances that used to maintain diversity may reduce it, and new regional conditions make old disturbances problematic.

Loss of grazing and a regional fire regime has increased the growth of woody vegetation. (Haying has somewhat substituted functionally for grazing and fire.) Restoration of bird populations (prairie chickens, for example) may be slowed by lack of vegetation structure maintained by grazing and needed at different points in a life cycle, or by wooded areas close to the restoration which harbor predators of ground-nesting birds.

Air pollution and wind erosion of soil are regional problems that may affect

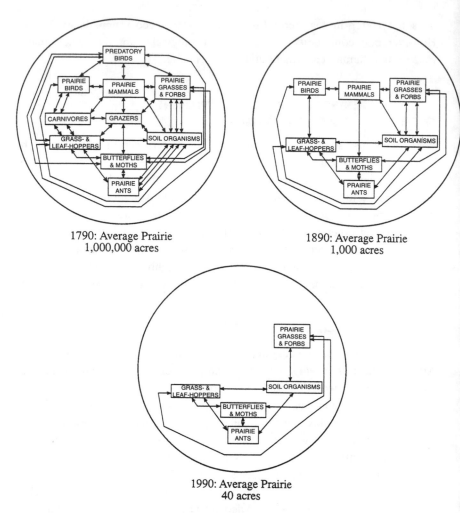

1790: Average Prairie
1,000,000 acres

1890: Average Prairie
1,000 acres

1990: Average Prairie
40 acres

Fig. 15.1. Changes in the prairie ecosystem, presettlement to present day.

restorations by favoring cool-season exotic grasses. Burning or haying helps to remove nitrogen, while grazing retains it.

Exotic species, such as leafy spurge, infest large regions in the Midwest. In such a region, small prairie restorations must withstand a rain of seeds from exotics unless the exotics are controlled in the larger region. Haying and grazing can introduce exotic plants.

Insects constitute the majority of prairie species. The abundance of each insect species depends on fire frequency and timing. Large prairie restorations allow for a diversity of vegetation patches that are managed to produce an abundance of the most fire-sensitive prairie insects somewhere every year (given good weather).

We distinguish between restoration of a process, and restoration of structure through plantings. Prescribed burning is a process introduced by managers because of its effectiveness and low cost. Mechanical removal (haying, brush-hogging) and grazing are also used to maintain treeless conditions. Removing trees and brush can bring back a diverse prairie groundlayer, especially on sandier soils.

Where there is no seedbank or vestiges of prairie plants, plantings are necessary. These re-create structure and reintroduce plant species more quickly than unassisted colonization from nearby prairie. For large-scale plantings, The Nature Conservancy uses combines to harvest good quality prairies and plants the uncleaned seed immediately in cropland treated earlier with glyphosate herbicide and light disking. Grasses form the basis for many large-scale restorations that serve as fair habitat for some birds and mammals. Most small-scale plantings take place in urban or home settings. These prairie restorations can provide prairie birds and mammals increased habitat opportunities.

Wetland Ecosystems

Wetlands have dynamic hydrology, chemistry, and biota. Normal hydraulic changes provide low-intensity disturbances that maintain diversity. The ecological processes that support wetlands (hydrological, nutrient, and sediment cycles) have been altered in agricultural and urban regions. Even wetlands that are unaltered are being degraded. A common cause of this is a change in the regional hydrological cycle, causing a rapid swing in water level from low to high to low as a result of precipitation. Consequently, native biological communities cannot establish or maintain themselves. Streams in urban and drained agricultural settings especially experience this problem. With hydraulic changes come increased sediments and altered nutrient status. These changes further alter the biological community. Along with hydrological changes, nitrogen and phosphorus adsorbed to soil particles carried into wetlands encourage development of monocultures of such aggressive, persistent plants as cattails (*Typha* sp., Apfelbaum 1985) and reed canary grass (*Phalaris arundinacea,* Apfelbaum and Sams 1987). Consequently, native species decline.

Restoration of wetlands requires establishing appropriate hydrological dynamics and fostering adequate water quality to maintain biodiversity. Channels must be stabilized against erosion, runoff reduced, and fire introduced to stimulate vegetation recovery. Often seedbanks in degraded wetlands respond quickly to restoration.

Oak Savanna Ecosystems

Oak savannas once dominated 50–100 million acres in North America but in the past 150 years have been dramatically reduced. Fire and browsing animals (for example, elk) maintained an open understory beneath a largely oak canopy; a savanna structure of grassland with scattered trees and brush; or prairie openings within largely forested areas. Plant and animal diversity was high often because of the proximity of three systems: woodlands, prairie, and savanna. A unique assemblage of species was associated with the savanna structure (Chapman 1984, Packard 1993).

Disruptions to this ecosystem included land clearing and fragmentation, fire suppression, elimination of browsing mammals, and introduction of permanent groundlayer grazing (first by pigs, for the oak mast, later by dairy cattle) and exotic plants.

Degradation also results from increasing shade by woody plants invading beneath the oak canopy. Oak branching permits relatively high amounts of light to reach the ground unless subcanopy and shrub layers intercept it. As shade increases, groundcover plants (grasses, sedges, and native forbs) decrease. These fine-rooted plants that hold the soil decline, and erosion begins on steeper slopes. (Tree and shrub roots do not hold the soil as well.) Runoff is not slowed sufficiently by plant stems, roots, and litter to allow it to infiltrate. Animal and plant diversity diminishes as the open structure is lost; decline of butterflies and breeding birds (Apfelbaum and Haney 1989; Haney and Apfelbaum 1990) is often rapid (fig. 15.2).

In contrast, a savanna ecosystem with abundant and diverse ground vegetation retains both water and topsoil after heavy snowmelt and rainstorms. When next to such a savanna, streams and wetlands exhibit more stable water levels and higher water quality.

Savanna ecosystem restoration requires an understanding of the potential for a site to recover following removal of the understory either mechanically or with fire. In some locations, this is sufficient to restore the groundlayer. Savannas are more easily restored on sands than on heavier soils. However, where shade suppression and erosion are long-standing problems, the soil seed bank should be tested for desirable species. It is critical to control exotic shrubs (European buckthorn, Asian honeysuckle, and others) and herbs (garlic mustard, for example) that invade savanna.

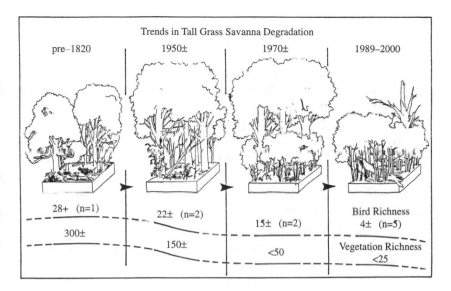

Fig. 15.2. One phase of presettlement savanna had a canopy of scattered oak, little to variable woody understory, and rich biotic diversity. With fire exclusion, trees and shrubs soon invade, eliminating many of the herbaceous species. Eventual closure of the subcanopy prevents oak regeneration and leads to loss of most herbaceous species and a remarkable decline in breeding avifauna richness. N is number of communities we have studied upon which richness data and degradation pattern is based. Time is estimated (Apfelbaum and Haney 1989).

Watersheds

In the Midwest, undeveloped watersheds retain water. This is achieved by (1) nearly continuous upland vegetation cover; (2) numerous isolated wetlands; and (3) headwaters that have no channels. These structures both store water and slow its flow through watersheds. Beaver dams also slow the movement of water through watersheds. Water in uplands and wetlands is lost through evaporation and evapotranspiration, to such an extent that in some streams little flow occurs during the growing season even in normal precipitation years. Because it is held back in the watershed, precipitation infiltrates and may more completely charge groundwater.

Altered midwestern watersheds eliminate water much more quickly. In the agricultural region of the Midwest, covering approximately a sixth of the continental United States, more than 75% of the original upland vegetation and more than 90% of wetlands are gone. Many modern streams were not observed by General Land Office surveyors during the 1830–1860 period.

These former wet swales were incorporated into a web of drains and ditches connected to tile networks in surrounding agricultural fields. The addition of impervious urban surfaces compounds the problems created by these changes.

The overall effect is a dramatic increase in the rate and volume of runoff in Midwestern streams (Apfelbaum in press, fig. 15.3). This destabilizes the soil around watercourses, creating stream channels where they may not have existed, increasing soil and bank erosion, and reducing water quality, vegetative cover, and diversity of aquatic plant and animal communities. Banks slump, and soil transported from uplands increases the scouring power of the flow, further entrenching channels downstream.

Watershed restoration and changes in surface and groundwater levels can threaten other land uses. Before undertaking a watershed restoration, know its impacts on neighbors. Vegetated filter strips along streams, graminoid cover on banks, revegetated uplands, and restored wetlands will stabilize stream banks and beds, reduce fluctuations in flow, stabilize and diversify the biological community, and retain more water, soil, nutrients, and contaminants in the uplands.

A BASIS FOR ECOLOGICAL RESTORATION

It is important to develop an appreciation for the necessity of restoration, for even after being given information about ecological degradation, some individuals are skeptical about the need to intervene. For a fuller discussion, consult essays by William Jordan and others in *Restoration and Management Notes* and *Environmental Ethics*. The objections to intervention with ''nature'' that we have experienced in our projects include:

People, not being part of nature, should leave nature alone. People are separate from nature and should not interfere with it. Ecosystems can take care of themselves. People should not become involved in ''natural'' processes and tinkering with ecosystems.

People are inherently flawed and can only destroy nature. Even if people are part of nature, their current numbers, technology, and lifestyle set them apart as nature's destroyer. Moreover, nature has an inherent sacredness which people should not interfere with because our ability to understand nature is fundamentally limited.

Nature is inherently flawed and will do what it wants to. Plants and animals have always gone, and will continue to go, extinct. The species and habitats in danger of extinction were on their way out anyway. Other plants and animals will arise and replace them. Therefore, why should we care about

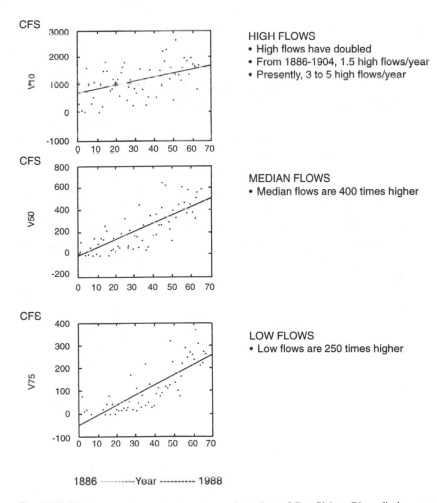

Fig. 15.3. Linear regression analysis and raw data plots of Des Plaines River discharge at Riverside, Illinois, 1886–1988. Low, median, and high flow data were derived from duration-flow curves for 75, 50, and 10 percentile annual flow levels (Apfelbaum, in press).

changes in the biota? Nature is dynamic, and the changes we are seeing now are part and parcel of its natural processes.

People cannot know enough to restore nature. People cannot know what ecological conditions to restore toward. People cannot know whether pre-settlement conditions or more recent conditions are most appropriate to model restorations after. Moreover, ecosystems are so complex, it is impossible to comprehend them.

We respond to these objections by explaining and implicitly defining the kind of restoration we advocate:

For aesthetic reasons (beauty). Nature is beautiful and we should value, preserve, and restore the beauty of the earth, as we would a beautiful or historical building or painting. I, my children, and my grandchildren deserve to see everything that nature has to offer, not just the remains of overused and abused ecosystems. Even though dinosaurs are extinct, what would you give to see one alive?

For the benefit of human use (the utilitarianism of preserving species and ecosystems). Extinction eliminates the potential for people to tap the genetic reservoirs of life. Nearly all our foods, medicines, and fibers come directly from, or are synthetically patterned after, natural materials. Providing future generations with the means of life, including the cure for disease, requires the preservation and restoration of ecosystems. If nature breaks down, then human survival is threatened, because we cannot substitute human labor and capital for such "free" ecosystem services as decomposition, pollination, oxygen production, and climate regulation. Why would we not intervene in nature if it meant saving ourselves?

For the reunification of people and their natural origins (human connection). The close relation that develops between people and the land they restore or manage is mutually beneficial. People restore their connection to nature and benignly transform their attitudes toward it, and nature is enhanced as restoration takes place.

For the benefit of the species and ecosystems themselves. Nature has an inherent right to exist. As human society has evolved, rights have been granted to an ever-widening circle beyond the individual, extended in recent times to include trees, birds, flowers, and all of nature. People have no right to destroy nature. Instead, people should do whatever they can to preserve and restore ecosystems to their most natural state.

Aldo Leopold in *The Land Ethic* expressed these ideas eloquently (Leopold 1966):

A thing is right when it tends to preserve the integrity, stability, and beauty of the biotic community. It is wrong when it tends otherwise. In short, a land ethic changes the role of *Homo sapiens* from conqueror of the land-community to plain member and citizen of it. It implies respect for his fellow-members, and also respect for the community as such. To sum up: a system of conservation based solely on economic self-interest is hopelessly lopsided. It tends to ignore, and thus eventually to eliminate, many elements in the land community [or ecosystem] that lack commercial value, but that are (as far as we know) essential to its healthy functioning. It assumes, falsely, I think, that the economic parts of the biotic clock will function without the uneconomic parts.

These reasons to intervene are based in human values, and because values differ among individuals and groups, not everyone is persuaded by them. In addition, each rationale, if acted on, leads to problems in the intervention itself (table 15.1). We believe that ecosystem management is the approach most likely to produce good restoration results. Knowing the shortcomings of the approach helps us to avoid pitfalls in designed restoration programs.

Restoration of ecosystems requires the maintenance or alteration of composition and structural components by introducing processes. In all cases, we argue that restoration should follow nature's lead, not in order to slavishly re-create an 1850s ecosystem but to restore an ecosystem's ability to respond to change.

In following nature's lead, we examine several ecological attributes of ecosystems for signs of deterioration from expected conditions (table 15.2). Such attributes as species number and relative abundance are predictable based on measurements of repeated patterns and trends in functional and degraded ecosystems. Energy flow in ecosystems determines the relations among trophic levels. Species-dominance studies document that richness and abundance are related in various ecosystems. Low-diversity natural systems aside, in the most degraded ecosystems, a few species dominate while few others are found at all (Whittaker 1975). Where species number and relative abundance differ from the patterns predicted by studies of unaltered areas, the variance can often be explained by historic events at a site, often caused by modern humans. At its simplest level, ecological restoration focuses on biodiversity by attempting to restore opportunities for species of plants and animals to perpetuate themselves.

Such processes of ecosystems (endogenous or internal) as forest succession or soil building in tallgrass prairie are the reactions of the vegetation and animal life to the environment. These functions are impaired when ecosystem composition and structures are severely altered, or if natural disturbances from outside the ecosystem (exogenous or external processes) are removed.

Loucks (1970) and Huston (1979) suggested that species presence and abundance and trophic relations can be structured by external processes. Natural disturbances maintain the diversity and productivity of ecosystems (Picket and White 1985), which creates resiliency to withstand large-scale

Table 15.1 The nature of human intervention and its shortcomings

Impaired attribute	Nature of the intervention	Shortcomings of the intervention
Beauty	Regulation (e.g., air and water emission standards, zoning, etc.) Landscaping (e.g., performance standards, etc.)	Aesthetic basis for restoration is strongly rooted in human value system and varies greatly with the individual. Consensus standards are difficult to develop.
Species diversity	Species recovery (e.g., endangered species recovery plans, transplants, and reintroductions, etc.) Restoration plantings	Costs of single-species recovery are high (e.g., eastern timber wolf), or wasted if species recovers on its own (e.g., bald eagle and osprey). Techniques for multiple-species recovery (e.g., plantings) are in infancy, and their level of functioning falls far short of natural ecosystems.
Processes	Ecosystem management (i.e., alteration or maintenance of the diverse compositional and structural components of ecosystems by mimicking or establishing the ecologically necessary processes)	Existing information can be incomplete, contradictory, or wrong. Difficult to apply at specific and local level. Approaches can run counter to accepted management practices. Approaches may appear to run counter to economic practice or appear politically infeasible.
Human connection	Volunteerism Experiential education	Lack of expertise can create credibility crisis. Private conservation initiatives may challenge agency authority and appear to question competence. Measured success is diffuse and long-term.

changes, such as those cause by climate. Thus, at its root level, ecological restoration seeks to reestablish an ecosystem's capacity to maintain species diversity, internal processes, and resiliency in the face of changing conditions. It is possible, as illustrated by the case studies above, to document some of the changes in ecosystems caused by humans, to measure the degree of change that has taken place, and to prescribe restoration programs that assist nature to reestablish composition, structure, and functions.

Table 15.2 Ecological indications that may require human intervention

Ecological attribute	Measured value of attribute	Indications for intervention
Total number of species	Overall species count by area Species counts in each habitat	Species count is lower than expected or has declined for an area or habitat(s). Species count includes high proportion of exotic species.
Relative abundance of species	Number of individuals of each species Proportion of community in different structural classes (e.g., canopy, large herbivore, etc.)	Number of individuals of each species relative to others is greater or less than expected or has significantly changed. The most abundant species are not those expected, or are exotic species. There are missing or additional structural classes.
Natural endogenous or internal processes	Rate of accumulative processes (e.g., soil development, succession, predation, etc.)	Relative to a fully functional ecosystem, the accumulative processes are reversed (e.g., soil erosion), accelerated (e.g., woody succession in prairies), or halted (e.g., predators removed).
Natural exogenous or external processes	Intensity and duration of disturbances (e.g., flooding, draw-down, fire, grazing, windstorm, drought, etc.)	Relative to a fully functional ecosystem, the disturbances are increased in intensity and duration (e.g., hydrological cycle, grazing), or decreased in intensity and duration (e.g., fire). Drought and other climatic disturbances result in simplification of such ecosystems.

SETTING RESTORATION GOALS

A successful restoration program begins by gathering information, forming hypotheses, and setting goals that can be changed in the face of new information and hypotheses:

(1) *Maps and descriptions of ecological conditions and history.* Complete an inventory of natural resources at the site and for adjacent lands. Document historical and postsettlement land uses from aerial photos, land survey records and maps, oral and written histories, and economic and

land-use records. Historical sources carry their own biases; cross-check information to establish a baseline. Map, describe, and sample the current ecosystem. Study and obtain data from nearby unaltered ecosystems for comparison. Field studies establish current conditions and also help confirm history.

(2) *Ecosystem hypotheses.* From the information, develop hypotheses about ecosystem composition, structure, and functions. Reviewing technical literature and visiting remnant natural lands helps to refine hypotheses. Once understood, the hypotheses (or model) for the ecosystem will help to explain the changes that created the current conditions and the significance of those changes to the ecosystem's future. Devise restoration units which encompass lands that have similar ecological problems and restoration solutions.

(3) *Restoration goals.* Develop goals for each management unit by assessing potential for restoration with reasonable effort and by specifying its desired future condition. Goals can be quantitative, or qualitative if achieving the desired conditions can be documented by the appearance or written descriptions of the habitat or landscape.

Goals can change as new information becomes available. We develop experiments that test our hypotheses about how species diversity, soil-holding capacity, and other desirable properties of ecosystems are maintained. We undertake small experiments to test a new technique and evaluate the costs of employing it in large-scale restorations.

If a goal cannot technically be achieved, except at high cost, it is better to discover that in a one-hectare test plot than in a 1,000-hectare project. Experimental treatments also determine the most effective approach to restore ecosystems.

We will never fully explain nature, and restoration strategies may not achieve all program goals. In these cases, achieving goals may require a larger-scale experiment based on regional ecosystem components and processes. For example, bison have been reintroduced at several Nature Conservancy preserves in the Great Plains and are being used to test the hypothesis that fire and grazing interacted to maintain a diverse prairie. If the ecosystem cannot be restored at a small site, the decision must be made to expand the site, change strategies, or abandon the goals. However, a decision to scale back restoration goals is a weighty one, since remnant populations of species, already disjunct and stressed, may further deteriorate.

(4) *Implementation plan.* The implementation plan and schedule converts goals into action. A plan often consists of several phases, specific to

management units. For example, degraded units may require drastic treatments that can be costly and time-consuming but necessary (remedial phase) before long-term and low-cost management can begin (maintenance phase). The remedial phase may be mechanical shrub removal in oak savanna, while the maintenance phase may consist of periodic low-intensity burning.

(5) *Monitoring program.* Monitoring programs are designed last and implemented first. Monitoring provides measurements of program effectiveness. All aspects of the restoration strategy should be known so that efficient measurements can document progress toward goals. Measurements of prerestoration conditions provide a baseline for comparisons with subsequent conditions. Collection of data continues with restoration; restoration methods or goals can be changed if goals are not being achieved.

IMPLEMENTING A RESTORATION PROGRAM

Carrying out a restoration program requires time, attention to detail, coordination with and education of landowners (especially adjacent ones), clear understanding of the physical limits and potentials of the site, and precise goals. The restoration will be strongly molded by the site—biological and physical conditions, site boundaries, and location. These must be understood in the physical and legal sense. For example, will a wetland restoration affect adjacent landowners, and how? Restoration goals flow from the information, and the program success hinges on how well the goals are formulated and work tasks are implemented.

When restoration sites are small, restoration opportunities are highly constrained. Site location and surrounding land uses determine to a large degree what is possible. On larger sites it is easier to consider the restoration of hydrological processes—for example, modifying flow regimes into and out of the site. We encourage the application of restoration programs over a large geographical unit, if possible. This can produce spinoff benefits, including lower costs and better restorations.

To organize thinking about restoration planning, we have prepared a worksheet (table 15.3) that provides a single location for recording important information required for restoration planning. (It is partly completed for an actual site.) The worksheet brings major issues and information for restoration planning into an accessible and easily used format.

We use maps, photographs, and resource reports that identify changes and external influences that damaged the ecosystems being considered, as well as

Table 15.3 Restoration and management planning worksheet

I. Management Unit: _____1_____ and Unit Name: __Ft. Snelling   Fen/Wetland__

II. Existing Vegetation Type (Reference or attach description): _____See Attached_____

III. Historic Vegetation Type (Reference or attach description): _____See Attached_____

IV. Plants present that may need focused management attention:

1. Reed canary grass   4. European buckthorn
2. Purple loosestrife   5. Phragmites
3. Cattails   6.

V. Significant changes in **physical** (hydrology, drainage, excavation, dredging, sedimentation, or erosion, etc.), **chemical** (contaminants, nutrients, erosion/sedimentation, agricultural/development, surface water loading, etc.), and **biological** (other problem species, seed bank depleted, shade suppression of ground cover, dominance by a few species, etc.) components of the unit:

1. Watershed changes-land development
2. Farming
3. Watershed acreage reduction-sewered
4. Fire suppression
5. Overland flow rather than ground water recharge

6. Urban deer and rare plant herbivory
7. Deicing materials/urban runoff and contaminants
8. Road bisects wetland offsetting hydrology
9. Wetland fill, ditching
10. Chemical changes in water

VI. Restoration/Management Goals: (Note priority or sequence of events)

( ) 1. Reduce hydrologic impacts
( ) 2. Provide upland biofilter to capture contaminated runoff
( ) 3. Reduce invasive shrub cover
( ) 4. Reduce sedimentation and nutrient loads

( ) 5. Remove road bisecting wetland
( ) 6. Increase groundwater recharge in developed uplands and reduce overland flow
( ) 7. Reduce exotic plant species
( ) 8. Remove fill and debris

VII. "Monitoring Attainment of Goals" tied to goals above:

Parameter to measure	Method to use	Timing/Frequency
1. *Surface water quality and quantity*	*field meters*	*quarterly/continuous*
2. *Ground water*	*piezometers (nested)*	*continuous*
3. *Vegetation changes*	*permanent transects*	*annually*
4. *Photographic (aesthetics)*	*permanent photopoints*	*quarterly*
5.		
6.		
7.		

VIII. Restoration/Management tasks* tied to goals above:

(*Property boundary surveying, landowner contacts, educational programs, brushing, noxious weed treatment, prescribed burning, install fences, install firebreaks, meetings, monitoring, photography, tours, volunteer efforts, urban wildlife management task, seed collection, planting, propagation and reintroduction of species, press conferences and PR, research, review and refinement of plan, etc.)

1.   A. *Brush management*

    B.

    C.

2.   A. *Road removal/revegetation*

    B.

    C.

3.   A. *Neighbor education*

    B.

    C.

(continued)

Table 15.3 (cont.)

IX. Phasing/Scheduling of tasks:

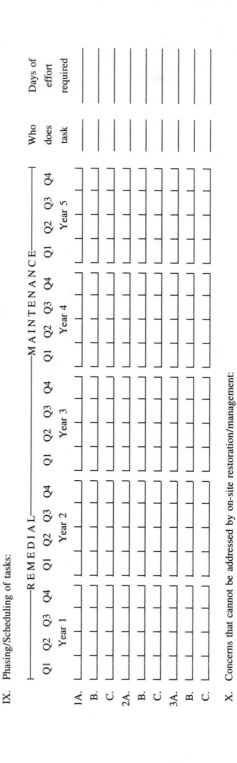

	REMEDIAL								MAINTENANCE								Who does task	Days of effort required
	Q1 Q2 Q3 Q4	Q1 Q2 Q3 Q4	Q1 Q2 Q3 Q4	Q1 Q2 Q3 Q4	Q1 Q2 Q3 Q4		
	Year 1	Year 2	Year 3	Year 4	Year 5		

1A.
B.
C.
2A.
B.
C.
3A.
B.
C.

X. Concerns that cannot be addressed by on-site restoration/management:

1. *Power line right-of-way interaction*   4.

2. *Urban water quality*   5.

3.   6.

XI. Site and off-site problems that need to be addressed or considered by others (i.e. land acquisition, property boundaries, wetland creation, permits, railroad, community outreach, volunteerism, etc.):

1. *Water quality*   6.

2. *Channel degradation*   7.

3. *Fertilizers/deicing materials*   8.

4. *Hydraulic volatility*   9.

5.   10.

XII. Contacted persons, agencies: Names, addresses, and telephone numbers:

1.

2.

3.

4.

5.

XIII. Running record of notes of correspondence on the above:

(date, person, point of communication)

1.

2.

3.

4.

5.

land ownership maps. Combined with specific information from field visits, the planning process progresses rapidly. A growing understanding of ecosystem composition and functions results from completing the worksheet. After developing an understanding of the ecosystem, we define and prioritize restoration and management goals and define and schedule specific tasks to accomplish the goals. Schedules can be based on quarterly budget cycles; detailed weekly schedules usually are required to implement restoration programs. We express the labor requirements in the schedule to directly calculate budgets.

The worksheet requires that off-site problems be identified. Contact persons and strategies for beginning to address these problems should also be documented. The prudent restorationist will write everything down; good records are required to keep the restoration process on track.

We use a "management unit" planning approach, even though this can be arbitrary and contrary to ecological conditions. Management units are useful as "work units"; that is, specific locations on the ground to which work crews are assigned a task. Burn units are often defined by straight-line firebreaks because igniting fires is safer and cheaper than along curving lines. Ideally, a site is managed as a whole, with management practices not reinforcing artificial human boundaries. Management should flow across units, as natural ecological forces did originally. Smaller sites will require a management unit approach, while larger sites may support an integrated landscape approach.

Finally, sharing successes and failures advances the science of ecological restoration. It is incumbent upon anyone undertaking a restoration program to keep good records, then publish in the appropriate journals or otherwise disseminate their results. As practical ecologists, restorationists above all should keep an open mind, despite their best laid plans. New information is guaranteed to come along and contradict preciously held ecosystem hypotheses and restoration goals. A truly successful restoration program adapts to changing conditions in the way that functional ecosystems do.

Many have shared over the years their technical knowledge of practical ways to view ecological restoration, and, perhaps more importantly, how to assist others in this thought process. We specifically appreciate the ongoing council of colleagues at The Nature Conservancy and Applied Ecological Services, Inc., and colleagues who have provided guidance, including J. Ludwig, A. Haney, L. Leopold, K. Wendt, J. White, D. Wedin, and others too numerous to mention. We thank Bill Jordan for his insightful review. The Nature Conservancy (Wisconsin) is acknowledged for supporting research on historic hydrology of the Des Plaines River; Cook and DuPage Counties Forest Preserve District (Illinois) and Illinois Non-game Funds (Illinois Department of Conservation) for supporting research on Oak Savanna systems; and the University of Minnesota Cooperative Extension and Department of Natural Resources, specifically B. Morrissey, D. Lime, K. Bolin, and D. Anderson, for fostering the course series Outdoor Recreation Management in the 1990s, where we had the opportunity to develop the tutorial information in this paper.

LITERATURE CITED

Apfelbaum, S. I. 1985. Cattail (*Typha* spp.) management. Natural Areas Journal 5(3): 9–17.

———. In press. The role of landscapes in stormwater management. Proceeding of a National Conference on Urban Runoff Management, March 30–April 2, 1993. Chicago, Illinois.

Apfelbaum, S. I., and A. Haney. 1989. Management of degraded oak savanna remnants in the upper Midwest: preliminary results from three years of study. Pages 290–291 *in* Proceedings of the Society for Ecological Restoration and Management, Oakland, California.

Apfelbaum, S. I., and C. Sams. 1987. Ecology and control of reed canary grass (*Phalaris arundinacea* L.). Natural Areas Journal 7(2):69–74.

Chapman, K. A. 1984. An ecological investigation of native grassland in southern lower Michigan. M.S. thesis, Western Michigan University, Kalamazoo, Michigan.

Haney, A., and S. I. Apfelbaum. 1990. Structure and dynamics of Midwest oak savannas. Pages 19–30 *in* J. M. Sweeney, editor, Management of dynamic ecosystems. North Central Section, The Wildlife Society, West Lafayette, Indiana.

Huston, M. 1979. A general hypothesis of species diversity. American Naturalist 13: 81–101.

Leopold, A. 1966. A sand county almanac. Oxford University Press, New York, New York.

Loucks, O. L. 1970. Evolution of diversity, efficiency, and community stability. American Zoologist 10:17–25.

Nature Conservancy. 1990. An approach to evaluate long-term survival of the tallgrass prairie ecosystem. The Nature Conservancy, Midwest Regional Office, Minneapolis, Minnesota.

———. 1994. The conservation of biological diversity in the Great Lakes ecosystem: issues and opportunities. The Nature Conservancy, Great Lakes Program, Chicago, Illinois.

Packard, S. Restoring oak ecosystems. Restoration and Management Notes 11:5–16.

Picket, S. T. A., and P. S. White. 1985. The ecology of natural disturbance and patch dynamics. Academic Press, New York, New York.

Whittaker, R. H. 1975. Communities and ecosystems. 2d Edition. Macmillan, New York, New York.

Wilson, E. O. (editor). 1988. Biodiversity. National Academy Press, Washington, D.C.

Yonzon, P. B., and M. L. Hunter. 1991. Cheese, tourists, and red pandas in the Nepal Himalayas. Conservation Biology 5:196–202.

FUTURE DIRECTIONS

A clear congressional commitment to ecosystem management would represent another bold step forward to perpetuate the nation's public land legacy.
—R. B. Keiter

In the long term, the issue ultimately boils down to one of human population growth. Even with the best management, growing demand for goods and services from Earth's ecosystems will continue to undermine sustainability. Conflicts over management policies and priorities, which already are burdensome, will increase. Issues of ecosystem management are most serious in developing countries. There exists concern that our conservation efforts will shift demand to tropical forests resulting in escalated extinction rates because species diversity is so much greater in the tropics.
—D. G. Sprugel

Overleaf: Highly mechanized methods for timber extraction are resulting in rapid rates of deforestation in many regions of the world.

Chapter 16 Implementing Ecosystem Management: Where Do We Go from Here?

Norman L. Christensen

From the presentations in this volume, as well as from policy statements by no fewer than nineteen federal agencies and departments (GAO 1994), it is clear that many decision makers view a shift to ecosystem management as a true revolution in management practice. Nevertheless, the phrase *ecosystem management* does not, by itself, convey much understanding of how such management differs from current or past practice.

An ecosystem is "a spatially explicit unit of the Earth that includes all of the organisms, along with all components of the abiotic environment within its boundaries" (Likens 1992). A. G. Tansley coined the word *ecosystem* in his 1935 exegesis on ecological terminology to emphasize the "chicken and egg" relation between organisms and their physical environment; while the distribution and abundance of organisms are determined by abiotic environmental factors, organisms play a significant role in determining the character of those factors. Thus, the composition and structure of forests are often influenced by the availability of water, and the hydrology of a watershed is very much a consequence of its vegetation.

The metaphor of the "superorganism" has been heavily used in ecosystem ecology, particularly in the definition of structure and function. Just as anatomical structure, such as the tissues constituting the digestive tract, is vital to such functions as digestion and assimilation, ecologists posited that structural attributes of ecosystems, such as species composition and diversity, were critical to such functions or processes as mineral cycling or primary production (Odum 1953). That ecosystem functions or processes are critical to the provision of "goods" such as food and wood fiber and "service" such as regulation of hydrological cycles and cleansing of water and air has long been recognized.

In a somewhat trivial sense, all management is ecosystem management. Virtually all human activities influence the structure and function of ecosystems. Ecosystem management has been characterized as management that focuses on the complexity and multiplicity of interactions that underlie function, emphasizes such large-scale management units as watersheds, acknowledges the importance of the dynamics of ecosystem processes, or recognizes the incongruity between the scale of ecological processes and patterns of land management jurisdiction (Christensen et al. 1996). Some view ecosystem management as focusing especially on societal impacts and involvement, as articulated in the Eastside Forest Health Assessment Team's (1993) definition: "optimal integration of societal values and expectations, ecological potentials, and economics." In contrast, others view ecosystem management as focused on "protecting native ecosystem integrity over the long term" (Grumbine 1994).

I agree with Jerry Franklin (Chapter 2) that ecosystem management can be defined in a more basic and far-reaching way as "managing ecosystems so as to assure their sustainability." In its recent report on this subject, the Ecological Society of America defined ecosystem management as "management driven by explicit goals, executed by policies, protocols, and practices, and made adaptable by monitoring and research based on our best understanding of the ecological interactions and processes necessary to sustain ecosystem composition, structure, and function" (Christensen et al. 1996). As Franklin pointed out, most, if not all, of the characterizations described above are likely to be integral strategies of an ecosystem management plan.

If sustainability is indeed *the* central goal or value of ecosystem management, it must be clearly defined in terms of what it is we wish to sustain and the time frame over which we wish to sustain it. Notions of sustainability and "sustained yield" are not absent from historic management schemes and practices. Furthermore, few have consciously advocated policies or activities that would diminish the capacity of ecosystems to provide goods and services. With regard to what it is we wish to sustain, ecosystem management acknowledges that the capacity of a system to deliver goods and services depends on ecosystem processes and that management should focus less on arbitrary goals for the yield of such "deliverables" as allowable cut or bag limits, and more on practices necessary to sustain the ecosystem structure and function necessary to deliver those goods and services. "Ecosystem managers should focus at least as much on what is left behind as on what is extracted" (Christensen et al. 1996).

In one sense, the issue of time frame is relatively simple. A commitment to sustainability is equivalent to a commitment to intergenerational equity; we should not deny future generations the resources and opportunities that we enjoy today. Thus, in this context, the time frame for management is "in perpetuity." Agreeing that such an open-ended commitment to sustainability is worthy, we must also admit that it is impossible to set operational goals for management without a more definite timeline. In practice, the time frame for implementation and evaluation of management practices should coincide with the temporal scales for the variation and behavior of vital ecosystem processes rather than the timing of fiscal planning or electoral cycles which have little relation to such processes.

ELEMENTS OF ECOSYSTEM MANAGEMENT

The Ecological Society of America Committee on Ecosystem Management identified the following elements as components of a comprehensive ecosystem management system: (1) clear operational goals, (2) sound ecological models and understanding, (3) an understanding of complexity and interconnectedness, (4) recognition of the dynamic character of ecosystems, (5) attention to context and scale, (6) acknowledgment of ignorance and uncertainty, (7) commitment to adaptability, and accountability, and (8) acknowledgment of humans as ecosystem components (Christensen et al. 1996; see also Grumbine 1994). Moving from concepts to practice will require attention to each of these elements.

Clear Operational Goals

Many critics fear that ecosystem management is simply a guise for expansion of wilderness preserves at the expense of extractive use of natural resources. The chapters in this book have demonstrated that the appropriate balance of land allocation among uses extending from "industrial strength" forestry and urban development to "reagent grade" pristine wilderness has long been, and will continue to be, a matter of debate among managers, decision makers, and various segments of the body politic. However, they also testify to the fact that ecosystem management is as applicable to intensive utilitarian objectives as it is to wilderness parks. So what is different?

While acknowledging the need to provide the goods and services for which demand from human populations continues to increase, the operational goals of an ecosystem management plan should not focus exclusively on such "deliverables" as board feet of timber, bag limits, or visitor days. "Ecosystem management does not deny the need or wish to harvest fiber from our forests, fish from our streams and lakes, or energy from rivers, it simply confronts the reality that in order to meet these needs or wants sustainably, we must value ecosystems for more than the goods and services we buy and sell in economic markets" (Christensen et al. 1996). Management goals must be explicated in terms of desired future conditions or desired future behaviors for the ecosystem structures and functions that are necessary to sustain those goods and services. Furthermore, these goals should be formulated in terms that are amenable to direct measurement and monitoring.

For example, we may wish to manage a large tract of forested land primarily for wood fiber production. An ecosystem management plan for such a tract would identify those ecological structures and processes such as woody debris, nutrient capital, and successional change upon which sustained fiber production depends, then set the goal of managing so as to sustain them (Kotar, Chapter 13). Similarly, goals for biological-diversity preserves should focus on the processes necessary for the maintenance of such diversity (Temple, Chapter 5). This is, of course, relatively easy to say, but in truth we are continually learning more and more about the complexity of ecosystems that underlies sustained function.

Among the most daunting realities for natural resource managers are the dynamics and increasing complexity of societal expectations for goods and services from ecosystems. While demand for fiber has increased significantly, so have public interests in the wildlife, high-quality water, recreation, and inspiration that forested ecosystems provide. We cannot anticipate with any certainty what priorities future generations will place on particular commodities or amenities. However, if goals and strategies are focused on the sustainability of ecosystem integrity and function, managers will most likely be able to respond to those changes.

Sound Ecological Models and Understanding

An understanding of basic ecological principles, processes, and interconnections is necessary to establish appropriate operational objectives. Sound ecological models are critical to cultivating that understanding. Models provide the means to organize information and express relations and interconnections. They may take the form of simple compartment diagrams or complex computer simulations that permit depiction of processes operating through time and across landscapes (Christensen et al. 1996). Models can be useful in identifying particularly sensitive ecosystem components or in setting brackets around expectations for the behavior of particular processes. Models are important in identifying indexes and indicators that provide a measure of the behavior of a broad suite of ecosystem properties. Finally, models often provide useful tools for exploring alternative courses of action (Lee 1993).

The term *ecosystem management* should not be taken to suggest that the only goals or science relevant to management are set or done at the ecosystem level. All levels of organization, from the morphology, physiology, and behavior of individual organisms, through the structure and dynamics of populations and communities, to patterns and processes at the level of ecosystems and landscapes, are potentially important. Models of one form or another are necessary to identify those features most critical to sustainable management.

Modeling is often criticized as a blatant attempt to simplify the complexity inherent in ecosystems, but, as Lee (1993) argued, that is indeed its virtue. Models are best thought of as maps to guide managers through the myriad complexity of ecosystems. It is, however, important to recognize that such models are relatively crude representations of the ecosystems they depict, just as map projections simplify and distort the complexity of the territory they represent. Lee (1993) argued that "models are indispensable and always wrong." The utility of models "comes from their ability to pursue the assumptions made by humans—assumptions with qualitative implications that human perception cannot always detect."

Complexity and Interconnectedness

The phrases "everything is connected to everything" and "everything goes someplace" have become clichés. If nothing else, a century of ecological research has taught us that ecosystem complexity and interconnections are fundamental to ecosystem function and, therefore, to the delivery of ecosystem goods and services. Human interests may focus on only a subset of the organisms or functions that constitute an ecosystem, but the overall complexity and diversity of ecosystems is critical to the maintenance of such key processes as nutrient cycles (David, Chapter 6), the resistance to and resilience from disturbances, and the adaptability of ecosystems to long-term environmental change (Wilson 1992, Christensen et al. 1996).

Intensive extractive management, such as agriculture or plantation forestry, involves purposeful simplification of ecosystems with the goal of concentrating production or yield in particular ecosystem components; there is no doubt that such management has greatly increased the availability of particular resources which humans need and want, and has thereby increased the carrying capacity for human beings worldwide. However, management practices that diminish ecosystem complexity and diversity are similar in some ways to financial investment in a portfolio focused on only one or two instruments with high growth potential; while the potential yield may be higher, so is the risk of failure. From a human standpoint, so long as populations numbered well below the carrying capacity, such risk may have been tolerable. However, worldwide and regional demand for many resources is approaching—or perhaps exceeds—capacity for sustained production, and failures can have catastrophic consequences in relation to human populations.

That each species that is part of the earth's diverse biota is the product of more than four billion years of evolution provides, for many of us, sufficient justification for preservation of biological diversity. Those who feel that human needs should be given a higher priority than diversity and complexity do not understand that, in the end, provision of those human needs is dependent on that diversity and complexity. That we do not understand the role of all ecosystem components is hardly justification for allowing them to disappear. As Aldo Leopold (1946) suggested, managers are tinkers, and the first rule of creative tinkering is "saving all the parts."

Recognition of the Dynamic Character of Ecosystems

Over their long-term and recent history, the earth's ecosystems have never been the same twice. Such natural disturbances as fire, windstorms, insect and pathogen epidemics, and floods are ubiquitous and, in many cases, critical to the maintenance of key ecosystem processes. For example, in many forested ecosystems, natural successional processes result in changes in the structure and distribution of trees and the woody debris they produce, which gradually increases the likelihood of fire. In such systems, fire redistributes nutrients, creates favorable seedbeds, and deflects successional change or restarts it altogether. Where fires have occurred regularly through evolutionary time, natural selection has favored life history adaptations such as serotinous cones or fire-dependent seed dormancy that have not only increased resistance to fire but resulted in dependency on it.

Much research, as well as mismanagement, has taught us the futility of equating sustainability with maintaining the status quo. Management determined to "freeze" ecosystems in a particular state has generally proven to be futile and unsustainable. For example, the suppression of wildfires in ecosystems as diverse as chaparral and mixed conifer forests has resulted in widespread accumulation of fuels and larger and more intense fires than would have occurred in the absence of such suppression. Thus, in seeking to define the "desired future condition" for an ecosystem, we must view "condition" as a dynamic concept or, better yet, think in terms of "desired future trajectory" (Christensen et al. 1996).

The past two decades have seen a significant shift in management to acknowledge the dynamic character of ecosystems, as witnessed by the widespread implementation of fire management programs. Nevertheless, such management is not without its pitfalls. Identification of optimal disturbance frequencies and intensities has proven difficult, given that historic patterns were highly variable and that this variability may be as important to the sustainability of ecosystems as the disturbance events themselves (Christensen 1994). Managing natural disturbance often involves significant risk to human property and life, resulting in considerable liability to managers (Christensen et al. 1989).

Simplistic views of such natural disturbance processes as fire, as well as undeserved confidence in our ability to manage such disturbances, have resulted in serious management dilemmas. The fact that we can artfully manage fire in such ecosystems as grasslands characterized by light fuels has no bearing on our ability to manage fire in the complex, heavy fuels of a mixed conifer forest. In such situations, we risk serious liability in implementing prescribed fire programs to diminish the risk to life and property of wildfires burning in excessive accumulations of fuel.

Given the liabilities inherent in simply allowing natural disturbances to run their course, we will need to use surrogates, such as prescribed fire or logging. However, we have little hard data to verify that such surrogates effectively simulate natural processes. Such interventions can actually increase rather than decrease flammable fuels. Certainly, such surrogates may have considerably different effects on processes like nutrient cycling or patterns of immediate postfire succession. This is not to argue against prescribed fire or logging as a means of managing forest fuels but rather to alert managers of the need for more study to be sure that such interventions are truly accomplishing long-term management goals.

Because the world is constantly changing, one might argue—indeed, some have argued—that human disturbances are in some sense "natural." So far as we can tell, at no time in the past has the earth experienced change at the rate at which it is currently occurring. Furthermore, many changes, such as extremes of land fragmentation and certain kinds of pollution, have no precedent in the earth's evolutionary history. It is a tremendous act of faith to suppose that ecosystems can and will respond to these novel insults as they would to natural disturbance processes to which they evolved over millions of years. If our faith is misplaced, the consequences may be disastrous, both in terms of the integrity of ecosystems and their ability to produce sustainably the goods and services upon which we depend.

Context and Scale

One of the most fundamental questions to emerge in ecology in the past decade is: At what distance do events or changes occurring at one location no longer have any meaningful effect on other ecosystems? Island biogeographic theory has taught us that the diversity of a particular location is very much influenced by the richness of nearby locations, the mobility of the organisms, and the hospitality of the intervening territory. Manipulations of forests in one area influence microclimates, even rainfall, in adjacent areas.

Clearly, ecosystem processes operate over a wide range of spatial and temporal scales, and their behavior at any given location is very much affected by the status and behavior of the systems or landscape that surrounds them. The importance of context in determining the behavior of ecosystems at a particular location is the basis for the "landscape approach" to management of terrestrial ecosystems (Noss 1983, Noss and Harris 1986).

So what is the appropriate scale or time frame for management? There is none! One the most frustrating characteristics of ecosystems is that they are open to inputs and outputs of materials, energy, and organisms at scales extending all the way from a small forest stand to the entire biosphere. Ecosystem scientists define the boundaries of the systems they study such that they can most easily monitor or manipulate the processes in which they are interested. Ideally, we would define the spatial and temporal scales of management in order to *manage* most easily the processes necessary to sustain ecosystem function. Instead, we have generally defined the boundaries of management jurisdictions with little or no reference to the behavior of ecosystem processes. Consider, for example, how often rivers form boundaries for management jurisdictions. Could a less appropriate border be selected if one wanted to manage processes driven by hydrology?

Even if we had the will and the means to rectify management jurisdictions with ecological processes, we would still face significant challenges. Our knowledge of the relevant scales for key processes is very limited. More important, we know that the appropriate scales for management of one process, say hydrology, may have little relevance to the management of other processes, such as the behavior of large mammals. "Because ecosystems exist at several scales, so, too, should efforts to coordinate activities that affect them" (GAO 1994). Successful management of ecosystems will depend on the development of mechanisms and institutions to reconcile conflicting resource goals across spatial boundaries.

Ignorance and Uncertainty

One often hears the aphorism, "You cannot manage what you do not understand." Yet in reality we must always manage with a knowledge base that is incomplete and with a limited ability to predict management outcomes. It is often the case that if we simply knew more, we would make better management decisions, but it is also true that there are limits to what we can know and to the precision of our predictions set by the complexity of the systems we try to manage. We must recognize that there will always be limits to the precision of our predictions set by the complex nature of ecosystem interactions and strive to understand the nature of those limits. For example, the notion that natural-disturbance dynamics can be thought of simply as an endless series of cycles characterized by amplitude and frequency neglects the obvious variability from one cycle to the next and the fact that such variability may be very high for some ecosystems and disturbance types.

Hilborn (1987) suggests that three types of uncertainty are inherent in ecosystem management: (1) "Noise" arises as a consequence of random variation and measurement errors. Variability is not only an inescapable part of ecosystem behavior and management, it contributes enormously to the diversity, complexity, and, therefore, sustainability of ecosystems. This sort of uncertainty can be statistically quantified and must be understood in setting goals and establishing monitoring programs. (2) Uncertainties due to insufficient knowledge can be reduced by adaptive management and additional research. Limits to knowledge may be governed by the complex nonlinear nature of ecological systems, by ethical constraints on research that risks human life or jeopardizes rare species or ecosystems, or by economic constraints. (3) Surprises are the most difficult category of uncertainty for managers. Examples include such events as catastrophic fires or storms that are outside the range of normal noise, or unforeseen connections between actions and effects, such as the relation between the release of CFCs and the depletion of stratospheric ozone.

To be sustainable, management must include a margin of safety for uncertainty. Ecosystem management is not a strategy to eliminate uncertainty. Rather it acknowledges its inevitability and accommodates it. The impacts of human activities make this uncertainty all the more daunting. In the words of C. S. Holling (1993), "not only is the science incomplete, the system itself is a moving target, evolving because of the impacts of management and the progressive expansion of the scale of human influences on the planet."

Public understanding and education on this matter is critical. Just as managers cannot claim perfect knowledge, the public should not expect it. Public acceptance of the nature of uncertainty in our science and management, as well as the potential risks from surprises, is a necessary prerequisite to informed involvement in decisions affecting the management of natural resources.

Adaptability and Accountability

Given the uncertainties described above, it is essential that management systems be adaptable. They must be adaptable to variations in environment (including the impacts and needs of humans) from location to location. They must also be adaptable to inevitable changes in those environments through time. Most important, management systems must acknowledge the provisional nature of our models and information base and be adaptable to new information and understanding (Holling 1978 and Walters 1986). To be adaptable and accountable, management objectives and expectations must be explicitly stated in operational terms, informed by our best models and understanding of ecosystem function, and tested by carefully designed monitoring programs that provide accessible and timely feedback to managers.

As Lee (1993) argued, we must see many of our management protocols and practices as experimental and devise the means to test the hypothesis that these interventions are, in fact, moving the managed system toward intended goals. In the context of ecosystem management, the question of whether management practices are truly sustainable should be central. To make management adaptive we must clearly explicate objectives in operational terms relevant to sustained ecological function, develop monitoring programs focused on data relevant to those operational objectives, provide for efficient analysis and management of data, and encourage timely feedback of information from research and monitoring programs to managers.

Clear operational objectives are prerequisite to an adaptive management program, and these objectives are not necessarily synonymous with overarching management goals. For example, a management goal of sustained provision of timber, recreation, and wildlife should translate into specific objectives for the ecosystem structures and functions that are most critical to providing this combination of goods and services. In this context, modeling can be especially helpful in identifying key ecosystem elements that are likely to be effective indicators of sustainable functioning.

It is particularly critical that managers not set such operational goals in terms of the management interventions themselves. For example, with the knowledge that fire plays a critical role in many ecosystems, managers have focused on its reintroduction by means of prescribed fire programs. However, the mere fact that managers apply prescribed fires to a grassland, shrubland, or forest does not certify that such fires have effects similar to historic natural wildfires. We do not manage areas in order to burn them; rather we suppress, augment, or encourage fire in order to achieve specific ecosystem goals (Christensen et al. 1989). In this case, operational goals for a fire management program should focus on such items as flammable fuels, hydrology, or biological diversity that are important to sustained ecosystem function.

Monitoring programs of one form or another are by no means new; however, three elements—focus, efficiency, and commitment—require special attention if they are to inform management in a meaningful way. Monitoring should be focused on management expectations (operational objectives) and designed to test the success and efficacy of management practices. Efficiency demands rigorous statistical sampling designs, with attention paid to issues of precision and bias in data gathering. Because management situations often offer limited opportunities for replication or are often biased by patterns of ownership and accessibility, sampling designs will often be flawed. Such flaws should not be taken as an excuse to avoid monitoring, but their likely impacts on data quality and uncertainty of conclusions must be explicitly evaluated.

A long-term vision and commitment to the development and maintenance of monitoring programs is critical. Such programs necessarily add cost and can be especially difficult to maintain where personnel turn over frequently. Thus, clear identification of target objectives for monitoring is important; ''shotgun'' approaches may miss key variables while incurring the unnecessary cost of gathering irrelevant data.

All too often monitoring programs have proven to have ''write only memories.'' Data may be gathered but not properly archived and analyzed, and data formats are often incompatible across management jurisdictions and between agencies. Hardware and software tools developed over the past decade provide great opportunity to facilitate data transfer, storage, and analysis, although considerable research is needed in the development of formats and archival systems that are likely to be accessible among management jurisdictions and into the future.

What is the role of research in an adaptive management program? We have much to learn regarding the factors critical to the sustained functioning of ecosystems, and more basic research in this area is clearly needed. Nevertheless, there is also a need to prioritize our research agenda. ''Research programs driven solely by the immediate needs of management risk overlooking new insights and opportunities; however, programs that are innocent of an understanding of management challenges and priorities risk being irrelevant. There is an infinitely large array of potentially interesting problems that the research community might wish to tackle, but only a considerably smaller subset of that array is relevant to compelling management questions'' (Christensen et al. 1996).

The community of academic scientists has much to offer the managers with regard to the design and execution of monitoring programs. Research is needed to refine models that will help identify key variables for monitoring, as well as in the design of cost-effective but efficient sampling protocols. Standards for data gathering are well developed in some areas (for example, hydrology and climate) and nonexistent in others (for example, biodiversity or site fertility). The development of such standards should be a high priority for agencies funding research relevant to management.

The adaptive management loop will be closed only if managers receive timely feedback of results of monitoring information, as well as new insights from basic research programs. Such feedback will require institutional change in public agencies charged with natural resource stewardship and may even require legislative initiatives to eliminate impediments to the exchange of information. Cultural barriers that have traditionally separated communities of managers from those of researchers must be broken down. This will require changes in modes of communication as well as changes in reward systems for both communities.

Humans as Ecosystem Components

Human impacts on ecosystems are ubiquitous; certainly no terrestrial ecosystem has escaped the effects of human activities. Whereas humans may present some of the most significant challenges to sustainability, they are also integral ecosystem components who must be part of any effort to achieve sustainable management goals. Given the growth in human populations, sustainable provision of ecosystem goods and services becomes an even more compelling goal.

As Christensen et al. (1996) have pointed out, ''many of our most celebrated 'environmental train wrecks' are not disputes between the 'rights of nature' versus the human demands for resources; rather, they are conflicts among competing human demands.'' Conflicts may result when practices associated with extraction of one commodity influence the availability of another, as, for example, in the effects of logging practices on breeding habitats for migratory salmon. Conflicts also result as a consequence of our limited ability to reconcile human wants and needs across scales of space

and time. Thus, the rights of individual property owners on increasingly fragmented landscapes may be in conflict with the collective vision of those same
property owners for the entire landscape, and our need to meet the constraints
of fiscal years and tax schedules may conflict with our wish to guarantee to
our children the same opportunities we enjoy.

Throughout this volume, the need for management at temporal and spatial
scales that match those of key ecosystem processes has been emphasized.
Nevertheless, the growth of human populations and the ever-increasing complexity and fragmentation of land ownership and resource stewardship responsibility make such management increasingly difficult. Although we have
not yet succeeded in placing a price tag on ecosystem processes and functions
that are not typically bought and sold in traditional markets, the value that
society places on such things is certainly influenced by the laws of supply
and demand. Wild places and the biological diversity they contain, air and
water quality, and the beauty of our landscapes are increasingly valued by
society, not only because of our increased awareness of their importance to
sustained ecosystem function and environmental quality but also because of
their diminished supply. Nevertheless, we are groping to reconcile these
large-scale values with aspirations and goals we set at smaller spatial scales.
The development of mechanisms and institutions devoted to this reconcilia-
tion must receive a high priority.

IMPLEMENTING ECOSYSTEM MANAGEMENT

This volume has, for the most part, emphasized the scientific basis for ecosystem management, and most authors have called attention to the inadequa-
cies of our knowledge base. It is tempting (not to mention self-serving) for
the scientific community to see implementation of sound management practices as being knowledge- or information-limited, and there can be no question that we would manage better in some cases if we understood more.
However, management questions are compelling; we cannot simply not manage until we have the complete picture, if indeed such a picture is possible.
Furthermore, as Vitousek (1994) suggests, we actually know a great deal.

In their opening chapter, Haney and Boyce argue that ecosystem management is about people more than anything else. Put another way, implementation of the principles of ecosystem management is limited more by our
social, political, and economic systems than by lack of scientific data. In this
context, four challenges to implementation are apparent: (1) information transfer, (2) uncertainty, (3) scales of space and time, and (4) public confidence.

Information Transfer

In truth, most of the principles outlined in this book and summarized in this chapter have been understood for many years—in some cases, many decades. The bottleneck to bringing the best information to bear on complex management issues is often not the generation of new knowledge but the incorporation of existing knowledge into management policies and protocols. Changes are needed in both the academic and management communities for such information transfer to occur in a more timely fashion.

The historic dichotomy between so-called basic and applied ecological research in academic institutions is becoming an anachronism. But scientists must understand that the information they generate will not have an impact on management if it is delivered only in the form of peer-reviewed articles. Direct interactions between the academic community and those who would use their products (the colloquium that produced this volume is an example) are necessary. The evolution of a cadre of scholar/managers dedicated to bridging this gap is encouraging (Christensen et al. 1996). By acknowledging that their existing knowledge base is provisional, managers will be eager for new information and receptive to modifications.

Uncertainty

Earlier, I discussed the need to treat management as an experiment and to acknowledge that surprises, however improbable, are certain to occur. We should not underestimate the challenges this uncertainty promises for managers and the public they serve. Managers not only face the challenge of developing research and monitoring programs to evaluate management goals, policies, and protocols, but they also must confront the risks of untried approaches and the inevitability of occasional management failures (Lee 1993). In many cases, frameworks must be developed for deciding whether a given management approach is indeed meeting objectives; the scientific community has much to contribute in this area.

If policy makers and mangers must be up front about their uncertainty, the public must be able to understand and accept it as well. Certainly, a more detailed public understanding of the nature and dynamics of ecosystems would facilitate such understanding and argues for a more prominent role for environmental education in formal and informal curricula. However, an even more fundamental understanding of the role of uncertainty and probability in the behavior of our world and in the conduct of scientific experiments is needed.

Scales of Space and Time

Disputes over the management of such natural resources as those in the Pacific Northwest have laid bare two critical truths. First, it is no longer possible for the ecosystems that provide goods and services to humans to meet the competing demands of different constituencies to the full expectation of each constituency. In most regions we can no longer simultaneously meet public expectations for scenery, biodiversity, wilderness, high-quality water, and fiber in historic amounts or at past rates. As indicated earlier, this is not a matter of "man against the rights of nature," it is simply a matter of conflicts in human wants. Very often such conflicts are a matter of reconciling management activities at one scale (say, the management of forest stands) with management objectives that are realized only on very large scales and over very long time periods (such as managing water quality in large rivers and streams or sustaining populations of large ungulates or top carnivores).

Second, institutions or mechanisms necessary to resolve conflicts across scales of time and space are imperfect at best and in some cases nonexistent. Silviculturalists can take pride in the improvement in forest practices at the stand level that have led to increased productivity and sustained site quality. However, the additive effects of even the best stand-level forestry practiced in an uncoordinated fashion across large landscapes are likely to conflict with other values that society has placed on those ecosystems.

Virtually every commentator on ecosystem management calls attention to the need for bottom-up rather than top-down approaches—that is, stakeholder involvement—to solving management problems. Experience to date suggests that much work needs to be done to implement such approaches. While FEMAT (Forest Ecosystem Management Team 1993) and similar large-scale projects have been criticized for focusing primarily on public ownerships, it is clear that leadership from the public sector will be needed to reconcile conflicts across complex ownership boundaries and among goals that are realized at different temporal scales. The exact role or roles for the public sector in the resolution of such conflicts is at the very core of arguments over private property rights and has yet to be articulated. This must be viewed as the grand challenge to implementation of ecosystem management on large spatial scales.

Public Confidence

Ultimately, our willingness to resolve conflict among goals and to accept management policies in the context of uncertainty depends less on a consensus on goals and agreement on policies than on our confidence in the institutions that will implement them. Perhaps the real environmental train wreck has been the deterioration of public confidence in those institutions. I suspect that by meeting the challenges outlined above, public confidence will be bolstered considerably. But, above all, what is needed among stakeholders is a shared understanding of what is at stake.

I have a favorite Gary Larson cartoon. A stegosaurus stands at a lectern addressing an audience of hadrosaurs, tyrannosaurs, velociraptors, and the like, and says, "Gentlemen, the picture is bleak, the climate is changing, the mammals are on the rise, and here we sit with brains the size of walnuts."

Debates over whether ecosystem management is or ought to be more biocentric than anthropocentric seem both tiresome and unproductive. Somehow, the earth's biota and ecosystems managed rather well for eons in the absence of *Homo sapiens*. Furthermore, from what we know of evolution and ecosystem resilience, if humankind disappeared from the earth tomorrow, ecosystems would recover in a matter of millennia. And, while I do lament human-caused loss of species, the geologic record provides testimony to the capacity for evolution to replenish the biota following catastrophes, albeit over very long spans of time. My point is simply that the challenges to natural resource managers result directly from the activities and desires of humans. So far as we are aware, no other organism in the history of life on earth has matched our species' powers of intelligence and self-awareness. Dinosaurs of one kind or another ruled the earth's ecosystems for more than one hundred million years; hominids have been around a few million years at most. Our greatest challenge is no less than to prove that intelligence and self-awareness will indeed prove to be adaptive traits in the long term.

The author was chair of the Scientific Basis for Ecosystem Management Committee of the Ecological Society of America, and some of the material presented here is derived from that committee's report (Christensen et al. 1996). The author wishes to thank the entire committee for many ideas that are presented here.

LITERATURE CITED

Christensen, N. L. 1994. Plants in dynamic environments: is "wilderness management" an oxymoron? Pages 63–72 *in* P. Schullery and J. Varley, editors, Plants and their environments: proceedings of the First Biennial Scientific Conference on the Greater Yellowstone Ecosystem. National Park Service, Mammoth, Wyoming.

Christensen, N. L., J. K. Agee, P. F. Brussard, J. Hughes, D. H. Knight, G. W.

Minshall, J. M. Peek, S. J. Pync, F. J. Swanson, J. W. Thomas, S. Wells, S. E. Williams, and H. A. Wright. 1989. Interpreting the Yellowstone fires of 1988. BioScience 39:678–685.

Christensen, N. L., A. M. Bartuska, J. H. Brown, S. Carpenter, D. D'Antonio, R. Francis, J. F. Franklin, J. A. MacMahon, R. F. Noss, D. J. Parsons, C. H. Peterson, M. G. Turner, and R. G. Woodmansee. 1996. The scientific basis for ecosystem management. Ecological Applications (in press).

Forest Ecosystem Management Team (FEMAT). 1993. Forest ecosystem management: an ecological, economic, and social assessment. U.S. Government Printing Office, Washington, D.C.

Government Accounting Office. 1994. Ecosystem management: additional actions needed to adequately test a promising approach. United States General Accounting Office Report to Congressional Requesters. GAO/RCED-94-111.

Grumbine, R. E. 1994. What is ecosystem management? Conservation Biology 8:27–38.

Hilborn, R. 1987. Living with uncertainty in resource management. North American Journal of Fisheries Management 7:1–5.

Holling, C. S. 1978. Adaptive environmental assessment and management. John Wiley & Sons, New York, New York.

———. 1993. Investing in research for sustainability. Ecological Applications 3:552–555.

Lee, K. N. 1993. Compass and gyroscope: integrating science and politics for the environment. Island Press, Washington, D.C.

Leopold, A. 1949. A Sand County almanac. Oxford University Press, New York, New York.

Likens, G. 1992. Ecosystem approach: its use and abuse. Ecology Institute Nordbute 23:W-2124, Oldendorf/Luhe, Germany.

Noss, R. F. 1983. A regional approach to maintain diversity. BioScience 33:700–706.

Noss, R. F., and L. D. Harris. 1986. Nodes, networks, and MUMs: preserving diversity at all scales. Environmental Management 10:299–309.

Odum, E. P. 1953. Fundamentals of ecology. W. B. Saunders, Philadelphia, Pennsylvania.

Tansley, A. G. 1935. The use and abuse of vegetational concepts and terms. Ecology 16:284–307.

Vitousek, P. M. 1994. Beyond global warming: ecology and global change. Ecology 75:1861–1902.

Walters, C. J. 1986. Adaptive management of renewable resources. Macmillan, New York, New York.

Wilson, E. O. 1992. The diversity of life. Belknap-Harvard Press, Cambridge, Massachusetts.

Contributors

STEVEN I. APFELBAUM, Applied Ecological Services, Inc., Brodhead, Wisconsin.

PETER E. AVERS, USDA Forest Service, Washington, D. C.

ROBERT G. BAILEY, USDA Forest Service, Land Management Planning, Fort Collin, Colorado.

MARK S. BOYCE, College of Natural Resources, University of Wisconsin, Stevens Point.

KIM ALAN CHAPMAN, The Nature Conservancy, Minneapolis.

NORMAN L. CHRISTENSEN, School of the Environment, Duke University, Durham, North Carolina.

DAVID T. CLELAND, USDA Forest Service, North Central Forest Experiment Station, Rhinelander, Wisconsin.

THOMAS R. CROW, USDA Forest Service, North Central Forest Experiment Station, Rhinelander, Wisconsin.

CLIVE A. DAVID, College of Natural Resources, University of Wisconsin, Stevens Point.

DAVID G. DAVIS, Office of Wetlands, Oceans and Watersheds, U.S. Environmental Protection Agency, Washington, D.C.

PHILLIP DEMAYNADIER, Wildlife Department, College of Forest Resources, University of Maine, Orono.

FRANK D'ERCHIA, National Biological Service, Environmental Management Technical Center, Onalaska, Wisconsin.

JERRY F. FRANKLIN, University of Washington, Seattle.

LEIGH H. FREDRICKSON, Gaylord Memorial Laboratory, University of Missouri, Puxico.

ERIC J. GUSTAFSON, USDA Forest Service, North Central Forest Experiment Station, Purdue University, West Lafayette, Indiana.

ALAN HANEY, College of Natural Resources, University of Wisconsin, Stevens Point.

MALCOLM HUNTER, JR., Wildlife Department, College of Forest Resources, University of Maine, Orono.

MARK E. JENSEN, USDA Forest Service, Regional Office, Missoula, Montana.

THOMAS KING, USDA Forest Service, Minerals and Geology, Washington, D.C.

JOHN KOTAR, Department of Forestry, University of Wisconsin, Madison.

W. HENRY MCNAB, USDA Forest Service, Southeastern Forest Experiment Station, Asheville, North Carolina.

DOUGLAS J. NORTON, Office of Wetlands, Oceans and Watersheds, U.S. Environmental Protection Agency, Washington, D.C.

REED F. NOSS, Department of Fisheries and Wildlife, Oregon State University, Corvallis.

NEIL F. PAYNE, College of Natural Resources, University of Wisconsin, Stevens Point.

WALTER E. RUSSELL, USDA Forest Service, Regional Office, Milwaukee, Wisconsin.

J. MICHAEL SCOTT, Department of Wildlife, University of Idaho, Moscow.

STANLEY A. TEMPLE, Department of Wildlife Ecology, University of Wisconsin, Madison.

JACK WARD THOMAS, USDA Forest Service, Washington, D.C.

INDEX